69-1986

TN263 69-1986
.B27 Bateman
The Formation of
Mineral Deposits.

**NO LONGER
PROPERTY OF
JEFFERSON
COLLEGE
LIBRARY**

Junior College District of
Jefferson County, Missouri
Library

The Formation of
MINERAL DEPOSITS

By ALAN M. BATEMAN

    THE FORMATION OF MINERAL DEPOSITS

    ECONOMIC MINERAL DEPOSITS

        Second Edition

Thirteenth century mining window in the Freiburg Cathedral, Germany, depicting olden mining in Schauinsland, in the Black Forest near Freiburg. (From print in Yale University library.)

The Formation of MINERAL DEPOSITS

ALAN M. BATEMAN

Silliman Professor of Geology
Yale University
Editor, Economic Geology

Junior College District of
Jefferson County, Missouri
Library

JOHN WILEY & SONS, *Inc.*

New York · London · Sydney

COPYRIGHT, 1951
BY
JOHN WILEY & SONS, INC.

All Rights Reserved

This book or any part thereof must not be reproduced in any form without the written permission of the publisher.

NINTH PRINTING, FEBRUARY, 1967

PRINTED IN THE UNITED STATES OF AMERICA

Preface

The growing realization by the general public that metals and minerals are the very basis of modern industrial development has incited much interest as to how these mineral substances occur in nature and just how they have been formed. I have often been asked by the scientist, the engineer, the industrialist, the layman, and the student about such matters. They have enquired also about books that might contain this information in language not too technical and that do not presuppose a basic knowledge of geology and mineralogy. There have been no such books embodying sound modern information.

This book is an attempt to supply in part answers to such questions. It is meant to convey fundamental information regarding the various means by which the deposits that yield our mineral wealth have been formed and how they occur. A knowledge of mineralogy is not prerequisite, and technical words are held to a minimum. A brief glossary in Appendix B explains the technical terms that are used in this book. This volume treats of the principles and processes of mineral formation and omits descriptions of individual deposits and the many industrial uses of the various metals and minerals, as such information may be found elsewhere. Tabular and statistical matter are likewise omitted, and there is no treatment of individual mineral commodities as such. It is hoped, however, that despite abbreviation and simplicity of treatment the quality of the content will stand unimpaired.

Much of the material contained in this book has been taken from the second edition of the author's *Economic Mineral Deposits* (1950), where a more expanded and more technical treatment may be found, as well as descriptions of deposits and the industrial uses of minerals.

The manner of treatment of the present volume is aimed to serve general readers interested in mineral substances, and scientists, engineers, industrialists, and others who deal with the mining profession and business, as well as students who desire collateral reading in other courses or in a brief course in mineral deposits.

ALAN M. BATEMAN

New Haven, Connecticut
July 24, 1951

Contents

1. INTRODUCTION   1

2. MATERIALS OF MINERAL DEPOSITS   5
Minerals and Ores, 5; The Formation of Rocks, 11; The Formation of Minerals, 12

3. BRIEF HISTORY OF EVOLUTION OF IDEAS ON THE ORIGIN OF MINERAL DEPOSITS   16
Ancient Times, 16; Egyptian, Greek, and Related Cultures, 17; The Beginning of the Scientific Eras, 19; The Seventeenth and Eighteenth Centuries, 21; The Later Centuries, 25

4. RELATION OF MINERAL DEPOSITS TO IGNEOUS ACTIVITY   30
Igneous Relations, 30; Magmas and Crystallization, 35; Products of Differentiation, 40; Summary of Relation to Igneous Activity, 42

5. THE PRIMARY MINERALIZING SOLUTIONS   45
Gases and Vapors, 47; Character of the Hydrothermal Solutions, 49; Evidence of Hydrothermal Solutions, 55

6. THE PROCESSES OF FORMATION OF MINERAL DEPOSITS   59
Summary of Origin, 62

7. MAGMATIC PROCESSES   66
Mode of Formation, 66; Early Magmatic Processes and Deposits, 68; Late Magmatic Processes and Deposits, 72

8. CONTACT-METASOMATIC PROCESSES   90
The Process and Effects, 92; The Resulting Mineral Deposits, 96

9. HYDROTHERMAL PROCESSES   99
Principles, 99–111; Rock Openings, 101; Movement of Solutions through Rocks, 104; Effect of Host Rocks, 105; Factors Affecting Deposition, 106; Wall Rock Alteration, 107; Localization of Hydrothermal Mineralization, 108; Mineral Sequence, 110

Cavity Filling, 111–126; Fissure Veins, 112; Shear Zone Deposits, 119; Stockworks, 120; Saddle Reefs, 120; Ladder Veins, 121; Fold Cracks, 122; Breccia Fillings, 122; Solution Cavity Deposits, 125; Pore-Space Fillings, 125; Vesicular Fillings, 126

Replacement, 127–140; The Process, 128; Resulting Mineral Deposits, 136

Contents

10. **SEDIMENTARY PROCESSES AND CYCLES** 142

   Some Principles Involved in Sedimentation, 143; The Cycle of Iron, 147; The Cycle of Manganese, 150; The Cycle of Phosphorus, 152; The Cycle of Sulfur, 154; The Cycle of Copper(?), 156; The Cycle of Uranium and Vanadium, 157; The Carbonate Cycle, 161; The Clay Cycle, 164; The Silica Cycle, 166; The Coal Cycle, 167; The Cycle of Petroleum and Natural Gas, 172

   The Cycle of Chemical Evaporites, 188–199; Deposition from Oceanic Waters, 189; Deposition from Lakes, 194; Deposition from Ground Water, 198; Hot-Spring Deposition, 199

11. **WEATHERING PROCESSES** 202

   Principles, 202

   Residual Concentration, 205–216; Residual Iron Concentration, 206; Residual Manganese Concentration, 208; Residual Bauxite Formation, 209; Residual Clay Formation, 213; Residual Nickel Concentration, 214; Other Residual Concentration Products, 215

   Mechanical Concentration—Placers, 216–226; Principles Involved, 217; Eluvial Placer Formation, 219; Stream Placer Formation, 220; Beach Placer Formation, 224; Eolian Placer Deposits, 225

   Oxidation and Secondary Enrichment, 226–245; Oxidation and Solution in the Zone of Oxidation, 228; Ore Deposition in the Zone of Oxidation, 235; Secondary Sulfide Enrichment, 238

12. **METAMORPHIC PROCESSES** 246

   Graphite Formation, 246; Asbestos Formation, 248; Talc and Soapstone Formations, 251; Sillimanite Group Minerals, 251; Other Metamorphic Products, 252

13. **GROUND-WATER PROCESSES** 254

   How Ground Water Occurs, 255; Withdrawal and Recharge of Ground Water, 259; Ground-Water Supplies, 261; Long Island, New York, Ground-Water Supply, 263; Florida Artesian Province, 264; The Dakota Sandstone Artesian Province, 265

14. **CONTROLS OF MINERAL LOCALIZATION** 267

   Structural Controls of Mineral Localization, 267; Sedimentary Controls of Mineral Localization, 268; Physical and Chemical Controls, 270; Igneous Rock Controls, 271; Localization of Ores in Parts of Batholiths, 272; Zonal Distribution of Mineral Deposits, 273; Metallogenetic Epochs and Provinces, 275

15. **EXPLORATION AND EXPLOITATION OF MINERAL DEPOSITS** 278

   Geological Exploration, 278

   Geophysical Exploration, 282–295; Magnetic Methods, 282; Electrical Methods, 284; Electromagnetic Methods, 288; Gravimetric Methods, 289; Seismic Methods, 291; Radiometric and Other Methods, 293

   Valuation of Mineral Properties, 295–296; Mining Properties, 295; Oil Properties, 296

   Extraction of Metals and Minerals, 296–310; Mining Methods, 297; Milling or Mineral Dressing, 307; Smelting and Refining, 309

Contents

16. MINERAL RESOURCES	312

Mineral Resources and Their Distribution, 313

Mineral Sufficiency and Deficiency of Leading Areas, 318–329; United States, 318; The British Commonwealth, 322; Russia, 323; Western Europe Bloc, 324; Atlantic Treaty Countries, 325; Eastern Asia, 326; South America, 326; Africa, 327; Australasia and Indonesia, 328; Surplus Minerals from Underdeveloped Areas, 328

Minerals in International Relations and War, 329–337; Distribution of Some Key Critical Minerals, 330

Conservation of Mineral Resources, 337–342

Some Economic Factors in the Development of Mineral Deposits, 340

Future Outlook for Minerals, 342–344

ADDITIONAL GENERAL REFERENCES	346
APPENDIX. GEOLOGICAL TIME TABLE	349
GLOSSARY	351
INDEX	361

CHAPTER ONE

Introduction

The materials of the mineral kingdom are wrested from the earth for man's necessities and comfort. Their search has given rise to voyages of discovery and settlement of new lands; their ownership has resulted in industrial development and in commercial or political supremacy or has caused strife or war. Their quest necessitates a knowledge of their occurrence, distribution, and mode of origin.

We have become aware that the development of the industrial nations has coincided with the utilization of their mineral resources and that those countries lacking them tended to become agricultural and handicraft nations. It is no accident that great manufacturing centers sprang up in central England, in the Ruhr, and around the Great Lakes of America, for there coal and iron met. The energy of coal and the solidity of metals spelled industrial growth and, in turn, industrial might, and other nations came to regard mineral self-sufficiency as one of the chief goals of economic nationalism. It is startling to realize that the insatiable demand for minerals to sustain modern industrialized economic life has caused the world to dig and consume more minerals within the period embracing the two World Wars than in all previous history. This made deep inroads into the storehouse of mineral resources—inroads heightened by war demands—and former bountiful supplies now show serious depletion. Moreover, it is disheartening to realize that the more we manufacture, the greater are the inroads upon the very basis of nearly all manufacturing, and the greater the depletion of unrenewable mineral resources.

This alarming consumption of mineral resources and the exhaustion of known reserves mean that new supplies must be discovered if industrial development is to persist unimpaired. With waning discoveries, search must be directed to less obvious occurrences, of which vast numbers must still lie hidden within the earth. Every phase of geology must be employed to this end. It is pertinent, therefore, to enquire into the geologic environment of mineral deposits, *why* they should be *where* they are, and where others might be found. This involves a knowledge of how mineral deposits are formed and the conditions that tend to localize them in one place and not in another. Enquiry into the origin of mineral deposits embraces a study of the primary sources of the mineral substances, how they came to be moved to their present sites, the agents of transportation, how they came to be deposited or concentrated into economic deposits, and how they became affected by the action of weathering or other subsequent geologic processes. Their study raises many interesting scientific problems that whet intellectual curiosity and promise opportunity for long-continued research.

Mineral deposits as they are found today are not, as commonly supposed, a part of the initial creation of the earth. They are exceptional features that constitute only an infinitesimal part of the earth's crust, but they assume an importance far in excess of their relative volume because of the highly valuable materials they supply to national wealth and industry. They have been concentrated in the rocks under peculiar and exceptional circumstances by various processes that have operated over a period of time in excess of 2,000 million years. They are thus later than the original earth formation, and the processes by which they have been formed are active even to the present day.

It will be shown that the many diverse materials of mineral deposits, which range from the noble metals and gemstones to fuels and iron, to the base metals that bulwark industry, to the chemical substances that serve the household, the apothecary, and the chemical industry, to the mineral substances that are the sinews of war, and to the lowly clay products that make commonplace buildings or beautiful ceramics, are equally the products of as many diverse geologic processes that operate within and on the earth's surface. Thus, we will delve into plutonic forces that give rise to fiery volcanism, to fumarolic gases and vapors that escape from buried molten lavas, to mineralizing fluids that escape from molten rocks, to surficial bodies of water that deposit economic sedimentary beds, to atmospheric weathering processes that create new substances or concentrate former disperse substances, and to the forces of metamorphism that have acted on all preceding bodies. Even the ground water will be considered in the part it plays in yielding desired minerals, and in life. These processes will be the main theme of the pages to follow.

Introduction 3

The study of mineral deposits falls within the sphere of *economic geology*, which in turn has long held kinship with *mining*. The two have grown up together since mining first became an art. Indeed, there is now a specialized division of mining known as *mining geology*. Broadly speaking, economic geology deals with the occurrence, genesis, composition, shape, size, localization, controls, and appraisal of mineral bodies within the rocks; mining deals with the exploitation and extraction of the minerals from the rocks; and *ore beneficiation* and *metallurgy* deal with the winning of the metals and minerals from the ores and making them into forms and compounds desired for industry. The relations between these studies may be understood better by realizing that the desired metals are locked up as chemical compounds in the form of ore minerals, which are admixed with undesired minerals or rock substance to form ores. The ores in turn occur in ore bodies or mineral deposits enclosed within the rocks of the earth's crust. The winning of metals, therefore, involves first an application of knowledge of geology to the appraisal of the ore deposits in the ground, next the technical engineering skills of mining to bring the ores to the surface, and next the techniques of ore dressing and metallurgy to separate the ore minerals from the waste minerals, and then the metals from the ore minerals. Of course, many earthy commercial minerals are not sought for metals that they might contain; they may not contain metals but are desired for multitudinous other uses as *nonmetallic* minerals. Petroleum, coal, and rocks are quite different mineral substances. Clay, for example, is not mined for its aluminum content nor asbestos for its magnesium; but clay is desired for pottery or ceramics, and asbestos for asbestos. These examples are not cited to show the properties and uses of mineral substances but rather to indicate the diverse materials that enter into a study of the occurrence and mode of formation of mineral deposits.

Consideration of mineral deposits also raises questions of mineral resources and particularly their international aspects and implications in national economy and security. Since minerals are not only the basis of industrial power but also of security, armament, and aggression, their distribution, ownership, and international movement become of paramount importance. An industrialized country cut off from adequate supplies of essential and critical minerals is shorn of its industrial power for peaceful economic development and for armament and defense because minerals are the very foundation upon which industrial strength depends. Therefore, mineral resources play a dominant role in aggression, defense, and security no less than in the development of national strength and international trade. It is fitting, therefore, to look into their occurrence, geographic sufficiency and deficiency, distribution, and political ownership, as is undertaken in Chapter 16.

First, however, it is desirable to consider briefly the materials of which mineral deposits are composed.

CHAPTER TWO

Materials of Mineral Deposits

MINERALS AND ORES

Mineral deposits for the most part are composed of desired and undesired materials. Some, like coal, oil, or clay, consist wholly or almost wholly of economic products, but mostly the valued substances are accumulations or concentrations of elements that are only sparsely distributed in the earth's crust.

Some 100 elements are known. Of these, only 8, according to Clarke and Washington, are present in the earth's crust in amounts exceeding 1 percent; and 99.5 percent of the outer 10 miles of the earth's crust is made up of the following 13 elements: oxygen, silicon, aluminum, iron, calcium, sodium, potassium, magnesium, titanium, phosphorus, hydrogen, carbon, and manganese. The remaining 0.5 percent includes all the precious and industrial metals such as gold, silver, copper, lead, zinc, tin, tungsten, nickel, chromium, and others, and these are the elements that are concentrated into economic mineral deposits.

Only a few elements occur in nature in elemental form. Mostly, they are bound together in combinations of two or more elements to form minerals, which are merely naturally occurring chemical compounds with specific properties and with definite arrangements of their atoms. Most rocks are aggregates of one or more minerals. Over 1,600 natural mineral species are known, but only about 50 of them are rock-making minerals, and the remainder come from mineral deposits. Only about 300 of these are classed as economic or industrial minerals. Thus, we speak of some as ore minerals (Table 2), gangue minerals (Table 3), nonmetallic minerals, or rock-forming minerals (Table 1), and others as simply minerals. Some of the substances of mineral deposits are not minerals at all but are rocks or gravels. Thus, in the usage of the terms mineral deposits or mineral

TABLE 1

Some Common Rock-Forming Minerals

Mineral	Composition
Feldspar (orthoclase, plagioclase)	Potassium, sodium, aluminum, silicon, oxygen
Quartz	Silicon, oxygen
Mica	Potassium, aluminum, magnesium, iron, silicon, oxygen
Pyroxene	Magnesium, iron, silicon, oxygen
Amphibole	Calcium, magnesium, iron, silicon, oxygen
Olivine	Magnesium, iron, silicon, oxygen
Magnetite	Iron, oxygen
Calcite	Calcium, carbon, oxygen
Dolomite	Calcium, magnesium, carbon, oxygen

resources, mineral is used in the broad sense of materials of the mineral kingdom.

Ore minerals are those that may be used to obtain one or more metals. For the most part they are what are spoken of as *metallic minerals*, or those that have a metallic luster; a few are *nonmetallic minerals*, or those without a metallic luster. A few of the ore minerals occur in nature as native elements, of which gold and platinum are examples, but mostly they are combinations of the metals with sulfur, arsenic, oxygen, silicon, carbon, or other elements. It does not follow that each metal is obtained from only one ore mineral. On the contrary, there are a dozen economic ore minerals of copper. Also, more than one metal may be won from a single ore mineral, as in copper-silver and copper-tin combinations. Further, several different ore minerals may be contained in the same mineral deposit. From this it follows that an individual ore deposit may yield several different metals.

Ore minerals are also commonly referred to as *primary*, or those formed during the original period or periods of mineralization, and *secondary* or *supergene* to designate those formed from the primary minerals as a result of weathering or other surficial processes resulting from the activities of surface agents.

The metals of commerce are derived from many metallic compounds. Most of the world's gold has been won from native gold; consequently, its removal from admixed minerals is a relatively simple process and offered no serious problems of extraction, even to the ancients. Silver, on the other hand, is obtained more from combinations with sulfur and other elements than from native silver. Most

Minerals and Ores

of the other metals of industry are obtained only from combinations with other elements, and their recovery in many cases involves complicated metallurgical processes of extraction. The iron that supplies the vast steel industry is obtained almost entirely from combinations with oxygen. It is from such simple combinations as these that the human race has been supplied with desired metals for over 2,000 years.

A list of the common ore minerals is given in Table 2.

TABLE 2
Some of the Common Ore Minerals

Metal	Ore Mineral	Composition	Percent Metal
Gold	Native gold	Gold	100
Silver	Native silver	Silver	100
	Argentite	Silver, sulfur	87
Iron	Magnetite	Iron, oxygen	72
	Hematite	Iron, oxygen	70
	Limonite	Iron, oxygen, water	60
Copper	Native copper	Copper	100
	Bornite	Copper, iron, sulfur	63
	Chalcopyrite	Copper, iron, sulfur	34
	Chalcocite	Copper, sulfur	80
	Malachite	Copper, carbon, oxygen, water	57
	Azurite	Copper, carbon, oxygen, water	55
Lead	Galena	Lead, sulfur	86
	Cerussite	Lead, carbon, oxygen	77
Zinc	Sphalerite	Zinc, sulfur	67
	Smithsonite	Zinc, carbon, oxygen	52
Tin	Cassiterite	Tin, oxygen	78
Nickel	Pentlandite	Nickel, iron, sulfur	22
Chromium	Chromite	Chromium, iron, oxygen	68
Manganese	Pyrolusite	Manganese, oxygen	63
	Psilomelane		45
Aluminum	Bauxite	Aluminum, oxygen, water	39
Antimony	Stibnite	Antimony, sulfur	71
Mercury	Cinnabar	Mercury, sulfur	86
Molybdenum	Molybdenite	Molybdenum, sulfur	60
Tungsten	Wolframite	Tungsten, manganese, iron, oxygen	76
	Scheelite	Tungsten, calcium, oxygen	80

Gangue minerals are the associated undesired materials of an ore deposit. They are mostly nonmetallic minerals or rock and are usually discarded in the treatment of the ore. Some of the gangue minerals may be collected as by-products and utilized. Indeed, sometimes the by-product gangue minerals have been instrumental in making marginal ore deposits profitable. Some of the common gangue minerals are given in Table 3.

TABLE 3

SOME COMMON GANGUE MINERALS

Name	Composition
Quartz	Silicon, oxygen
Limonite	Iron, oxygen, water
Calcite	Calcium, carbon, oxygen
Dolomite	Calcium, magnesium, carbon, oxygen
Rhodochrosite	Manganese, carbon, oxygen
Barite	Barium, sulfur, oxygen
Fluorite	Calcium, fluorine
Feldspar	
Garnet	
Rock matter	

Ore is the mixture of ore minerals and gangue from which metals may be extracted at a profit, thus excluding noncommercial occurrences. Profit depends upon the amount and price of the metal and the cost of mining and placing it on the market. This in turn depends upon the geographic location of the deposit, which involves the cost of mining and treatment and the availability of transportation. A good quality iron deposit located in the Arctic might not be ore because the cost of extraction and transportation might exceed the value of the iron, but given transportation and a nearby market it would become ore.

Mostly, the geographic factors of location and transportation are obstacles to be overcome, rather than deterrents, provided the deposits are of sufficient size and grade to withstand the costs involved. Few large mineral deposits have remained undeveloped because of geographic factors. Rather, the incentive to develop mineral bodies into commercial ores has stimulated the construction of transportation facilities and opened up the hinterland areas of the western Americas, northern Canada, and central Africa. Agriculture and industries followed in the wake of mining, and the transformation of former submarginal mineral deposits into wealth-producing ores has brought about the change of sparsely settled areas into flourishing states.

The relative proportion of ore minerals and gangue minerals in ores varies greatly. In general, gangue predominates, and whether material may or may not constitute ore may depend upon our technical ability to remove the gangue. It is costly to smelt valueless gangue

Minerals and Ores

in order to obtain the enclosed metal, so that it is customary to mill the ore to concentrate the ore minerals and discard the waste. Thus, if a ton of copper rock contains $6.00 worth of metal, and mining costs $3.00 and freight and smelting $3.00, it is obviously unprofitable and is not ore, but if 10 tons of rock were susceptible of concentration to yield one ton of concentrates, then there would be freight and smelting charges on only 1 ton of concentrates instead of on 10 tons, which would be equivalent to about 30 cents per ton of rock mined, and the rock would be profitable ore. Consequently, the proportion of ore minerals to gangue is of vital importance in determining whether a material is ore; the gangue may play fully as important a part as the ore minerals.

The amount of ore minerals present varies considerably in ores of different metals, and also in ores of the same metal. High-grade iron ores range from 80 to 100 percent hematite; copper ores range from a few percent up to 75 percent metallic minerals, the lower ones being concentrated and those with higher percentages being smelted directly. In contrast, gold ores may contain only an infinitesimal amount of gold. For example, the Alaska Juneau gold mine for years yielded profitable gold ore carrying only 0.00016 percent of gold, or about $4/100$ ounces per ton. The amount of metal that must be present to constitute ore obviously depends upon the value of the metal. A gold ore containing ½ ounce of gold per ton of ore would be a rich ore, but copper rock carrying 5 pounds of copper per ton would be valueless. During periods of low metal prices, only rock with a metallic content higher than normal can be classed as ore.

The metal content of ore is referred to as *grade* or *tenor*. The tenor of ores of some of the commoner metals, and their units and price range, are given in Table 4.

Ores may yield a single metal (simple ores) or several metals (complex ores). Those commonly worked for only a single metal are iron, aluminum, tin, chrome, mercury, manganese, tungsten, and some ores of copper. Gold ores may yield only gold, but silver is a common associate. Much gold, however, is extracted as a by-product from other ores. Most silver is a by-product. Ores that commonly yield two, three, or more metals are gold, silver, copper, lead, zinc, nickel, cobalt, and antimony. Many of the minor metals are obtained only as by-products. Where precious metals accompany base metals, their presence may make profitable ore of otherwise unprofitable material. Some of the common associations of metals in ores are: gold and silver; silver and lead; lead and zinc; lead, zinc, copper; copper,

TABLE 4

DATA ON METALS AND THEIR ORES

Metal	Unit of Measure	Tenor Low	Tenor Average	Common Associates	Commercial Unit	Price Ranges 1925–40	Price Ranges 1941–55
Gold	oz/ton	0.15	0.2–0.3	Ag	oz Troy	20.67–35.00	35.00
Silver	oz/ton	10	12–30	Au, Pb	oz Troy	0.25–0.70	0.35–0.90
Platinum	oz/ton	0.1	0.3	Pt group	oz Troy	31.00–67.65	36.00–93.00
Iron	% Fe	30	40–60	Mn	ton iron	15.00–24.00	20.00–56.00
Copper	% Cu	0.7	1–5	Au, Ag	lb Cu	0.05–0.21	0.12–0.33
Lead	% Pb	3	5–10	Zn, Ag	lb Pb	0.03–0.09	0.06–0.215
Zinc	% Zn	3	10–30	Pb	lb Zn	0.03–0.08	0.075–0.175
Tin	% Sn	0.5	1–5	W	lb Sn	0.22–0.65	0.52–1.28
Nickel	% Ni	1.5	1.5–3	Cu, Pb	lb Ni	0.35–0.39	0.31–0.64
Aluminum	% Al_2O_3	30	55–65	....	lb Al	0.17–0.24	0.15–0.22
Antimony	% Sb	20	40–60	Ag	lb Sb	0.05–0.17	0.14–0.417
Bismuth	% Bi	BP*	40–60	W	lb Bi	0.85–1.30	1.25–2.25
Beryllium	% BeO	8	10–12	....	unit BeO	30.00–35.00	26.00–47.00
Arsenic	% As_2O_3	BP	......	....	lb As_2O_3	0.01–0.03	0.04–0.06
Cobalt	% Co	5	8–10	Ag, Cu	lb Co	1.10–3.00	1.50–2.60
Chromium	% Cr_2O_3	32	35–50	....	ton Cr_2O_3	17.00–47.00	34.00–48.00
Cadmium	% Cd	BP	......	Zn	lb Cd	0.55–1.42	0.75–2.25
Manganese	% Mn	35	45–55	Fe	units/ton	0.18–0.55	0.70–0.80
Mercury	% Hg	0.5	1–3	....	flask-76 lb	58.00–202.00	76.00–324.00
Molybdenum	% MoS_2	0.4	1–3	....	lb MoS_2	0.34–0.45	0.45–1.05
Titanium	% TiO_2	3	4–40	Fe	lb TiO_2	....	4.50–6.00
Tungsten	% WO_3	.....	60–70	....	unit WO_3	9.20–20.61	24.00–28.50
Vanadium	% V_2O_5	2	3–8	....	lb V_2O_5	0.20–1.05	0.27–0.31

* BP = by-product

gold, molybdenum; iron and manganese; titanium and iron; chromium and platinum; tin and tungsten; and zinc and cadmium.

Nonmetallic materials consist of solids, liquids, and gases in the form of minerals, rocks, hydrocarbons (oil, gas, and coal), and brines. The term ore or ore mineral is not applied to such substances; they are referred to by the name of the substance itself, such as asbestos, mica, coal, or petroleum, and the undesired associated material is called *waste*.

Since nonmetallic materials are, in general, bulky common substances, their price is correspondingly low as compared with that of metals. Except for a few substances, such as gemstones, the deposits consist predominantly or entirely of industrial materials, and there is generally little or no waste.

The nonmetallic materials consist of a vast array of substances utilized in modern civilization. They include such substances as fuels.

The Formation of Rocks

clays, rocks, sands, sulfur, salts, mica, fluorspar, feldspar, gypsum, barite, talc, chemicals, fertilizers, graphite, and gemstones. The associated waste consists mostly of rock material, undesired nonmetallic minerals, or unfit products. The desired materials are separated from the undesired materials by *processing* mechanically or by hand.

The determination of what is economic material is not so dependent upon the associated gangue as it is in ores. Rather it is dependent upon the price and the physical and chemical properties of the products themselves. A metal is always a metal, but clay, for example, to be usable must meet definite specifications regarding plasticity, fusibility, strength, and other properties. Different requirements exist for each nonmetallic product and for its uses.

THE FORMATION OF ROCKS

All mineral deposits lie in or on solid rocks, and, except for a thin skin of soil or water, rocks compose the earth's crust. Rocks may be divided into three groups: igneous, sedimentary, and metamorphic.

Igneous rocks are those that consolidate from molten material that originates within the earth. This molten material, called *magma*, is molten because of high temperature, but it is really a high-temperature chemical solution that obeys chemical laws. It has become familiar through lava flows. The molten material contains the constituents of rock-forming minerals, gases, much water vapor, and small quantities of the metals of commerce. When it cools, loses its gases and vapors, and freezes, igneous rocks result. Lava rock is some of this residuum. During cooling crystallization occurs and gives rise to the rock-forming minerals. Most magmas do not reach the surface, and cooling occurs slowly deep within the earth, with the result that the crystals grow large and form *coarse textures*. If cooling is rapid, as on the surface, then *dense* or *aphanitic* textures result, and the rock has a stony appearance. If cooling is very rapid, no crystallization visible to the naked eye occurs and the resulting rock is a glass, with a *glassy texture*. The contained gases and vapors may erupt to the surface with explosive violence and blow the magma to dust, giving *volcanic tuff*, or it may blast the previously consolidated portions into fragments, large and small, creating *volcanic breccias*. In any of the solid rocks certain crystals may grow larger than the surrounding *groundmass*, and the result is a *porphyritic* texture.

Sedimentary rocks are those laid down in water or air from (*a*) detrital particles of rock and mineral fragments, or (*b*) chemical deposition of substances contained in solution. The constituents of course are derived from former rocks and minerals. The settling of particles through a fluid brings about a layering or stratification, and because of this, sedimentary rocks are often called stratified rocks. The sediments are unconsolidated when first laid down and later become compacted or cemented into hard rocks. Accumulations of gravels give rise to *conglomerates* or *breccias;* sands to *sandstones;* clay and silt to *shales;* and calcium and magnesium carbonates precipitate to *limestone* and *dolomite.* Other special substances, of particular value to man, that are laid down as sedimentary rocks are coal, iron ore, manganese ore, phosphate beds, sulfur, magnesite, oil shale, diatomaceous earth, and clays.

Metamorphic rocks are igneous or sedimentary rocks that have undergone change as a result of elevated temperature and pressure. They are squeezed, the minerals are flattened and elongated, and the rocks are foliated, cemented, or recrystallized. Thus, igneous rocks become *gneisses* and *schists,* sandstones become *quartzites,* shales become *phyllites* or *schists,* and carbonate rocks become marbles. Some new minerals have been developed from former ones, such as white mica and garnet from shales. Most metamorphic rocks become crystalline, and are sometimes called crystalline schists.

A few of the common rocks and their composition are given in Table 5.

THE FORMATION OF MINERALS

It is commonly but mistakenly thought that the minerals of economic mineral deposits were formed at the same time as the earth itself. This is not true. All the constituents of mineral deposits have been formed later than earlier rocks and minerals, and from periods of a few thousand years to hundreds of millions of years ago. The minerals, therefore, have been introduced by some means into or onto pre-existing host rocks, and the mode of formation of individual minerals is essential to an understanding of the formation of the deposits of which minerals are the constituents.

Minerals are formed in several different ways of which the chief are: crystallization from magmas; deposition from gases and vapors; chemical and organic deposition from liquid solutions; weathering; and metamorphism.

CRYSTALLIZATION FROM MAGMAS. Crystallization takes place in the manner described for the formation of rocks. Since a magma is a molten silicate solution it follows the laws of aqueous solutions. When it cools and the saturation point is reached for any given min-

The Formation of Minerals

TABLE 5
A Few Common Rocks and Their Chief Constituents

Name	Group *	Color and Texture	Chief Constituents †
Granite	I	Light, coarse	Quartz, orthoclase, mica, Pl, A, P
Syenite	I	Light, coarse	Orthoclase, mica, Pl, A, Py
Granodiorite	I	Medium, coarse	Quartz, Or, Pl, mica, A, P
Quartz monzonite	I	Light, coarse	Plagioclase, orthoclase, quartz, mica, A, Py
Diorite	I	Dark, coarse	Plagioclase, Or, mica, A, Py
Gabbro	I	Dark, coarse	Plagioclase, Py, A, olivine
Peridotite	I	Dark, coarse	Olivine, Py, Pl
Rhyolite	I	Light, dense	Same as granite
Andesite	I	Dark, dense	Same as diorite
Basalt	I	Dark, dense	Same as gabbro
Tuff, breccia	I	Fragmental	Same as rhyolite, andesite, basalt
Conglomerate	S	Pebbly	Gravel
Sandstone	S	Granular	Quartz sand grains
Shale	S	Clayey	Clay, silt
Limestone	S	Dense	Calcite, (dolomite)
Gneiss	M	Crystalline	Same as igneous rocks
Schist	M	Crystalline	From igneous or sedimentary rocks
Quartzite	M	Vitreous	Quartz grains
Marble	M	Crystalline	Calcite, (dolomite)
Slate	M	Crystalline	From shale

* I, igneous; S, sedimentary; M, metamorphic.

† Or, orthoclase feldspar; Pl, plagioclase feldspar; A, amphibole; Py, pyroxene; subordinate constituents are abbreviated; any igneous rock may be porphyritic.

eral, that mineral will crystallize out as a solid. Thus, certain minerals of economic interest, such as magnetite, chromite, or apatite, may be formed. This complicated process is discussed further in Chapter 7.

DEPOSITION FROM GASES AND VAPORS. Hot gases and vapors given off from magmas may contain many of the elements and compounds that enter into the composition of minerals. The fascinating Valley of Ten Thousand Smokes in Alaska throws much light on this process, for there E. G. Zies found that the highly heated gases and vapors rushing up the fumaroles carried large quantities of acids and mineral constituents. Deposition of many minerals took place around the fumaroles by chemical reactions between different gases and vapors. For example, native sulfur, iron ore minerals, and many other minerals were formed by such reactions. Similar reactions are considered to have taken place in the past at depths below the surface.

Other reactions between gases and vapors and solids (rocks and minerals) have taken place on a vast scale where magma intrusions have occurred, and have formed a large variety of high-temperature minerals of economic importance. Still other reactions with liquids also bring about deposition of minerals.

CHEMICAL DEPOSITION OF MINERALS FROM HOT SOLUTIONS OF MAGMATIC ORIGIN. This is one of the common methods of formation of minerals of industry. The hot waters, called *hydrothermal solutions,* are given off in the late stages of magmatic cooling, and they course through openings in the rocks seeking places of lower pressure. They carry vast quantities of mineral matter in solution and give rise to most of the minerals in mineral deposits. As the temperature and pressure drop in their ascent into cooler and shallower rocks, the least soluble minerals become deposited first and the most soluble ones later. These hydrothermal minerals may fill openings in the rocks such as fissures, caves, pore spaces, or breccia openings. The deposition from solution may be aided by contact with other solutions of different chemical character or by catalytic action.

The most important method of deposition of minerals from hydrothermal solutions, however, results from reactions between wall rock and solutions, whereby ore and gangue minerals take the exact place of rock minerals, the former being deposited from solution and the latter being taken into solution with no change in volume. This is known as *metasomatism* or *replacement,* and the process gives rise to huge mineral deposits.

DEPOSITION FROM SURFACE SOLUTIONS. Deposition may take place by the action of bacteria, shell life, or other organic substances, giving rise to rocks and minerals such as beds of limestone, iron ore, and manganese. Simple oxidation of mineral-bearing surface waters may also give rise to deposition of iron and other minerals. Also, evaporation of brines, marine or continental, brings about deposition from solution of many salines and salts, such as common salt, gypsum, potash, borax, and sodium carbonate. Nitrates are formed through evaporation of ground water in Chile, and the greatest copper deposit in the world (Chile) has resulted from the deposition of copper minerals by evaporation from surficial copper solutions.

WEATHERING. Weathering results in the mechanical and chemical breakdown of rocks and minerals. Few substances escape the unrelenting attack of oxygen, water, and carbon dioxide, and the soil cover the world over attests to such activity and resulting decay of rocks. As a result of weathering many new minerals are created, and some of these are commercial products. Bauxite, the ore of aluminum, is created by the weathering of rocks high in alumina, and valuable porcelain clays, as well as many other mineral products, are formed in this manner. Weathering acts to change minerals that are unstable under surface conditions into other minerals that are stable.

METAMORPHISM. Metamorphism at elevated temperature and pressure acts to change former minerals into new minerals that are stable under the new conditions. Thus clays born under the conditions of weathering, when buried deeply by later rocks, may be converted to mica schist or garnet; and talc, sillimanite, or graphite may similarly be formed from pre-existing substances.

The above remarks indicate that many kinds of minerals may be formed in many different ways and that the processes of formation are generally not simple ones. The formation of deposits of these minerals is also not simple, as will be seen in later chapters.

Next, we will trace briefly the development of knowledge regarding the evolution of ideas as to the origin of mineral deposits.

SELECTED REFERENCES ON MATERIALS OF MINERAL DEPOSITS

Minerals and How to Study Them. The late E. S. Dana, revised by C. S. Hurlbut, Jr. John Wiley & Sons, New York, 1949. A modern elementary textbook on minerals.

Rocks and Rock Minerals. 3rd Edit. L. V. Pirsson and Adolph Knopf. John Wiley & Sons, New York, 1947. Elementary reference book on rocks and rock minerals.

Industrial Minerals and Rocks. 2nd Edit. Committee on Industrial Minerals, S. H. Dolbear, chairman. Am. Inst. Min. Met. Eng., New York, 1949. Comprehensive, authoritative volume giving descriptions, occurrence, uses, preparation, and marketing.

CHAPTER THREE

Brief History of Evolution of Ideas on the Origin of Mineral Deposits

ANCIENT TIMES [1]

The early utilization of mineral products by primitive man must have persisted long ages before the crude knowledge developed into a craft and later into a science. The early incentive for the acquisition of knowledge of minerals was undoubtedly utilitarian, but later the Greek philosophers raised it to an intellectual plane.

The first materials used by primitive man were nonmetallic substances—flint, chert, quartz, and certain hard and soft stones such as obsidian, quartzite, soapstone, and limestone—sought for their use as weapons, implements, household utensils, and for carving. Clay was widely and extensively used, first for pottery and later for bricks. Unquestionably clay represented the first large-scale mineral industry, an industry that has persisted continuously through the ages. Burned clay figures believed to be Aurignacian (30,000–20,000 B.C.) have been discovered in Moravia, and excellent Paleolithic pottery of the Solutrean period ($+10,000$ B.C.) has been found in Egypt. Brick, tile, and clay tablets were extensively used by the Chaldeans, Babylonians, and early Egyptians for building cities and irrigation works and making writing tablets. Early Asiatic and African dwellings were built of clay bricks, and building stones were extensively used later. During the building of the Pyramids (2980–2925 B.C.) the stone industry must have been on a grand scale, as the Pyramid of Gizeh contains 2,300,000 blocks of stone averaging $2\frac{1}{2}$ tons apiece.

Paleolithic man between 100,000 and 7000 B.C., according to S. H. Ball, used 13 varieties of minerals—chalcedony, quartz, rock crystal,

---

[1] Several of the following paragraphs have been extracted from *Economic Mineral Deposits* by the author.

Egyptian, Greek, and Related Cultures

serpentine, obsidian, pyrite, jasper, steatite, amber, jadeite, calcite, amethyst, and fluorspar. He also utilized ochers or mineral paints. At about the time Neolithic man became acquainted with gold and copper, he also used nephrite, sillimanite, and turquoise. Most of these minerals are common ones that probably were found by accident and whose quest neither greatly stimulated human curiosity nor created knowledge regarding their origin. They were accepted as found and utilized. Enquiry into the origin of mineral deposits had not yet been made; it was the pre-dawn stage.

EGYPTIAN, GREEK, AND RELATED CULTURES

As the desire for gemstones and metals, which represented wealth, became more urgent, facts of occurrence were recorded and crude theories of origin were evolved. Expeditions were even organized for the discovery and working of deposits, and ownership and barter of these minerals became an important part of the life of the people, even more important relatively than it is today. The mining and use of gemstones reached a high art among the early Egyptians, Babylonians, Assyrians, Indians, and Chinese. Gemstones were greatly prized, and, living or dead, the Egyptian was bedecked with jewels, which attained important significance among a people obsessed with mysticism. In pre-Dynastic times (+3400 B.C.) it was the color rather than the substances that the Egyptian prized most. The Theban craftsman created pleasing color schemes by utilizing the azure of the lapis lazuli, the red of the carnelian, the purple of the amethyst, the green of the malachite, the yellow of the jasper, and the blue of the turquoise. He also used agate, beryl, chalcedony, and garnet, and shaped and polished hard stones, producing not only ovoid but also faceted beads. Even in those remote times international barter is recorded because the lapis found in Egypt was probably obtained from Afghanistan, some 2,400 miles away.

According to Ball, other stones are known to have appeared, such as the onyx in the 2nd Dynasty, azurite and jade in the 3rd Dynasty, and amber in the 6th Dynasty (2625–2475 B.C.). The stele of Nebona (18th Dynasty) reads: "I have consecrated numerous gifts in the temple of my father Osiris in silver, in gold, in lapis lazuli, in copper, and in precious stones." Later, under Greek influence, in the time of the Ptolemies, several other stones were introduced, including the Indian gems sapphire, zircon, and topaz.

The oldest mining was for gems and decorative stones, and for over 2,000 years the Pharaohs dispatched expeditions including engineers and prospectors to the Sinai Peninsula for turquoise, and into the Sudan. Ball identifies as the first economic geologist the Egyptian Captain Haroeris, who about 2000 B.C. led an expedition to Sinai and after 3 months' prospecting discovered and extracted large quantities of turquoise. The ancient Egyptians (from 1925 B.C.) sank literally hundreds of shafts for emeralds on the Egyptian coast of the Red Sea. Some of these excavations are said to have been 800 feet deep and large enough to permit 400 men to work at a time therein.

The first metals used were probably gathered as native metals from streams by primitive man. Gold is presumed to have been used before copper, and copper is considered by some to have been discovered 18,000 years before Christ; certainly copper was known to the Egyptians in 12,000 B.C. and was widely used in Europe about 4000 B.C. Strabo relates that "in the country of Saones, where is Colchis, by the Euxine, the winter torrents bring down gold which the barbarians collect in troughs pierced with holes and lined with fleeces." Hence the legend of the Golden Fleece. The lining of fleece skins entangled the gold particles. The coarse gold was shaken out but the fine adhered to the wet wool, and the fleeces, hung on the trees to dry so that the fine gold could be beaten out of them, spurred Jason and the Argonauts in the ship *Argo* to seek the Golden Fleece near the shore of the Euxine. This is the earliest record of placer gold mining and a poetic expression of the first gold-mining rush. Even today, cowhides are inserted in wooden troughs to collect fine placer gold in certain parts of South America.

At the ancient mines of Cassandra, Greece, which Sagui estimates to have been mined from about 2500 to 360 B.C., the skilful extraction of gold-silver ores was based upon a knowledge of their localization at the intersection of fissures, toward which tunnels were run to hit the intersections beneath the zone of weathering. A beginning had been made in the understanding of the occurrence of ores.

Some knowledge of how ores occur and the beginning of curiosity regarding their genesis are shown in the writings of Greek and Roman philosophers. Herodotus (484?–425 B.C.) told of the occurrence of gold in the quartz veins in the Krissites district, Greece, later described by Diodorus. Theophrastus (372–287 B.C.), a pupil of Aristotle, in his *Book of Stones*, the first mineralogy textbook, described 16 minerals, grouped as *metals, stones,* and *earths*. Strabo, writing in A.D. 19, says in reference to alluvial mining in Spain: "Gold is not only dug from mines, but likewise collected; sand containing gold being washed down by the rivers and torrents . . . at the present day more gold is produced by washing than by digging it from the mines." (H. C. Hamilton and W. Falconer). Many descriptions of ore occurrences in Spain are given in Pliny's elaborate technical descriptions. He tells also that Hannibal had a silver mine, named the Baebulo, in southern Spain, in a mountain that had been penetrated 1,500 paces. Pliny stated it yielded 300 pounds of silver daily. The production of silver-lead ores was an important industry in Attica at a remote period, the famed mines of Laurium having been worked long before the days of Xenophon, who reported upon them in 365 B.C. The ancients sank here more than 2,000 shafts, one of which is 386 feet deep, and their locations disclose an accurate knowledge of the occurrence of ores. Throughout the Dark Ages little appears to have been added to the knowledge of the early philosophers, except by Avicenna (980–1037), the Arabian translator of Aristotle, who grouped minerals as *Stones, Sulfur minerals, Metals,* and *Salts* (Crook), thus definitely recognizing the sulfide group of minerals.

THE BEGINNING OF THE SCIENTIFIC ERAS

The first reasonable theory of ore genesis was formulated by Georgius Agricola (Bauer, 1494–1555, Fig. 3–1). Born in Saxony amidst the mines of the Erzgebirge he became a keen observer of minerals and a careful recorder. He drew heavily from an earlier work, von Kalbe's *Bergbüchlein* (*Little Book on Ores,* about 1518), which is stated to be the first printed work in the field of mining. Although some of Agricola's views were fantastic, he showed in his *De Re Metallica* that lodes originated by deposition of minerals in "canales" (fissures) from circulating underground waters, largely of surface origin, that had become heated within the earth and had dissolved the minerals from the rocks. He made a clear distinction between *homogeneous minerals* (minerals) and *heterogeneous minerals* (rocks). The former he divided into *earths, salts, gemstones,* and *other minerals.* He also classified ore deposits genetically into veins (vena profunda), beds, stocks, and stringers. Prior to Agricola, most writers thought that mineral lodes were formed at the same time as the earth, but he recognized clearly that they were of different age from the enclosing rocks, as he states: "To say that lodes are the same age as the earth itself is the opinion of the vulgar." (*De Re Metallica,*

1556). This knowledge of earth minerals led him to be the first to refute vigorously the efficacy of the forked hazel stick (Fig. 3-2) commonly used at that time in the effort to find metals and water

FIGURE 3-1. Portrait and signature of Agricola, "The Father of Mineralogy," about 1550. (From Hofmann; F. D. Adams, *Birth and Development of the Geologic Sciences.*)

(and still mistakenly used to find water). He made accurate observations on the weathering of rocks and the surface decomposition of metallic sulfide ores. Many of his careful observations are quaintly portrayed by interesting woodcuts (Fig. 15-13).

Agricola's writings are among the most original contributions to the study of the origin of ores. They were a marked advance in scientific thought and influenced greatly the thought of later writers.

FIGURE 3-2. Medieval miners employing the divining rod. (Agricola, *De Re Metallica*.)

THE SEVENTEENTH AND EIGHTEENTH CENTURIES

The extraordinary mining atmosphere of the Erzgebirge and Harz mountains in Germany must have stimulated knowledge of rocks and minerals, but little information appears to have been recorded from the time of Agricola to that of Descartes, whose *Principia Philosophae* was published in 1644. His conception of the earth as a cooled star with a hot interior led him to suggest that the ore minerals were driven upward, from a deep metalliferous shell, by interior heat (a theory revised by J. S. Brown in 1948) in the form of exhalations and resurgent surface waters, to be deposited as lodes in the fissures of the outer stony crust. This conception is clearly the forerunner of some of the ideas held today and exhibits a canny insight into earth processes.

In the eighteenth century the accumulated factual knowledge further aroused curiosity as to the origin of ore materials. Under the

FIGURE 3-3. Exhalations of vapors and hot waters from the bowels of the earth, giving rise to fumaroles and hot springs and forming veins of copper by reaction with stony materials. Copper ore is being mined, refined, and shipped. (From John Friedrich Henkel, *Pyritologia*, Leipzig, 1754, Yale Univ. library.)

Seventeenth and Eighteenth Centuries

stimulus of inspiring leaders, hypotheses of origin, and resulting controversy, burst forth. Most of the theories emanated from German mining districts, but early contributions were also made by the Swedes, who were much interested in their own mineral deposits. Becher (1703) and Henkel (1725) attributed the origin of ore veins to the action on stony materials of vapors arising from "fermentation" in the bowels of the earth (Fig. 3-3). Henkel's idea of "transmutation" had in it the germ of modern replacement, discussed in Chapter 2. In 1749 Zimmermann also anticipated the idea of replacement when he ascribed the origin of lodes to the transformation of rocks into metallic minerals and veinstones by the action of solutions that entered through innumerable small rents and other openings in the rocks. Another early idea that escaped attention is that of von Oppel (1749) to whom belongs the credit for having shown that ore veins were mainly the filling of fault fissures and that their formation preceded the circulation of the ore-depositing solutions. Lehman, in 1753, explained that the upward branching of veins indicated deposition from exhalations and vapors that emanated from the earth's interior and rose through the crust, like sap rising from the roots into the branches of a tree. Such were the theories of origin of mineral deposits before 1756.

In that year an event occurred that profoundly affected all later ideas of the origin of mineral deposits. The famous Mining Academy of Freiberg, Germany, was founded in the midst of the varied mineral deposits of the Erzgebirge. Its famed teachers conducted excursions to study the nearby ores and enclosing rocks, and extensive mineral collections were made and studied in the academy. Scholars flocked to study under its famed masters. Here the newer geology flourished, and for over a century and a quarter its teachings in ore deposits influenced world thought.

Throughout most of the eighteenth century the prevalent view was that mineral deposits were formed by substitution of rock matter or by exhalations from the earth's interior that brought up the metals from depth and deposited them in fissures. In the latter part of the century, however, Lassius, Delius, and Gerhard explained ore solutions as diffused ascending water that dissolved scattered grains of metals from the rocks through which they passed. Thus, most of the germs of theories of mineral deposits had been considered, even if unscientifically, before the controversial time of Hutton and Werner.

When Abraham Gottlob Werner (Fig. 3-4) (1749-1807) became professor of mineralogy and geology at the Freiberg Mining Academy in 1775, he discarded the early theories of an interior source for the metals and became an insistent advocate of the theory that mineral veins were formed by descending percolating waters derived from the primeval universal ocean, from which, according to his views, not only sediments but all the igneous and metamorphic rocks were precipitated. These waters, he said, descended from above into fissures

FIGURE 3–4. Portrait of Werner. (Beck; F. D. Adams, *Birth and Development of the Geologic Sciences.*)

and there deposited the vein materials by chemical precipitation. His stimulating personality and fiery lectures caused students from all Europe to flock to him and to return as zealous disciples to defend his Neptunist theories. His thought then dominated in all matters relating to the origin of ores, particularly after the publication, in 1791, of his classic treatise on the origin of veins. Werner's enthusiastic lectures perhaps carried conviction to his hearers more by his personality and oratory than by the soundness of his facts. In one sense his leadership actually retarded the advancement of thought regarding mineral origin, but at the same time his dogmatic statements aroused vigorous opposition and thus stimulated wider consideration of other ideas, of which the most noteworthy was the Plutonist or Vulcanist school headed by Hutton.

Hutton, a quiet Scotsman, was a careful observer and investigator. In his *Theory of the Earth* (1788) he first defined the true origin of magmatic and metamorphic rocks, and his proponents waged bitter controversy with the Neptunist school. Hutton also applied his magmatic theory not only to rocks but to all mineral deposits. He claimed that ore minerals were not soluble in water but were igneous injections. In the words of his advocate, Playfair, "The materials which filled the mineral veins were melted by heat and forcibly injected into the clefts and fissures of the strata." Hutton's correct conclusions of the magmatic origin of igneous rocks made him go too far in attributing all veins to melted injections and discarding water as a possible agent. Werner's correct conclusions regarding the role of water in the formation of veins made him go too far in ascribing granites and basalts to water deposition. The Plutonists won out with respect to the rocks; the Neptunists prevailed with respect to the dominance of water in the formation of mineral veins, although, owing to the disrepute of Werner's explanation of rocks, his ideas regarding ores were overlooked for some time.

THE LATER CENTURIES

The Plutonist and Neptunist controversy quickened observations regarding the occurrence of minerals and brought forth many new data. Hutton's igneous injection theory of mineral veins was quickly forgotten. The early nineteenth century writers reverted to the pre-Werner ideas of mineral formation by exhalations from the interior of the earth but did recognize the significance of water in the formation of veins. Gradually, water of igneous derivation was considered to play the important role. This view received confirmation by the work of Necker (1832), who demonstrated the close relationship in many regions between igneous rocks and mineral deposits.

The connection of the mineral-forming solutions with magmas was given further emphasis by the French geologists Gay Lussac, Daubrée, Scheerer, and Elie de Beaumont; the brilliant Daubrée introduced early experimental methods into the study of mineral deposits in 1841. Scheerer in 1847, following Scrope, who had concluded that magmatic water played a part in the formation of igneous rocks, stated clearly that water was an important constituent of granite magmas and that mineral veins were formed by exudations of aqueous solutions from granite intrusions.

A few months later an important paper by Elie de Beaumont appeared, which Thomas Crook stated was perhaps the "most important and influential paper ever published on the theory of ore deposits." Elie de Beaumont might well be called the father of modern thought on the formation of mineral deposits. He was the first to show that most mineral deposits must be regarded as just one phase of igneous activity. He recognized that water vapor was an essential feature in volcanic activity and that mineral veins were formed as incrustations upon the walls of fissures from hot water of igneous origin. He distinguished such veins from dikes injected in a molten condition. He also cited occurrences of segregations of magnetite and chromite in basic igneous rocks, which he considered had crystallized out during the cooling of the magma. Elie de Beaumont thus recognized clearly the igneous affiliation of many types of mineral deposits—that some were formed as segregations during crystallization of the magma and that others were formed from hot aqueous emanations that escaped upward from the igneous intrusion. It is rather surprising that essentially similar views are held today, but for many years the clear statements of Elie de Beaumont were overlooked.

The conflicting opinions of these times were carefully weighed by von Cotta, whose excellent treatise on ore deposits appeared from Freiberg in 1859 and remained a standard textbook on mineral deposits for two decades. He affirmed judiciously the various theories of mineral genesis and correctly concluded that no one theory was applicable to all deposits. In his concluding observations he remarks (Prime's English translation, 1870), "Thus the formation of lodes shows itself to be . . . very manifold; and appears to have always stood in some connection with neighboring . . . eruptions of igneous rocks. The local reaction of the igneous-fluid interior of the earth created fissures, forced igneous-fluid masses into many of the same, caused gaseous emanations and sublimations in others; and in addition, during long periods of time impelled the circulation of heated water, which acted, dissolving at one point and again depositing the dissolved substances at another, dissolving new ones in their stead. The whole process is not confined to any particular geologic period, or any particular locality." Such statements might well be a part of a modern textbook. He shows clearly that most mineral deposits were the products of deep-seated igneous action. Von Cotta's balanced treatise exerted a profound influence upon the theories of the origin of mineral deposits.

A trend back toward the Hutton views of an igneous origin for mineral deposits may be noted in the writings of Fournet (1844, 1856) and

Belt (1861), who considered that veins and many other mineral deposits were the result of an igneous injection into fissures in a molten state, and were thus *ore magmas,* to use the term later proposed by Spurr (1923).

In the meantime these views of hydrothermal and igneous action in the formation of mineral deposits were partly obscured during the advocacy of another startling hypothesis of the origin of ores.

Bischoff in 1847 put forward a theory of lateral secretion by waters of meteoric origin. This theory was supported by some data regarding the dissemination of ore minerals in superficial rocks. Somewhat similar views were advanced in America by T. Sterry Hunt in 1861 and in England by J. A. Phillips in 1875, but they were advocated in an extreme form in Germany in 1882 by Sandberger, who sought to establish (1) that the gangue of ore veins corresponded with the wall rocks, and (2) that traces of the heavy metals occurred in the wall rocks. Meteoric waters were supposed to dissolve these substances from the wall rocks and deposit them in the fissures. Daubrée had concluded that hot waters were the most important agent in the formation of mineral deposits and that these waters, for the most part, were not magmatic but were meteoric waters that had become heated in depth and had risen again. These were the views of Hutton and Phillips, and likewise later of S. F. Emmons, who in 1886 explained the origin of the ore deposits of Leadville, Colorado, by surface waters that had leached metallic ingredients from the wall rocks. Similar views were elaborated in 1901 by C. R. Van Hise, who concluded that magmatic waters played a minor role and that most mineral deposits resulted from surface waters that had descended to depths where they became heated, dissolved metals from the rocks, and rose again to deposit their metallic content in fissures—a circulation resembling that of a hot-water heating system.

This gave rise to a series of animated controversies on the respective merits of the descensionist, ascensionist, and lateral-secretionist theories. The last lost ground rapidly under attacks by numerous geologists who contended that the mineral substances were not dissolved from the surrounding rocks by meteoric waters but instead were deposited there by ascending hot water that arose from deep-seated sources. Pošepný and De Launay, following the earlier ideas of Elie de Beaumont, sought a source for the mineral solutions in deep-seated eruptive rocks or still deeper sources in the barysphere. They were hot-water ascensionists, although Pošepný clearly distinguished types of mineral deposits formed by the circulation of meteoric waters. The vigorous discussions that followed in the beginning of the twentieth century directed attention once more to the close association between mineral deposits and volcanism, and the idea of an igneous origin for mineralizing solutions became rather generally accepted.

In the meantime J. H. L. Vogt of Norway had been laying the foundations of our present-day conceptions of the origin of mineral deposits. He delved into the source of ascending hot solutions. Careful field studies of deposits of magnetic iron, chromite, and nickel lead him to conclude that these substances were igneous injections of magmatic concentrations. He also concluded that the hot mineralizing waters were likewise derived by magmatic differentiation. Thus he linked, more clearly than his predecessors, the processes of the formation of igneous rocks and ore formation and recognized that magmatic differentiation gave rise to (1) ore segregations or injections, (2) mineralizing gases and vapors, and (3) hot mineralizing waters. These ideas have been elaborated, clarified, and experimentally verified by work of investigators, particularly in the United States and Germany. Such conceptions are the ones generally current today to explain the origin of most primary mineral deposits.

A later contribution to the theory of the origin of mineral deposits is the extreme magmatic view of J. E. Spurr (1923) who, following Thomas Belt (1861), postulates that many or most ore deposits have resulted from the injection and rapid freezing of highly concentrated magmatic residues, for which he proposed the terms "ore magmas" and "vein dikes." Spurr's ideas are not generally accepted. The latest concept on the origin of mineral deposits is that proposed by J. S. Brown in 1948, which he terms the metallurgical interpretation, an alternative to the hydrothermal theory. This hypothesis discards the action of hot water in ore formation and advocates that, during the former molten stage of the earth, the metallic minerals collected in deep zones according to their specific gravity and that they were later brought to the surface, the upper layers first and the lower layers later, in the form of vapors, from which the metals and minerals were deposited. This theory does not deny that water has taken some part in mineral formation but holds that the waters are not magmatic but are meteoric waters heated by the vapors. This theory is yet too new to be sufficiently proved or disproved, but it appears to have few adherents.

The advance in the study of mineral deposits has in large part centered around the development of theories of origin, and each new theory has quickened field observations directed to prove or disprove it. Consequently these various theories have given rise to a great mass of information regarding the character, distribution, and localization of mineral deposits. In this review of the development of ideas concerning the origin of mineral deposits it will be noted that most theories held today have been advanced in previous periods and that at present we are largely elaborating upon or amplifying earlier conceptions.

SELECTED REFERENCES ON HISTORY OF THEORIES OF ORIGIN

History of the Theory of Ore Deposits. Thomas Crook. Thos. Murby and Co., London, 1935. An interesting, brief, historical treatise.

Man and Metals. T. A. Rickard. McGraw-Hill, New York, 1932. Scattered parts of general historical interest.

The Romance of Mining. T. A. Rickard. The Macmillan Co., Toronto, 1944. Many references to historical development.

"Historical Notes on Gem Mining." S. H. Ball. *Econ. Geol.* 26:681–738, 1931. The most thorough treatment of this subject.

The Birth and Development of the Geologic Sciences. F. D. Adams. Williams and Wilkins, Baltimore, 1938. An exhaustive treatment dealing with pre-nineteenth century development of ideas; particularly Chap. 9, "The Origin of Metals and Their Ores."

Minerals in World Affairs. T. S. Lovering. Prentice-Hall, New York, 1943. Chap. 3, "Minerals in History Before the Industrial Era." Early use, distribution, and trade in minerals.

Bergwerk- und Probierbüchlein. Translation of U. R. von Kalbe's German edition of 1518 by A. G. Sisco and C. S. Smith. Am. Inst. Min. Met. Eng., New York, 1949. Translation of first known treatises on technology of minerals, metals, and veins.

De Re Metallica. Georgius Agricola; translated into English from the first Latin edition of 1556 by Herbert C. Hoover and Lou Henry Hoover, 1912. Reprinted by Dover Publications, New York, 1950. Best record of ancient knowledge of ores, mining, and metallurgy; illustrated with many quaint woodcuts.

CHAPTER FOUR

Relation of Mineral Deposits to Igneous Activity

IGNEOUS RELATIONS

The history of the development of ideas of the origin of mineral deposits, given briefly in Chapter 3, indicates a growing conviction that most mineral deposits are in some way related to igneous activity. It is now pertinent to enquire further into this relationship and the underlying reasons for it.

World-wide observations of countless mineral occurrences yield cumulative evidence that most metalliferous deposits occur in areas of igneous activity. Their habitat is dominantly in regions where deep-seated igneous intrusives have cut disturbed rocks and where erosion has bitten deeply. They are mostly in mountains or in the eroded roots of former mountains. The few exceptions to such areal associations, such as the zinc-lead ores of the Mississippi Valley region, are so notable as to have stirred much wonder and controversy. These deposits need not necessarily be considered as exceptions, however, since there may be underlying igneous bodies that have not yet been exposed. The widespread and repeated association cannot be coincidence. It must indicate kinship between some igneous activity and the associated ores; both must be offspring of the same parent. Moreover, the ores are later than the associated igneous rocks, suggesting further that they are after effects of the igneous intrusions and were generated as an end phase of the igneous activity.

Let us examine more closely the evidence of these associations to determine if they are assumed or actual relationships and to see what light they may throw on the origin of the ores.

IGNEOUS ROCKS AS ORES.   Some igneous rocks are themselves bodies of ore, such as some magnetic iron deposits, and bodies of chrome, titanium ore, and even diamonds. The ore minerals are igneous minerals and some are common constituents of igneous rocks. Since such bodies are merely igneous rocks, although of an unusual composition that happens to be of value to man, they indicate one direct relationship between magmas and mineral bodies.

RELATIONSHIP BETWEEN CERTAIN METALS AND SPECIFIC ROCKS.   Field observations disclose such a general and widespread association of certain minerals with specific rocks as to preclude coincidence. Such associations establish a definite relationship between the ore minerals and rocks, indicating thereby a magmatic source for both. For example, primary platinum deposits occur only in ultrabasic rocks, such as dunite or peridotite; diamonds in kimberlite; chromite in peridotite or serpentine; ilmenite and titaniferous magnetite in anorthosite or gabbro; nickel sulfides in norite or gabbro; tin in granites; and beryl in granite pegmatite. These associations are so universal as to render inescapable the conclusion that each of these rocks and its associated ore sprang from a particular magma source.

RELATION TO VOLCANOES.   Volcanoes afford opportunity to observe relationships between mineral deposits and magmas. Their throats may be or have been occupied by magma, and around the crater of a volcano one can see sublimations of sulfur, tin, cobalt, lead, zinc, copper, bismuth, and phosphorus. Around Vesuvius have been noted hematite, tenorite, boracic acid, and compounds of sodium, iron, and copper; other minerals have been noted elsewhere. Such minerals clearly establish a magmatic source.

FUMAROLES.   Even better than the throats of volcanoes, whose violence is not congenial for deposition of minerals, is the evidence of deposition around fumaroles. Those of the Valley of Ten Thousand Smokes in Alaska are particularly illuminating. Their gases arise from an underlying magma, and nothing but gases and vapors have coursed their conduits. Quantities of such minerals as magnetite, hematite, pyrite, and lead, zinc, copper, and molybdenum sulfides have been deposited around the orifices in the porous ash beds. There are also salts and some metals. They are being deposited today from the gaseous emanations, and these have come out of the magma and demonstrate not only a magmatic source for such substances but also that the substances are being carried in a gaseous phase.

HOT SPRINGS.   Hot springs associated with volcanism have long enticed the geologist and chemist because they contain many mineral substances and prove that such waters can dissolve, transport, and deposit the ingredients of minerals.

It should not be assumed that all hot springs consist of magmatic waters; most consist entirely of surface waters; others are mixtures of both. Temperature alone does not serve to distinguish magmatic from surface waters, because surface waters may be heated by hot rocks. However, if the waters contain such distinctively magmatic substances as boron, arsenic, chlorine, fluorine, and other substances characteristic of fumaroles, these indicate that the waters, in part at least, are of magmatic derivation. The hot springs of Yellowstone National Park and of Steamboat Springs, Nevada, have been shown to be of this type. Depositions of gold, silver, copper, lead, zinc, and other metals take place almost before our eyes at Steamboat Springs and indicate, therefore, a magmatic source for these substances. These and many other hot springs of similar type that are clustered around recent volcanic centers strongly indicate expiring magmatic activity.

It used to be thought, after the work there of Arnold Hague, that the hot waters of the Yellowstone Park hot springs and geysers were down-seeping surface waters that received their heat and mineral content from the underlying hot rocks. Later, Allen and Fenner made some boreholes and studied the waters, gases, and rocks. They found that silica and potash have been added and soda abstracted from the underlying rocks and that among the many dissolved constituents there were six elements (carbon, sulfur, boron, arsenic, chlorine, and fluorine) that are found only in traces in the underlying rhyolite and that are typical magmatic substances in the gaseous state in fumaroles; also, they found that superheated steam is present. This study led to the conclusion that the hot springs are caused by the heating of surface waters from the condensation in them of magmatic vapors that contributed some water and the magmatic constituents.

Most hot-spring waters are largely meteoric, with minor magmatic contributions. One cannot assume, however, that such hot-spring waters are characteristic of most of hydrothermal mineralizing solutions, although they may be characteristic of solutions that form some shallow-seated deposits within the zone of ground water. However, I believe that most mineral deposits are formed at depths far below the downward limit of ground water.

MINERAL ZONING. It has been observed in a great many mineral districts, such as Butte, Montana, that the ores are arranged zonally outward from a hot center, with minerals that have been formed at higher temperature nearest the igneous body or hot center and those at lower temperature farther out. Naturally, the temperature would have been highest nearest a molten mass and lowest farther out. However, one cannot assume that temperature change has been the only factor in bringing about deposition of minerals from solutions. But, if the minerals are definitely of igneous affiliation and are zonally arranged with respect to a hot center of mineralization, it does constitute a clear-cut indication of kinship with a magma.

IGNEOUS CONTACT DEPOSITS. There are certain distinctive types of mineral deposits that are found only at or close to the contacts of intrusive igneous rocks. These are referred to as contact-metasomatic or contact-metamorphic deposits and are discussed in Chapter 8. These deposits are characterized by an extraordinary assemblage of minerals, most of which are formed only at high temperatures and presumably by action of gases and vapors on calcium carbonate rocks. They are developed to the fullest next to the igneous body and disappear outward from it. Surrounding the igneous body is an aureole of intense high-temperature rock alteration in which the intruded rocks are baked, hardened, and recrystallized and in which many new minerals are formed by reactions between emanations from the magma and the constituents of the intruded rocks. Many of the common and rare ore minerals are deposited in this altered zone. Their localization near the igneous intrusive and their high temperature of formation are conclusive of their relationship to the intrusive igneous body.

COINCIDENCE OF AGE OF DEPOSITS AND OF INTRUSIONS. In some areas as, for example, at Bisbee, Arizona, porphyry intrusives cross stratified rocks of a certain age and are earlier than overlying stratified rocks. Also, nearby, ore deposits occur within the same stratified rocks and are earlier than the same overlying rocks. Where several such associations are repeated mere coincidence is eliminated, and kinship between intrusions and ore deposits is indicated.

SPATIAL DISTRIBUTION AND INTRUSIVES. In many regions it is found that there is a spatial distribution of specific types of ore deposits with respect to one or several igneous intrusives. For example, in Cornwall, England, in the region of the tin veins that attracted the Romans, the veins are clustered around or in intrusives of granite. This arrangement is repeated throughout the entire Cornish tin belt and definitely indicates a close relationship between the granite intrusions and metallization.

THE "PORPHYRY COPPERS." The "porphyry copper" type of ore deposits (see Chap. 9) are localized in the tops or margins of porphyry intrusives or in the immediately adjacent rocks. This general association and position implies that the intrusives and the copper deposits that nestle in their crackled borders arose through the same channelways from the same magma reservoir beneath. This implication is even more conclusive where a contact-metamorphic aureole of baked and altered rocks surrounds the intrusive and contains similar copper minerals, as at Ely, Nevada.

METALLOGENETIC PROVINCES AND EPOCHS. A broad coincidence, both in time and place, of mineral deposits with widespread igneous activity and mountain building has long been noted. Thus, regions of thick accumulations of sedimentary rocks that have become folded and broken by mountain-building forces and intruded by deep-seated igneous bodies are commonly the sites of mineral deposits.

In each of the continents there are certain time periods that recur in different geologic ages where there were widespread igneous intrusions and accompanying formation of mineral deposits. These are called *metallogenetic epochs* (see Chap. 14). In North America the Precambrian geologic era is characterized by widespread deposits of iron, copper, zinc, gold, silver, and other metals. Wide metallization at this time also occurred in Scandinavia, Africa, Australia, and parts of South America. Another notable metallogenetic epoch was in the late Mesozoic period, when great igneous intrusions occurred in the Rocky Mountain states and along the Pacific Coast from Alaska to Mexico. A wealth of minerals was formed at the same time. Another outstanding metallogenetic period occurred in the early Tertiary epoch, when great porphyry intrusions throughout the North and South American cordillera gave rise to prolific deposits of copper, silver, gold, molybdenum, and other metals.

Specific regions characterized by relatively abundant mineralization, dominantly of ore type, are referred to as *metallogenetic provinces*. In Arizona and adjacent regions is a great copper province where copper metallization occurred in four different geologic periods, extending over a period of some 600 to 800 million years. Each period was also accompanied by igneous intrusions with which the copper deposits are spatially connected. Also in the Precambrian Canadian shield, stretching 2,000 miles from Great Slave Lake to eastern Quebec, is a great gold province, where the deposits are likewise associated with igneous activity. A mercury province embraces the coastal ranges of North America from British Columbia down into Mexico.

Many other examples exist of metallogenetic epochs and provinces where igneous activity is accompanied by associated ores. Some of these are treated in Chapter 14.

MAGMAS AND CRYSTALLIZATION

The masses of molten silicate solutions within the earth that are called magmas are really high-temperature solutions of silicates, silica, oxides, and dissolved volatile substances that follow the laws of ordinary chemical solutions. Their temperatures range from 600° C up to 1,250° C, and their composition is variable. The volatiles consist chiefly of water, carbon dioxide, sulfur, chlorine, fluorine, and boron, and these are largely expelled when the magma consolidates.

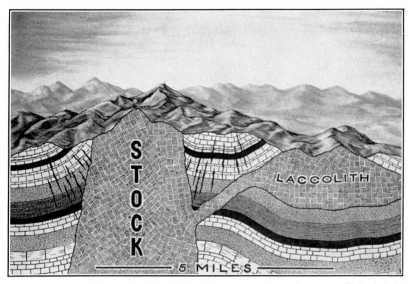

FIGURE 4-1. Diagram of a laccolith and accompanying dikes and sills. (K. K. Landes, *Physical Geology and Man.*)

Magmas are temporary features within the earth's crust. They form in pockets or reservoirs, are forced upwards to places of lesser pressure, and then consolidate. It is known that the temperature of the earth increases with depth, since such increases can be measured in mines and deep boreholes. It is also well known that most materials expand when they melt and if expansion is prevented melting is also prevented. Deep in the earth the temperature must be above the melting point of rocks, else we would not have volcanoes. But the enormous pressure of the weight of overlying rocks would prevent expansion and melting. However, should the pressure be relieved by buckling, faulting, or removal of the overlying load by erosion, then melting would occur and magma result. As for the source of the heat, the known radioactivity of the rocks is more than sufficient to yield enough accumulated heat, which escapes only slowly, to melt rocks. The general occurrence of igneous intrusions and lava outpourings along great zones of weakness in the earth's crust where crustal disturbances and deep erosion have occurred indicates that such places favor magma formation and its upward movement.

Liquid magma, like any liquid, tends to move to the place of least pressure. Hence, its movement is dominantly upward. It may be

FIGURE 4–2. Diagram of a batholith with associated dikes and sills. (K. K. Landes, *Physical Geology and Man*.)

squeezed upward to force overlying rocks apart, or it may eat its way upward by melting and prying off blocks of overlying rock that sink into the liquid, giving rise to *magmatic stoping*. The magma may wedge apart weak rocks or it may be squirted into fractures or along bedding planes of sedimentary rocks, giving rise to dikes, sills, or laccoliths (Fig. 4–1). It may be expelled to the surface, giving rise to volcanism, or it may solidify at great depth, forming huge intrusive bodies such as stocks and batholiths (Fig. 4–2), which are revealed at the earth's surface only by erosion of the cover rocks.

CRYSTALLIZATION. The familiar incident of water losing its heat and crystallizing to the mineral, ice, also extends to magmas, except that they are much more complex solutions of several substances, some of

Magmas and Crystallization

which are less soluble than others. As a magma slowly loses its heat, it also begins to solidify or crystallize when the temperature drops below the saturation point of its constituents. Following the laws of solutions, the least soluble substances are the first to form little crystals of minerals, which continue to grow larger as cooling proceeds, until all of that substance is crystallized. The most soluble substances crystallize last and mesh in around the earlier-formed crystals. The volatiles accumulate toward the end of solidification and of course do not crystallize but form the mother liquors or mineralizing solutions. Thus, there is a definite order of crystallization of the individual minerals, and in an acidic magma the basic minerals crystallize first and the acidic or siliceous ones crystallize last.

MAGMATIC DIFFERENTIATION. It has been noted that in many igneous bodies the rocks in one part differ in composition from those of another part, with gradational types between them. Also, it may be observed that the same volcano has yielded successive outflows of unlike lavas. One cannot conceive that a single conduit has tapped different reservoirs of unlike magmas, and the conclusion is irresistible that originally homogeneous magmas have split up into unlike fractions, a process called *magmatic differentiation*. This process is extremely important in supplying source materials for the formation of mineral deposits, and a proper understanding of how it operates is essential. By means of differentiation, rocks of different kinds, pegmatites, bodies of iron ore, chrome or nickel, vapors and gases, and mineralizing solutions are formed and constitute *magmatic products of differentiation*. The individual products may crystallize or be tapped off at different times.

Most magmas have probably undergone some differentiation, and this may have occurred in one or more stages. In an original magma chamber deep within the earth, consolidation of the differentiates may take place at the site where they were formed; or some of the differentiates may be squeezed out before consolidation to form intrusive or extruded igneous bodies and then crystallize directly as already differentiated bodies; or such separate fractions may undergo further differentiation in their last resting place and upon consolidation give rise to smaller, distinct but related bodies of rocks. Before complete consolidation, some of these later differentiates may again be tapped off to form satellitic dikes or other minor bodies. Last of all, the aqueous solutions that have accumulated with the residual fractions and have gradually become more and more concentrated with the metals originally present in the magmas are tapped off and supply the constituents of mineral deposits.

During the progress of differentiation some metallic substances, such as ore minerals of iron, chromium, nickel, and copper, may be collected into fractions concentrated in these substances and consolidate to form *magmatic ore deposits*. Thus, differentiation yields not only different rock types from a common magma but also magmatic mineral deposits of great commercial importance and in part the mineralizing solutions that form the vast majority of mineral deposits.

Although the fact of differentiation is clearly recognized the *modus operandi* is imperfectly understood, and several ideas have been advanced. Most geologists regard crystallization as the prime cause of magmatic differentiation, but it seems probable that other processes may also enter in, particularly where bodies of minerals are concerned. The end result of differentiation is to give rise to separate fractions of different composition. We will now consider briefly how differentiation proceeds.

DIFFERENTIATION BY CRYSTALLIZATION. At the commencement of crystallization in a magma the first solids to form are generally the heavy basic minerals such as magnetite. These, being heavier than the lighter liquid magma, will sink in it, and the liquid portion of the magma thus becomes depleted in the constituents that enter into the composition of such early-formed crystals. As crystallization progresses all of the early-forming minerals will have settled out, and a liquid of different composition, or a residual magma, will remain above. Should the early-formed minerals be lighter than the remaining liquid, as in certain basic magmas, they would rise to the top, and a similar separation of liquid and solid would occur. This process is called gravitational crystallization differentiation.

With further crystallization, still other minerals will similarly separate and collect, again depriving the remaining magma of those constituents and leaving a residual magma of still different composition. Finally, the residual magma, enriched in the minerals latest to crystallize, and generally more silicic, will consolidate to form rocks high in silica, such as granite, of quite different composition from the earlier crystallizations. By this process, early crystals of magnetite or chromite may accumulate to form economic deposits of iron and chrome ores.

In some basic magmas the order of crystallization is reversed; basic feldspars may be the first to crystallize and iron becomes concentrated in the residual magma. This iron-rich residuum may also be concentrated sufficiently to form economic iron ore deposits.

A variation of crystallization differentiation occurs where a magma has partly crystallized and consists of a mush of crystals with residual liquid in the interstices. If subjected to pressure by earth movements or other means, the crystals may be mashed together and the residual liquid squeezed out, like water out of a sponge, thus forming fractions of different composition. Field studies by Osborne and others in the Adirondack Mountains have led to the conclusion that many of the magmatic iron deposits there represent just such a squeezed-out residual liquid, highly concentrated in iron. Also, in such a mush of crystals, heavy iron-rich residual liquid may drain out and gravitate to the bottom of the magma chamber and give rise to a gravitative liquid accumulation.

Another variation of crystallization differentiation is that advocated by Balk, in which a mush of crystals and interstitial liquid is forced through a restricted opening. The suspended crystals because of their greater surface friction will be relatively retarded. This causes them to group into clusters and coalesce, like a log jam in a river; the liquid in the interstices will then drain out and thus bring about separation of liquid and solid.

LIQUID SEPARATION DIFFERENTIATION. Most of the constituents of a magma are soluble or miscible in each other, and thus no separation, as liquids, of unlike fractions takes place. However, Vogt has shown that in some basic magmas nickel-copper sulfides may separate as immiscible liquid droplets, like oil in water, and percolate downward to collect as a liquid at the bottom of the magma chamber. Such a melt will remain liquid after the silicate minerals have crystallized and then consolidate in place or be squirted by pressure into adjacent fractured rocks. The accumulation of liquid melt is analogous to the accumulation of molten copper beneath the slag in a copper smelter.

OTHER PROCESSES OF DIFFERENTIATION. Other theories about processes of differentiation that have been advanced are: (1) convection and diffusion, by which crystallization is presumed to commence at the cooler margins of the intrusive where early-forming minerals will deposit and by which new supplies of constituents would be moved to the site of deposition, thus adding continuous new supplies and continuous deposition; marginal deposits of magmatic iron ore have been considered to have been formed in this manner; and (2) assimilation of foreign materials by fusion of engulfed blocks of overlying rocks, which lighter melt would rise to the top of the magma chamber and there solidify.

PRODUCTS OF DIFFERENTIATION

From the foregoing it will be seen that by differentiation a magma tends to separate into (1) immiscible sulfide melts that settle and form magmatic sulfide bodies, (2) silicate mineral crystals that form igneous rocks, (3) crystals of metallic oxides that may form metallic mineral deposits, (4) gases and vapors, and (5) residual liquids. Each of these magmatic products is formed at different stages of the differentiation, some early and some late, and the residual liquids are the last end products.

IGNEOUS ROCKS. In the course of differentiation, the aggregation of early-formed silicate crystals in general gives rise to the more basic igneous rocks, which are mostly heavy and dark and are composed of silicate minerals high in iron and magnesia, such as pyroxenite or gabbro. Igneous rocks of intermediate composition arise from the next stage of differentiation after most of the early-crystallizing constituents are removed from the magma. The residual magma may then consist chiefly of the constituents of feldspar, quartz, mica, and a few other rock-mineral constituents, along with most of the water that was originally present, gases, other vapors, and any metals present. The crystallization of this stage gives rise to granitic rocks. The last siliceous minerals to crystallize are feldspar and quartz. The final residuum is chiefly water, enriched in gases, and metals that may originally have been present. Some common igneous rocks and rock-forming minerals are given in Chapter 2.

PEGMATITES. Before final crystallization, the late residual liquid of a magma from which granite, for example, crystallizes is made up chiefly of low-melting silicates and considerable water, along with other low-melting compounds and volatiles, and a relative concentration of many of the substances that enter into mineral deposits of igneous derivation. In addition to water, the volatile substances consist of compounds of boron, chlorine, fluorine, sulfur, phosphorus, and other rarer elements. They affect crystallization by decreasing the viscosity of the magma and by lowering the freezing point of minerals. This is an igneo-aqueous stage—a transition between a strictly igneous stage and a hot-water or hydrothermal stage, tending more to the igneous—and is spoken of as the *pegmatite stage*. If the cooled outer portion of the igneous body is fractured, some early pegmatite liquid

Products of Differentiation 41

may be tapped off and give rise upon crystallization to simple *pegmatite dikes* (Fig. 4–3), which are varieties of igneous rocks. They

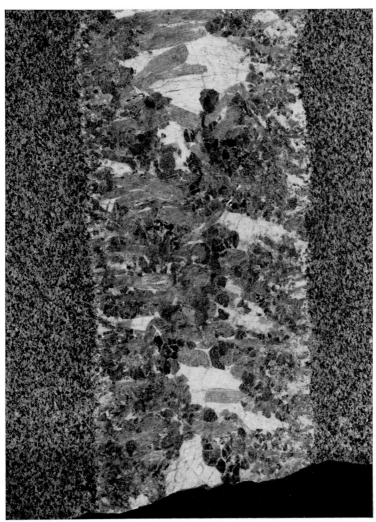

FIGURE 4–3. Small pegmatite dike cutting granite. White, feldspar; gray, spodumene crystals (lithium mineral); dark, quartz. Long Creek Church, North Carolina. (F. L. Hess, *Econ. Geol.*)

consist, chiefly, of large crystals of feldspar, quartz, and mica, with minor rare minerals. Later withdrawals of a more aqueous stage in the differentiation yield pegmatites, commonly characterized by small openings that may be lined with rare crystals and containing compounds of uranium, tin, tungsten, titanium, beryllium, boron, fluorine, chlorine, other elements, and some of the minerals of ore deposits; sulfides are uncommon. These pegmatites are closer to the aqueous than to the igneous end and have often been referred to as pegmatite veins rather than dikes. They are a transition from a magmatic melt to a hydrothermal vein. Some indeed are valuable mineral deposits and have yielded tin, tungsten, tantalum, columbium, uranium, lithium, and other elements of commerce.

As the withdrawn pegmatite liquid cools and crystallizes, the earliest-formed minerals will be those of latest crystallization in granite, such as potassium feldspar, quartz, and mica. These abstractions enrich the residual pegmatite liquid in water, soda, lithium, and other substances. In this liquid the potash feldspar becomes unstable and may be replaced by soda feldspar (albite). Similarly, other replacements of the original pegmatite minerals occur, both by their own residual liquid and by other hot waters that carry in additional substances. In consequence, many of the minerals of pegmatite dikes are of secondary replacement origin. Transitions have been noted from igneo-aqueous pegmatite dikes to pegmatitic quartz and then to hydrothermal quartz veins carrying ore minerals.

In addition to the metals mentioned above the pegmatites also yield many nonmetallic industrial minerals such as feldspar, quartz, mica, lithium minerals, beryl, and gemstones.

GASES, VAPORS, AND LIQUIDS. The water, gases, and other vapors of the magma gradually accumulate as residual fluids during the progress of differentiation and may be tapped off during late stages of crystallization. Upon final solidification all the fluids are of course expelled except some water that may be incorporated in the crystallization of water-containing minerals. These fluids, by themselves or mingled with meteoric waters, may constitute the *mineralizing solutions* or *hydrothermal solutions* that have given rise to so many kinds of metallic mineral deposits. They will be considered in more detail in Chapter 5.

SUMMARY OF RELATION TO IGNEOUS ACTIVITY

The magma is the source of magmatic mineral deposits and most of the materials of those mineral deposits that are formed later than the enclosing rocks. Through crystallization and differentiation some constituents of the magma collect as crystal aggregates or molten liquids within the magma chamber to form magmatic deposits of oxide or sulfide minerals.

Progressive differentiation of silicic magmas results in a residual liquid that becomes more and more enriched in the volatiles and other constituents, including metals, that formerly were sparsely dispersed throughout the magma. Before final consolidation some or all of this liquid may be tapped off to form pegmatites, whose crystallization leaves an aqueous residue that may give rise to one type of hot-water mineral deposit, possibly barren or low-grade veins. Continued crystallization of the magma yields a final aqueous residue enriched in volatiles, metals, and other constituents. When solidification is complete or nearly complete, this mobile, nonviscous, aqueous residue may be expelled as the hydrothermal mineralizing fluids that later deposit their load to form various types of mineral deposits. It must not be assumed, however, that every magma contains sufficient metals to yield commercial ore deposits, else deposits would not be such uncommon features.

If the consolidating magma is rather close to the earth's surface, the volatiles under slight pressure will boil off and may reach the surface as acid gases, giving rise to fumaroles or gas vents, such as those of the Valley of Ten Thousand Smokes in Alaska. They transport and dissipate the metals they formerly collected, and no appreciable metallization results.

At less shallow depths boiling may occur, giving rise to an acid distillate that rises through fractured rocks, and reactions with the wall rocks will cause deposition of some ore minerals. The vapors condense to form hydrothermal liquids from which minerals may be deposited, or they may mingle with meteoric waters and emerge as hot springs.

SELECTED REFERENCES

Rocks and Rock Minerals. L. V. Pirsson and A. Knopf. John Wiley & Sons, New York, 1947. Pp. 138–148. Brief description of magmatic differentiation and of the origin of igneous rocks and pegmatites.

"Geyser Basins and Igneous Emanations." E. T. Allen. *Econ. Geol.* 30:1–13, 1935. Derivation of geyser waters from meteoric and magmatic contributions.

Hot Springs of the Yellowstone National Park. E. T. Allen and A. L. Day. Carnegie Institution, Washington, D. C., 1935. A comprehensive discussion of spring waters and their origin.

The Evolution of the Igneous Rocks. N. L. Bowen. Princeton Univ. Press, Princeton, N. J., 1928. A thoughtful discussion of differentiation and evolution of igneous rocks.

Igneous Rocks and the Depths of the Earth. R. A. Daly. McGraw-Hill, New York, 1933. Origin and differentiation of igneous rocks.

Ore Deposits of the Western United States—Lindgren Volume. Am. Inst. Min. Met. Eng., New York, 1933. Chaps. 3, 4, 5, 7, and 8.

"Nature of Ore-Forming Fluid." L. C. Graton. *Econ. Geol.* 35:Supp. to No. 2, 1940. Lengthy discussion of origin, state, composition, and migration of fluids; ore fluids leave magma as alkaline liquids.

Eruptive Rocks. S. J. Shand. John Wiley & Sons, New York, 1947. Chaps. 1, 3–7, 10–12. Magmas—temperature and pressure conditions, fugitive constituents, crystallization; late-magmatic and post-magmatic reactions; the genesis of pegmatites; and the relation between eruptive rocks and ore deposits. Good bibliographies.

"On Hydrothermal Differentiation." Heinrich Neumann. *Econ. Geol.* 43:77–83, 1948. Hydrothermal solutions are formed as a separate phase later than the pegmatitic stage.

Economic Mineral Deposits. 2nd Edit. Alan M. Bateman. John Wiley & Sons, New York, 1950. Chapters 4, 6.

Internal Structure of Granitic Pegmatites. E. N. Cameron, R. H. Jahns, A. H. McNair, and L. R. Page. Econ. Geol. Pub. Co. Mono. 2, 1949. A profusely illustrated monographic work on the occurrence, character, mineralogy, and origin of pegmatites.

CHAPTER FIVE

The Primary Mineralizing Solutions

In the preceding chapter it was mentioned that the crystallization of a magma leaves behind excess water, which may form a mineralizing solution. It is pertinent now to consider the character of such solutions and the steps by which they are accumulated, become charged with mineral content, and are expelled.

For simplicity, consider a magma granitic to dioritic in composition, for such are the most common parental magmas of the hydrothermal types of deposits. Its original water content would perhaps be about 1 percent. As crystallization proceeds the earlier-formed crystals exclude water, which would become relatively concentrated in the remaining molten magma. Similarly, the sparse content of metals and volatiles originally dispersed throughout the magma would likewise become concentrated in this remaining fluid. At this stage there would exist a solid fraction composed of the rock minerals already formed and a fluid phase consisting of the remaining magma and contained water. Thereafter, further crystallization would result in still greater enrichment in water. Indeed, the accumulation of water, as pointed out by H. Neumann, may even proceed to a point where the content becomes greater than the remaining molten magma can dissolve. In other words, a saturation point may be reached, which the investigations of R. W. Goranson indicate will be about 10 percent water under favorable pressure conditions. Under such conditions there would then be three coexisting phases, namely, the solid phase of rock minerals, a molten phase of magma saturated with water, and a gas phase of excess water.

It is not yet established definitely that a separate water phase exists at this stage of crystallization. If it does and were forced out into fractures it might give rise to mineralizing solutions, which, however, might not yet be sufficiently enriched in metals to form workable mineral deposits. For one ore vein that is rich enough to be profitably mined there are hundreds that are low grade or barren. Such solutions, drawn off rather early in later phases of crystallization, may be the explanation of these numerous barren and low-grade veins. Most mining districts contain both barren and workable veins, and the difference between them may depend upon the stage of crystallization at which the solutions are tapped off from the magma chamber.

Just before final crystallization of the magma the concentration of water is almost at the maximum and so is the enrichment of metals and volatiles, provided of course that the water has not been lost to the surrounding country rock as rapidly as it became excluded by rock solidification. This then is almost the optimum stage for the generation of the mineralizing fluids, enriched as they would be to nearly the maximum with the soluble metallic and nonmetallic compounds that may have been present in the magma. Of course, all magmas may not have been originally charged with the ore metals, so that the final mineralizing fluids may be barren or almost barren. Just why some mineralizing solutions contain gold, others copper, and others practically no metals is an unsolved problem.

At this late stage of crystallization the residual fluids, perhaps enriched in metallic and nonmetallic compounds, are being circumscribed by the ever-crystallizing rock minerals. The space left for them is encroached upon by growing crystals. The fluids become more and more confined and consequently are subjected to greater and greater pressure. Here then lies an expelling force for the residual fluids, perhaps greater than the confining strength of the enclosing solidified walls of the igneous body and sufficient to impel them upward towards places of less pressure. The final act of crystallization will expel under great pressure (as Goranson has pointed out) all excess water that does not enter into the rock minerals. Thus are generated the "mother liquors" or mineralizing fluids that later may give rise to mineral deposits.

It will be noted that simple crystallization by itself can generate mineralizing solutions, and in Chapter 4 it is pointed out that magmatic differentiation also yields a final product of water with dissolved constituents. Since magmatic differentiation is brought about chiefly through progressive crystallization, the same end accumulation of mineralizing fluids results. Differentiation where it has occurred probably favors a more thorough concentration of the excess water and its dissolved constituents. All magmas, however, do not undergo differentiation, particularly those that lose their heat quickly and so crystallize rapidly.

GASES AND VAPORS

Enormous quantities of gases and vapors escape from underlying magma into the atmosphere during volcanism. In the great eruption of Vesuvius of 1906, vividly described by eyewitness F. A. Perret, a first phase of four days of lava flows was followed by a second phase of paroxysmal gas ejection during which gases and vapors under stupendous pressure were shot up as a continuous blast at high velocity to a height of 8 miles. So immense was the force of this colossal gas jet that it enlarged the crater to a width of 2,200 feet. This giant gas column continued for a day and then tapered off into great expulsions of gas and ash, the whole eruption lasting 18 days. This gigantic gaseous eruption is a realistic manifestation of the amount and power of the gaseous emanations of a magma. Fortunately, few magmas reach the surface.

The gaseous constituents originally were in solution in the magma, and as long as the magma remained liquid and the pressure unchanged, they remained in solution. Thus, in a closed magma chamber the volatiles would be excluded during crystallization and tend to rise to the highest part of the magma chamber, there to accumulate as a part of the residual magma. A rise of the magma to levels of lesser pressure would tend to bring about a separation of the gaseous phase. In traversing the magma both gases and liquids collect metals and other minor constituents present, and these are concentrated in the residual mother liquors that become the mineralizing fluids. If the magma chamber becomes ruptured some of these constituents may escape before final consolidation occurs. Generally this happens.

The character of the gaseous emanations from deep-seated magmas cannot be determined directly. It can only be inferred from the constituents they carried, the nature of the reactions with country rock, and observations on present-day emissions and laboratory experiments. Obviously, it is not feasible to be close enough to collect gaseous emanations during violent eruptions like that of Vesuvius (Fig. 5–1). However, the numerous fumaroles or gaseous jets that cluster around areas of recent volcanism or unexposed igneous activity do afford precise information regarding the constituents of the escaping emanations. Fumaroles commonly accompany volcanism and are found in areas of active and recently active volcanism such as those around Vesuvius, Etna, and Stromboli, and volcanic areas in Japan, the Pacific islands, Mexico, Central America, the United States, and Alaska. They also occur in dormant periods between eruptions of volcanoes, and in some places they are the only surface manifestation of shallow underlying igneous activity. The Mediterranean fumaroles have been extensively studied, but the new group that came into being in the Valley of Ten Thousand Smokes in Alaska, where Katmai volcano erupted in 1912 and blew dust clouds over vast areas of North America, afforded unusual opportunity for scientific investigation shortly after their birth and during a succeeding check-up period.

FIGURE 5–1. Ancient woodcut of man killed by poisonous exhalations given off from rich mineral veins during a thunderstorm. (From *The Historia of Olaus Magnus*, A.D. 1567.)

The gaseous emissions consist chiefly of steam, which may constitute from 90 to 99 percent of the whole. Other included gases are carbon dioxide, carbon monoxide, hydrogen sulfide, hydrogen, hydrochloric acid, hydrofluoric acid, methane, sulfur, and others. Those that give off sulfurous gases are called solfataras. Other contained substances are chlorine, fluorine, boron, nitrogen, and many volatile and other compounds, such as the chlorides of hydrogen and of sodium, potassium, calcium, and ammonium and other metals, and sulfides, fluorides, tellurides, and arsenides.

The fumaroles of the Valley of Ten Thousands Smokes yield information not only about the gases but also about the minerals deposited from the gases in the volcanic ashes around the vents. These vents have always been gas conduits, and no liquid solution has ever been active in them. Gas temperatures up to 645° C were measured.

The gaseous substances, of course, consist chiefly of water, but some estimation of their acid content may be gained from Zies' estimate that

Character of Hydrothermal Solutions

the fumaroles emit annually 1,250,000 tons of hydrochloric acid and 200,000 tons of hydrofluoric acid. Also, there were deposited sodium chloride, potassium chloride, sulfur, selenium, tellurium, and evidences of arsenic, bismuth, thorium, and beryllium. Great quantities were found of such metallic ore minerals as magnetite, hematite, pyrite, molybdenite, galena, zinc blende, and covellite. Zies found that the magnetite also contained lead, copper, zinc, tin, molybdenum, nickel, cobalt, and manganese. The various metals have been made volatile by the presence of chlorides and fluorides in the magma and were thus able to leave the magma in the volatile state. By reaction with other volatile compounds of the fumaroles metallic minerals were deposited in the porous volcanic ashes. Hematite, an ore mineral of iron, is commonly formed in this manner. During one eruption of Vesuvius a fissure 3 feet wide was filled with hematite in a few days. The Katmai and other fumaroles demonstrate the ability of gases and vapors to collect, transport, and deposit minerals that are the common constituents of mineral deposits. Such observations, in fact, early led to the concept of a genetic relationship between igneous activity and the origin of mineral deposits.

Gaseous action that gives rise to the deposition of minerals is called *pneumatolytic*, and some geologists consider that gaseous emanations are "the agents best adapted to effect primary separation of metals from the magma and transport them outward into the surrounding rocks" (Fenner). Fenner is inclined to assign to gaseous emanations the major role in such work. As will be elaborated in Chapter 8, certain types of ore deposits, such as *contact-metasomatic deposits*, have long been considered as due to gaseous action.

When the gaseous emanations cool and condense, their content of mineral matter is transferred from the gaseous to the more familiar watery environment, and hydrothermal liquids result. The ultimate fate of all magmatic gaseous emanations that travel far through cool rocks is to change over to liquid solutions.

CHARACTER OF THE HYDROTHERMAL SOLUTIONS

It will be seen that there are two concepts as to the preliminary character of the hydrothermal solutions: one, that they left the magma chamber as hot liquids and remained as liquids; the other, that they left the magma chamber as gaseous emanations and later condensed to hot liquids. Both are probably correct, but differences of opinion exist among geologists. In either case, the later phase is the same, i.e., a liquid hydrothermal phase, and the early phase is dependent upon the physical conditions in the magma chamber at the time of exit. If the external pressure is less than the vapor pressure of the residual liquid and the enclosing rock is somewhat permeable, a vapor phase would probably be present. If the reverse pressure conditions exist, a liquid phase would exist. It is demonstrated that a vapor phase can and does collect, transport, and deposit metals. Likewise it is known that hot liquids transport and deposit metals. Those who contend for a gaseous phase think that this is the only means by which escape from a magma chamber might be effected. Those who contend for a continuous liquid phase think a gas phase impotent to transport the hundreds of millions of tons of varied ore minerals that make up single deposits. I am not troubled by the transporting ability of a gaseous phase, nor by the ability of liquid solutions to escape from the magma chamber as such, under conditions of high confining pressure and the high impelling pressure provided in the later stages of magma solidification. In fact, some ore deposits give the suggestion of there having been two phases of metallization, an early phase of rock alteration and deposition of silica, potash, and sulfur without the useful metals, and a later phase or phases of ore deposition.

The composition of the mineralizing solutions can be arrived at only by inference since it is not susceptible of direct observation as are gaseous emanations. It is commonly thought that hot mineral springs are the direct emission of magmatic liquids comparable to fumaroles and are a true sample of hydrothermal solutions. There are hot springs *and* hot springs. Some have no magmatic affiliations whatever; others have only in part. Hot springs of original magmatic derivation may now represent in part condensations of magmatic vapors, or they may have undergone great change in composition in passage through the rocks, or they may have lost by deposition most of their original dissolved content, and they unquestionably represent intermingling with near-surface meteoric waters. Thus hot springs cannot be true samples of original hydrothermal mineralizing solutions. Some of them do, however, give some clue as to the constituents carried out from the magma chamber.

HOT SPRING WATERS. Hot springs originate in different ways. Some have no connection with volcanism whatever, such as those at Hot Springs, Arkansas, Saratoga in New York, Vichy in France, or Bath in England. These are surface waters carried in deep aquifers warmed by the internal heat of the earth. They are, of course, mineral springs because they carry mineral matter dissolved out of the underlying rocks.

Hot springs of volcanic regions are clustered about areas of active or quiescent volcanism, and their distribution bespeaks volcanic association. However, it is now considered that the waters of such thermal springs are made up chiefly of ground water heated by magmatic steam. The waters issue as springs because the weight of a downward-moving column of heavier cold surface water forces a lighter column of hot water to rise upward and issue as a hot spring.

Formerly it was thought, after the work of Arnold Hague in Yellowstone National Park, that the hot springs were entirely surface water that sank and became heated by contact with hot lava rock below and rose again, like the circulation of a hot-water heating system in a house. A careful restudy was made by Allen, Day, and Fenner. They sank deep boreholes and analyzed the waters, contained gases, and the alteration of the underlying rock. They found: (1) that the waters carried silica, sodium, potassium, calcium, chlorine, borates, fluorides, arsenates, pyrite, and hematite; (2) that soda predominated over potash; (3) that silica and potash have been added to the underlying rock and that soda has been abstracted; (4) that hydrogen sulfide and carbon dioxide are abundant; (5) that the dissolved substances include six elements of which there are only traces in the underlying lava rock, namely, carbon, sulfur, boron, arsenic, chlorine, and fluorine, and that these six elements are typically magmatic substances found in the gaseous state in fumaroles; and (6) that superheated steam was present—steam at a temperature higher than that necessary to yield steam under the existing pressure. The superheat must have come directly from the magma. These findings led to the conclusion that hot springs are caused by the heating of ground water of surface origin by condensation in it of magmatic vapors and gases that contributed some water and magmatic constituents. About 10 to 15 percent of the water is considered magmatic. Similar findings were arrived at for the thermal waters of Lassen Volcanic National Park in California and Steamboat Springs in Nevada. This conclusion is strengthened by the fact that in some volcanic regions hot springs change to fumaroles during the dry seasons when the surface waters become less. Thus, volcanic hot springs consist largely of surface waters, and therefore, as previously pointed out, cannot be considered representative of hydrothermal solutions. They can, however, be studied as diluted hydrothermal solutions, such as may have given rise to some types of shallow-seated deposits.

It should be pointed out, however, that most primary hydrothermal mineral deposits have been formed at depths far below the reach of ground water, where magmatic waters dominate.

COMPOSITION OF HOT-SPRING WATERS. In addition to the common hot-spring calcium carbonate and silica deposited as travertine and geyserite, hot springs also carry and yield many of the minerals that go to make up economic mineral deposits. The two chief types of waters are sodium chloride-silicate and sodium carbonate. The famed hot springs of Carlsbad in Czechoslovakia carry:

Antimony	Copper	Phosphorus
Arsenic	Fluorine	Selenium
Boron	Gold	Strontium
Bromine	Iodine	Tin
Chromium	Lithium	Titanium
Cobalt	Nickel	Zinc

The contents of the Yellowstone Park waters have previously been mentioned. The geyser waters of New Zealand are somewhat similar and carry also gold, silver, and mercury. Steamboat Springs in Nevada have been intensively investigated of late. There almost before our eyes occur depositions of gold, silver, mercury, antimony, arsenic, iron, copper, lead, molybdenum, and chromium, as well as other metals and nonmetals. The waters are dominantly alkaline and moreover are superheated, indicating a magmatic source. Many other examples could be cited of hot-spring waters that carry metals commonly found in economic mineral deposits along with elements of strictly magmatic derivation.

The hot-spring waters, although only in small part of igneous derivation, do show that they are capable of transporting and depositing the various minerals found in hydrothermal mineral deposits.

COMPOSITION OF THE MINERALIZING SOLUTIONS. Deep within the earth where the hot mineralizing fluids are generated there is no dilution by surface waters, as they do not reach these depths. The residual liquid of a differentiating magma must be alkaline, according to N. L. Bowen, owing to the interaction between water and silicates. Also, the waters of thermal springs that transport and deposit metals at the surface are alkaline, and some of the metals have been carried as alkaline sulfides. Consequently, it is considered that the mineralizing solutions also are alkaline. This is supported by examination of liquid inclusions commonly formed in some of the minerals of ore deposits, and these are also alkaline.

Character of Hydrothermal Solutions

The gas phase that escapes from a magma chamber, however, must be acid because it will contain an excess of hydrochloric, hydrofluoric, sulfuric, carbonic, and other volatile acids. Bowen states that there should also be present hydrogen, oxygen, chlorine, sulfur, fluorine, boron, potassium, sodium, iron, titanium, and aluminum, along with tin, copper, lead, zinc, gold, silver, and other metals. We have already seen that direct observations of the hot gases escaping from the underlying magma at Katmai, Alaska, corroborate this. They are acid and yield vast quantities of hydrochloric and other acids.

The hot solutions formed from the condensation of the gaseous emanations would readily attack most minerals except quartz and alkali feldspars. Carbonate rocks and basic igneous and metamorphic rocks would be particularly susceptible to attack. Certain oxides of the heavy metals would be readily transported in the acid solutions. However, reactions of the solutions with the rock minerals would render the solutions alkaline, and the dissolved oxides of the heavy metals would then be deposited. Bowen states, "At greater distances the [acid] solutions will be alkaline, for this is the ultimate fate of hot waters in contact with silicate and other rock minerals." Thus, whether the solutions were initiated as liquids or became liquid by condensation from gases and vapors, they would be alkaline carriers of metals and minerals.

The attack of acid solutions upon silicates would make available an ample supply of free silica in colloidal form and capable of being transported as such. This may well be a source of much of the quartz that is an ever-present gangue mineral of ore deposits. Many geologists now hold that silica has been carried in colloidal form and precipitated as a gel that crystallizes to chalcedony or quartz. Lindgren and others have also advanced the idea that the metals may have been transported largely as colloidal solutions, in which colloidal silica may have helped keep the metals in a dispersed state. The simplicity of transportation and particularly of deposition renders the idea intriguing.

The character of the hydrothermal solutions must be interpreted largely by inference and by analogy with hot springs and laboratory experiments. Their action is visible in the minerals deposited from them and in the nature of the alteration of the containing rocks.

TEMPERATURE AND PRESSURE CONDITIONS. The temperatures of hydrothermal solutions cannot be measured directly like hot springs or fumaroles; they have to be inferred. As the term hydrothermal implies, they are hot waters that probably range in temperature from 500° C down to 50° C. The temperatures are arrived at by minerals, called *geologic thermometers,* that yield information as to the temperatures of their formation. Some minerals yield only the maximum temperature of formation, others minimum, others exact temperature, and still others only approximate temperatures. The more important methods of geologic thermometry are:[1]

Melting points, e.g., stibnite (antimony ore) at 546° C or bismuth at 271° C. There are many others. This gives only the maximum temperatures at which they can crystallize, and they may form at any temperature below the melting point.

Dissociation, e.g., pyrite dissociates into pyrrhotite and sulfur above 615° C at atmospheric pressure.

Inversion points, e.g., high-quartz changes over to low-quartz at 573° C or the mineral argentite changes over to acanthite at 179° C. The inversion is accompanied by a change in the internal crystal arrangement or symmetry of the atoms. There are many pairs of inversion minerals.

Exsolution, e.g., a solid solution of chalcopyrite and bornite unmixes at 475° C into the individual minerals. This is indicated by their textures. There are many minerals whose exsolution point has been determined.

Recrystallization, e.g., native copper at 450° C.

Liquid inclusions in minerals. Heating of minerals containing cavities partly filled by liquid causes the liquid to expand, and the point at which the liquid just fills the cavity indicates the temperature of formation. For example, sphalerite liquid inclusions from the Tri-State zinc district indicate a temperature of formation of 115° to 135° C.

Changes in physical properties, e.g., amethyst loses its color between 240° C and 260° C.

Even changes in temperature during crystal growth have been indicated. An interesting case is a fluorite crystal from New Mexico, described by W. S. Twenhofel, whose center started growth at 202° C and whose outside finished growth at 150° C. The above illustrations are only samples of the many geologic thermometers that disclose the temperature of the mineralizing solutions at the time the minerals were formed.

[1] This subject is treated more fully in *Economic Mineral Deposits,* 2nd Edit., 1950, by Alan M. Bateman, on pp. 36–43, and lists of many thermometers are given on pp. 41–43.

Evidence of Hydrothermal Solutions

The pressure on the solutions increases with depth. Generally the higher the temperature present, the higher the pressure.

EVIDENCE OF HYDROTHERMAL SOLUTIONS

Up to this point discussion of the mineralizing solutions has hinged around their igneous source, how they are generated, how they obtain their metallic load, and their character and composition. It has been mentioned several times that they give rise to hydrothermal ore deposits, but little evidence has been adduced to show that mineral deposits actually result from these solutions. It is pertinent now to consider this evidence further.

The concept of the formation of hydrothermal mineral deposits from water solutions has evolved slowly as supporting evidence has accumulated. Most of the support, however, has come by analogy with hot-spring deposition and laboratory experiments. It was only natural that early investigators turned to familiar chemical reactions in solutions to account for the natural depositions of ore minerals in the rocks. If copper or lead minerals can be deposited from solutions in the laboratory, then, ran the argument, does not this demonstrate that the natural minerals have been formed similarly? It does not, of course, demonstrate that natural deposition *has* taken place in a similar manner. It only shows that natural deposition *could* take place in this manner, following general chemical laws. However, it has now been shown that most of the minerals of hydrothermal deposits have been produced artificially and, as pointed out previously, natural deposition of antimony, mercury, silver, copper, lead, and others can almost actually be seen taking place at Steamboat Springs, Nevada. Since these thermal waters are in part magmatic, it seems a fairly logical deduction that similar deposition has taken place from hydrothermal solutions to form similar minerals in vein deposits.

Again, when one examines crystals of minerals from the hydrothermal deposits small, or even large, liquid inclusions can be seen within the crystal. These certainly indicate, if they fill the cavity, that the crystal grew in a liquid environment. Moreover, drops of the liquids have been extracted and analyzed and found to consist largely of water with magmatic constituents in solution. As mentioned under geologic thermometers, these inclusions also can be made to reveal the temperature of the solution at the time the crystal was formed. Such tests have been made on innumerable crystals, and temperatures that range from a few tens of degrees to several hundreds of degrees have been found. Most work on liquid inclusions has been done on gangue minerals that are transparent in thin slices under the microscope. Until recently, similar tests could not be made on opaque metallic minerals, but recently H. S. Scott determined that the opaque minerals are also filled with minute inclusions. He heated mineral fragments until the bubbles expanded and cracked the fragments with microscopic explosions, which can be recorded by radio amplification. Thus, he demonstrated that the metallic minerals also contain liquid (and gas) inclusions and could record the temperature of explosion, which fixes the upper limit of the temperature of formation.

The studies of inclusions have shown remarkably consistent results for the temperatures of formation for minerals of the same age. More enlightening, however, is the fact that earlier minerals of a single deposit show higher temperatures of formation than the later minerals of the same deposit, indicating a drop in temperature of the solutions during the period of building up the deposit.

Another bit of evidence is *crustification* (Fig. 9–2) in ore veins. This consists of a layer or crust of one mineral on either wall of an open fissure, with crystal faces developed toward the open space. On top of layer 1 there may be a succeeding crust of a different mineral, and layers 3, 4, 5, and perhaps more may similarly be deposited successively on layer 2. This is interpreted to indicate slow growth of minerals from a solution that coursed along the fissure. Parts of rock fragments may likewise be coated with successive layers of different minerals, each layer following precisely around irregularities of the fragment. This gives rise to familiar *cockade ore* (Fig. 9–16).

Again, one often finds in vein and cave deposits beautiful crystals of minerals with well-developed faces upon which are perched delicate clusters of other perfect crystals of mineral 2 that delight the heart of the mineralogist. These are interpreted to represent undisturbed quiet growth by slow accretion from a solution that fills the open space in which they grow. In places the later crystals are perched on the earlier crystals entirely in one direction, indicating flow of the solutions and the up-current side. The shape of some crystals, growing most against the current and least on the lee side, also indicates direction of flow of the solutions that supplied them with the constituents for their growth. This growth arrangement in flowing solutions has been demonstrated in the laboratory.

Another indication of deposition from liquid solutions is that of low-temperature replacement of one mineral by another. A solid cube of pyrite, for example, may be almost completely replaced by covellite, starting from outside and progressing toward the center. Some medium is necessary for the transfer inwards of copper ions and outwards of iron and sulfur ions; water solution is the obvious medium at temperatures lower than that of vapors and gases. Such changes actually take place today in the weathering and oxidation of ore deposits, where downward-trickling cold surface waters containing copper sulfate react with iron sulfide (pyrite) and replace it by copper sulfide (covellite). Moreover, this process, known as *secondary (supergene) sulfide enrichment,* has operated extensively to enrich the upper parts of copper or silver deposits the world over.

The above features, among others, supply abundant evidence as to the availability, ability, and actuality of mineralizing solutions in giving rise to the minerals of hydrothermal mineral deposits.

SELECTED REFERENCES ON THE MINERALIZING SOLUTIONS

"On Hydrothermal Differentiation." Heinrich Neumann. *Econ. Geol.* 43:77–83, 1948. Hydrothermal solutions are formed as a separate phase later than the pegmatitic stage.

Ore Deposits of the Western United States—Lindgren Volume. Am. Inst. Min. Met. Eng., New York, 1933. Chap 3, Part I. "Pneumatolytic Processes in the Formation of Minerals and Ores," by C. N. Fenner. A scholarly discussion of magma end phases and of the dominant role played by gaseous emanations in transporting minerals from a magma. Part II. "Magmatic Differentiation Briefly Told," by N. L. Bowen. A concise summary of differentiation and pegmatite evolution. Part III. "Differentiation as a Source of Vein and Ore-Forming Materials," by C. S. Ross. A thesis that hydrothermal mineralizers leave the magma as liquids. Part IV. "Pegmatites," by W. T. Schaller. Compositional characteristics and origin.

Economic Mineral Deposits. 2nd Edit. Alan M. Bateman. John Wiley & Sons, New York, 1950. Pp. 35–44, Geologic thermometers.

"Transport and Deposition of Nonsulphide Vein Minerals. III. Phase Relations at the Pegmatitic Stage." F. G. Smith. *Econ. Geol.* 43:535–546, 1948. Experiments show that a granitic rest-magma during crystallization would separate into two liquid phases between 290° C and 550° C, forming two immiscible residual solutions, one constituting the hydrothermal solutions.

"Decrepitation Method Applied to Minerals with Fluid Inclusions." H. S. Scott. *Econ. Geol.* 43:637–654, 1948. Opaque minerals with fluid inclusions are heated until they burst, thus giving audible record of temperature of formation.

"The Pyrite Geo-thermometer." F. G. Smith. *Econ. Geol.* 42:515–523, 1947. Determination of temperature of formation of pyrite by measuring thermoelectric potential.

"Liquid Inclusions in Geologic Thermometry." Earl Ingerson. *Am. Mineral.* 32:375–388, 1947. Discussion of determining temperatures of crystallization by liquid inclusions and pressure corrections; quartz determinations.

"Geyser Basins and Igneous Emanations." E. T. Allen. *Econ. Geol.* 30:1–15, 1935. Derivation of geyser waters from meteoric and magmatic contributions.

Hot Springs of the Yellowstone National Park. E. T. Allen and A. L. Day. Carnegie Institution, Washington, D. C., 1935. A comprehensive discussion of spring waters and their origin.

"Preliminary Geochemical Results at Steamboat Springs, Nevada." W. W. Brannock, Philip F. Fix, Vincent P. Gianella, and Donald E. White. *Am. Geophys. Union Trans.* 29:211–226, 1948.

CHAPTER SIX

The Processes of Formation of Mineral Deposits

Chapters 4 and 5 give some indication of the part played by igneous activity and the products that spring from the magma. They convey that magmatic mineral deposits arise through magmatic differentiation, and hydrothermal deposits arise through the activity of hydrothermal solutions, the offspring of magmas. The prominence given these processes in the forepart of this book might give the impression that they are the all-important processes in the formation of mineral deposits. This is far from the truth. Igneous activity does play an extremely important role in supplying directly or indirectly the ingredients of most mineral deposits, but the vast bulk of the minerals of commerce spring from the operation of processes other than magmatic or hydrothermal. A sedimentary bed of iron ore, a bed of salt formed by evaporation of sea water, or a body of aluminum ore resulting from rock weathering, for example, all result from processes other than magmatic, although igneous activity supplies most of the ingredients to start with. The realization that many different processes operate singly or jointly to yield different kinds of mineral deposits has led to a clearer understanding of the origin of mineral deposits.

The formation of mineral deposits is complex. There are many types of deposits, generally containing several ore and gangue minerals. No two are alike; they differ in minerals, texture, content, shape, size, and other features. They are formed by diverse processes, and more than one process may enter into the formation of an individual deposit. Their origin is often puzzling or baffling, and the geologist, like a detective, must search for clues with which to piece the whole together. Often the picture cannot be fitted together until a single last clue is deciphered. The modes of formation of minerals by themselves, discussed in Chapter 2, although many, are but part of the larger processes that build up mineral deposits in their entirety.

Among the agencies that enter into the formation of mineral deposits, water in one form or another plays a dominant role. It may be in the form of water vapor, hot magmatic water, cold meteoric water, ocean, lake, or river water, ground water, or atmospheric water.

Temperature also plays an important part in all deposits formed by magmatic agencies, but many processes operate at surface temperatures and pressures, as do sedimentation, evaporation, and weathering. Other factors that enter into the genesis of mineral deposits are various chemical actions, catalytic agents, the action of life and sunlight, the action of streams, currents, and waves, the effect of bodies of standing water or of ground water, and metamorphic effects.

Different processes may operate to produce different and distinct types of deposits of the same metal. For example, deposits of iron ore may be formed by magmatic differentiation, by the action of hot gases and vapors, by hydrothermal solutions, by processes of sedimentation, and by weathering processes. The distinction among them is vital, particularly from an economic viewpoint. To illustrate, an iron ore deposit formed by magmatic processes may be expected only in association with certain kinds of igneous rocks and be relatively restricted in extent. One formed by weathering processes will carry many impurities and will be restricted in areal extent and be of shallow depth. Those formed by sedimentary processes, on the other hand, would have wide areal extent, would be characterized by general uniformity of thickness and quality of ore, and would be involved in any folding affecting the enclosing rocks. To a certain extent, conversely, a single process may give rise to similar types of deposits of many different metals. Thus, iron ore, gold, copper, or fluorspar may be formed from the action of hydrothermal solutions, or common salt, or potassium salts by evaporation of sea water.

Again, a single deposit may originate through the action of several different processes acting simultaneously or at different times. For example, sedimentation may give rise to a bed of low-grade iron ore too poor to be economic; weathering may enrich it to commercial grade, and later metamorphism may enrich it still further.

It is desirable, therefore, to distinguish the various processes that operate to give rise to mineral deposits and to arrange them in logical grouping. Then, broad generalizations, deductions, and conclusions can be applied to the various kinds of deposits and metals that originate from that process. Such a grouping follows:

Processes of Formation of Deposits

Process	Resulting Deposits
1. Magmatic concentration	Magmatic deposits
2. Sublimation	Sublimates
3. Contact metasomatism	Contact-metasomatic deposits
4. Hydrothermal action	Fillings of open cavities
	Replacement of rock masses
5. Sedimentation	Sedimentary beds
	Evaporation beds (evaporites)
	Coal beds, oil pools
6. Weathering	Residual concentrations
	Placers
	Oxidized and supergene sulfide deposits
7. Metamorphism	Metamorphic deposits
8. Hydrology	Ground-water supplies, brines, caliche deposits

Recognition of a mineral-forming process and the deposits resulting therefrom permit deductions applicable to ore finding and appraisal. The mode of occurrence of one representative given in the table above may indicate favorable places to search for similar, unknown deposits. The characteristics of one well-explored representative can be applied to slightly explored members of the same group, and conclusions can be drawn regarding the probable size, continuity in depth, and probable tenor, based upon experience gained from other deposits formed by the same process. Past experience with deposits formed by one process can be applied quickly to similar new deposits under consideration. When, for instance, the iron ore deposits of the Birmingham steel district of Alabama were determined to be sedimentary beds, and not shallow deposits of weathering origin, a potentially productive area could be quickly outlined by sparse exploration; dependable assumptions of continuity of ore between distant drill holes could be drawn; possible tonnages could be readily calculated; and there was confidence in risking venture capital for exploitation; thus a great steel industry arose. While the weathering origin of the iron ore held sway, confidence was absent, exploration was desultory, and development was retarded.

The eight groups of mineral-forming processes enumerated above will be considered in detail in successive chapters. Each of the constituents that make up economic mineral deposits have for the most part been derived either directly or more remotely from magmatic sources. This applies not only to hard minerals but to the constituents of sedimentary rocks, to the salts and waters of the sea, and even to the constituents of plant and animal life that gave rise to the mineral fuels of coal and oil. The magmas are the parents; the mineral deposits, the offspring. Some are immediate descendants that originated during or soon after the consolidation of the parent magma. Others, immature at birth, have been built up during one or more periods of growth by some of the processes listed above to yield subsequently fully developed deposits. A placer gold deposit, for example, may have been originally a lean magmatic offspring which first suffered disintegration and later became more concentrated in a sedimentary bed; later reworking by stream action may have reconcentrated this diffused sedimentary product into an economic placer deposit eagerly sought by the wily prospector. Still other lean magmatic offspring may not have matured until they were exposed to the beneficent influence of the atmosphere (also of magmatic derivation), whose leaching action has removed the metals from undesired companions and concentrated them into cleaner, fatter, richer deposits of supergene origin, but of magmatic ancestry nevertheless.

SUMMARY OF ORIGIN

Before we proceed to a more detailed consideration of the various mineral-forming processes outlined in the preceding table, a brief summary of the outstanding features of each is presented in order that their relation to each other may be better understood.

The initial stages of magma crystallization are attended by separation of metallic oxides, sulfides, and native elements. Some of these crystallize and may become segregated into mineral deposits early in the magmatic stage; others solidify later than the rock minerals and either accumulate at the original site or are injected into the cooled intrusive or into the surrounding rocks to form late magmatic deposits.

During the crystallization of the magma, the abstraction of the early-crystallizing rock minerals leaves a residual fluid that gradually becomes enriched in gases and vapors. These contain and in part consist of compounds of the metals and other substances that formerly were sparsely distributed throughout the magma. If the enclosing pressure is relieved they escape into the enclosing wall rocks, and under favorable circumstances contact or pneumatolytic metamorphism takes place, giving rise to contact-metasomatic deposits. Should some of these gases and vapors reach toward the earth's surface, active volcanism or fumaroles may be formed from which sublimates may be deposited, forming sublimate deposits of sulfur or other substances.

During the late phases of the solidification of the magma, silicic residual liquors may be tapped off to form pegmatite dikes, many of which carry valued economic minerals.

Summary of Origin 63

Just before and at the time of final consolidation of the magma, the residual aqueous solutions in the form of liquids or gases, or both, are expelled toward places of less pressure. These constitute the hot mineralizing fluids or hydrothermal solutions and may be charged with the constituents of mineral deposits. In their upward journey the solutions seek lines of easiest flow and follow fissures, cracks, joints, bedding planes, permeable beds, rock pores, and other openings in the rocks. Their mineral content may be precipitated from solution and fill these openings to form cavity-filled deposits, or they may replace the rock matter to form replacement deposits. The deposition may occur at high, medium, and low temperatures and pressures, imparting in each situation different characteristics to the resulting mineral deposits. These offspring of the magma may be deposited in sufficient concentrations to constitute economic mineral deposits, or they may be sparsely deposited, requiring other and later means of concentration to enrich them into commercially workable deposits.

Secondary processes may then operate to form still different kinds of mineral deposits. Weathering and land wastage yield vast quantities of mineral substances that may be transported mechanically or in solution to basins of sedimentation, there to give rise to sedimentary beds of industrial rocks and minerals and also to bedded deposits of metallic ores such as those of iron and manganese. Organic processes also enter into the growth of plants from which sedimentary beds of coal may result, or into mud deposits high in organic matter from which petroleum may be generated. Soluble substances released during land denudation may become concentrated in basins of water such as parts of the sea, lakes, or ground water. The salts of the oceans are considered to have accumulated through geologic time mostly from the dissolved substances discharged by the rivers of the lands. Moisture from the oceans precipitates upon the lands as rain and snow. This run-off water dissolves substances from the surface mantle and passes through rivulets to master streams that enter the oceans or enclosed inland basins, there discharging their dissolved load. Each year, the Mississippi River alone discharges some 135 million tons of dissolved salts. Under conditions of aridity, cut-off bodies of saline waters undergo evaporation, which causes precipitation of the salines to form bedded deposits of common salt, gypsum, potash, and other substances, called *evaporites*.

Under the continuous and relentless attack of weathering, rocks and enclosed mineral deposits succumb to chemical decomposition and mechanical disintegration. Under warm humid conditions, decay and accompanying solution take place. The soluble parts of rocks are removed; the insoluble may accumulate as residues. If the residues happen to be of value, commercial deposits may result. Many great deposits of the ores of aluminum (bauxite), iron, manganese, and numerous other substances such as china clays result from this gradual accumulation, in place, of insoluble residues through the removal of undesired substances. If the products of decay and disintegration are washed away, as they generally are, the strong sorting power of water (and also of wind) enters in. Heavy substances are separated from lighter ones, and those that are hard or durable as well as heavy may be concentrated into workable placer deposits of gold, tin, diamonds, and other minerals.

The processes of weathering also affect pre-existing ore deposits, as well as rocks, and cause them to undergo profound changes. Some metals are dissolved and carried down below in cold dilute solutions and are there precipitated, supplying additional metal to that already existing and thereby bringing about supergene enrichment. Uneconomic material may thus be made into economic ore. Many great copper and silver deposits owe their economic life to this process. In other cases ores are oxidized to new oxygen-bearing compounds, generally with removal of waste minerals, and are reprecipitated in the zone of oxidation, resulting in an increase in metal content of the ore. There are many great deposits of copper, lead, zinc, and other metals composed of such oxidized ores.

Many minerals that are stable under surface or near surface conditions of formation become unstable under metamorphism involving elevated temperatures and pressures. Common clay, born on the earth's surface, undergoes transformation by metamorphism into slate or a rock composed of white mica. Similarly, economic minerals may be formed the first time through metamorphosis from something else. Thus, graphite may be formed from coal or other carbonaceous materials; talc from magnesian minerals; asbestos from serpentine; abrasive garnet from rock minerals; or the refractories, kyanite and sillimanite, from rocks high in aluminum and silica.

Finally, economic deposits of mineral substances are formed through the action of ground water. Ground water itself, in vast regions of the world, is one of the most vital mineral supplies, essential for the support of life. Also, it dissolves common salt, forming natural brines. Other salts dissolved in ground water may be deposited at the surface as caliche deposits, such as those of natural nitrates or of lime.

From this brief review it will be seen that economic mineral deposits are many and varied and that numerous and even unrelated geologic processes must be invoked to explain their origin. In the deciphering of the genesis of mineral deposits and in the applications of the deductions drawn, it is clear that more than one hypothesis of origin must be considered. Each of the various processes of genesis will now be considered.

SELECTED REFERENCES ON THE PROCESSES OF FORMATION OF MINERAL DEPOSITS

Economic Mineral Deposits. 2nd Edit. Alan M. Bateman. John Wiley & Sons, New York, 1950. Chap. 4, "Magmas, Rocks, and Mineral Deposits," 45–67; Chap. 8, "Classifications of Mineral Deposits," 355–365.

"Classification of Magmatic Mineral Deposits." Alan M. Bateman. *Econ. Geol.* 37:1–15, 1942. A new classification of magmatic deposits and processes.

CHAPTER SEVEN

Magmatic Processes

Several of the minerals of igneous rocks are desired for economic purposes. Normally such minerals are sparsely scattered throughout the igneous body in amounts too small to be utilized commercially, and some process of concentration is essential to collect the economic minerals into workable deposits. Such concentrations occur during the magma stage and are termed, therefore, *magmatic processes*. There are some exceptions, of course, in which concentration is not necessary and simple crystallization will yield an economic product, as with diamonds, which are so sparsely scattered that there may be only a single crystal in a hundred tons of rock.

Magmatic processes give rise to many valuable mineral deposits. Some of these are large and rich; some, such as diamonds, constitute an important part of world production; others, such as chromite, are the sole source of chromium; many, however, are of greater scientific than commercial interest. Representatives of magmatic concentration are many and widespread, although there are relatively few types. Their mineralogy is simple, and the minerals yielded are few. Although individual deposits are of great value, magmatic deposits are as a whole greatly overshadowed in importance by deposits resulting from other processes to be considered farther on.

Magmatic deposits are characterized by their close relationship with intermediate and deep-seated intrusive igneous rocks and particularly with basic ones. Actually, they themselves are igneous rocks whose composition happens to be of particular value to man. They constitute either the whole igneous mass or a part of it, or lie as adjacent bodies. They are magmatic products that crystallize from magmas.

MODE OF FORMATION

Magmatic mineral deposits result from simple crystallization or from concentration by differentiation of intrusive igneous masses. The proc-

Mode of Formation

esses and stages of crystallization and differentiation have already been described in Chapter 4. It used to be thought that all magmatic deposits were formed by the earliest crystallization of economic minerals from a magma. These minerals were supposed to have become collected together and aggregated into mineral deposits during the early stages of magma crystallization. Also, it was formerly considered that this early segregation took place by the forming of individual crystals and that the deposits were the early segregation of such crystals. This is still held for many magmatic deposits, but it later came to be realized that in sulfides of metals little molten droplets of metallic minerals would separate out like the droplets of lead in a smelting furnace and settle to the bottom and collect there as a melt. In this instance the separation from the magma and the segregation do not take place by the forming of crystals that settle out of a molten magma. Subsequently it was learned that in many magmatic deposits the ore minerals actually crystallized later than the associated rock minerals and, therefore, could not have been formed by early crystallization and settling. For such deposits the former simple concept had to be modified. It is now realized that there have to be several modes of formation of magmatic deposits and that the deposits originate during different periods of magma crystallization. In some instances the ore minerals crystallized early and remained as separated crystals scattered through the parent rock, as in the example of diamonds or corundum. In others, they crystallized early and settled through the molten magma to collect into deposits. In still other instances the ore constituents became concentrated in a late residual magma, and the ore minerals were the last to crystallize. Before final solidification the ore magma may have been collected by draining or may have been pressed out into the outer solidified rock shell. Again, metal sulfides may have separated out as molten droplets and settled through the still fluid magma to collect on the bottom of the magma chamber.

Thus, we must recognize in magmatic concentration an early stage and a late stage, with suitable subdivisions under each. The classification in Table 6 is proposed.

Such a classification of magmatic processes and magmatic deposits should eliminate much of the confusion that has hitherto existed with respect to this controversial group of deposits. A clear distinction between the processes that operate early and those that operate late in the magmatic period will serve to clarify understanding of the disposition of magmatic mineral deposits and their relation to the parent and

TABLE 6

CLASSIFICATION OF MAGMATIC PROCESSES AND DEPOSITS

Processes	Examples
I. Early Magmatic:	
A. Disseminated crystallization without concentration	Diamond pipes; some corundum deposits
B. Crystallization–segregation	Some chrome deposits; platinum
II. Late Magmatic:	
A. Residual liquid accumulation and/or injection	Titanomagnetite and some chrome, titanium, and platinum deposits
B. Liquid separation and accumulation	Some nickel–copper deposits
C. Pegmatites	

intruded rocks. Each of these magmatic desposits will now be considered in more detail.

EARLY MAGMATIC PROCESSES AND DEPOSITS

Early magmatic deposits arise from processes that operate at the beginning of the magma period. They involve crystallization with or without accompanying differentiation and concentration into ore segregations. The ore minerals crystallize earlier than the rock minerals and remain within, and are an integral part of, the parent igneous rock. Their position is thus distinctly localized. This conclusion leads to the practical consideration that search for early magmatic deposits will be fruitless outside of the parent intrusive igneous body and must be restricted to the igneous body itself. The deposits are formed by (*a*) simple crystallization without concentration of the early crystals, and (*b*) crystallization and concentration of the early-formed crystals before solidification of the rest of the magma.

DISSEMINATED CRYSTALLIZATION. Crystallization of a deep-seated magma *in situ* will yield a granular igneous rock with interlocking grains in which early-formed crystals are disseminated throughout it. If such crystals are valuable and sufficiently abundant to permit the rock to be mined, the result is an igneous rock that is a magmatic mineral deposit, which may prove of economic importance. The whole rock mass, or just a richer part of it, may constitute the deposit.

More specifically, the diamond pipes of South Africa illustrate this type of early magmatic deposit (Fig. 7–1). Plugs of a basic igneous rock, called kimberlite, have been pushed up from the depths into overlying rocks and there cooled slowly enough that they crystallized into granular rocks. The rising magma brought up fragments and boulders of underlying deep-seated rocks and also crystals of diamonds, whose crystallization had been initiated in a deeper magma chamber. Upon solidification of the magma the diamond crystals be-

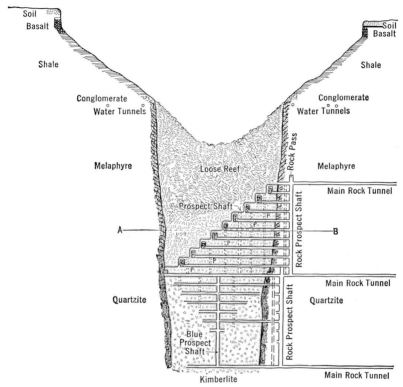

FIGURE 7–1. Plan view and cross section of the Kimberley diamond pipe to a depth of 3,000 feet. Kimberley, South Africa. (After du Toit, *15th Intern. Geol. Cong.*)

came sparsely disseminated throughout the igneous plugs, and the whole kimberlite pipe is the mineral deposit. There was no concentration of diamonds before the final magma consolidation. Disseminated crystals of corundum, the next hardest mineral to the diamond and also used as an abrasive, in nepheline syenite in Ontario is another example of an early magmatic dissemination. Disseminated crystallization also yields noncommercial bodies of other desired minerals, which, however, may undergo secondary concentration by weathering processes to yield economic residual or placer deposits.

In this manner placer deposits of platinum, ilmenite, monazite, and some gemstones have been formed.

The deposits of this class have the shape of the intrusive, which may be a dike, pipe, or small stocklike mass. Their size is large.

CRYSTALLIZATION-SEGREGATION. In certain magmas crystallization generally commences first with any accessory ore minerals that may

FIGURE 7–2. Chromite bands (black) in stratiform igneous rock (anorthosite). Bushveld Igneous Complex, Dwars River, Transvaal. Note convergence of some chromite bands and inclusion of anorthosite in chromite. (Photo by author.)

be present, and is followed by a sequence of minerals from basic to less basic to silicic. The solid crystals that form are mostly heavier than the molten magma from which they spring and naturally sink in the melt, growing the while during their downward progress. It has been noted that where layers of magma have been intruded between horizontally bedded rocks, forming sills, the lower parts of such sills, as in the Palisades of the Hudson, are enriched in the heavier basic minerals that settled down during early crystallization. Similar settling has likewise taken place on a larger scale within large bodies of magma. If the magma contained, to start with, constituents that would grow to form the mineral chromite (chromium, iron, oxygen), for example, the chromite crystals would form early, settle downward, and aggre-

Early Processes and Deposits 71

gate into a body consisting largely of chromite. An early magmatic segregation of chromite would result (Fig. 7-4, E3). If the segregation were composed largely of chromite and were of sufficient size, the result might be a commercial deposit of chrome ore, a substance desired to make strong, tough, stainless steels, chrome alloys, or refractory bricks. This is the manner in which most chrome deposits are considered to have originated. They are thus a phase of, and a result of, crystallization differentiation.

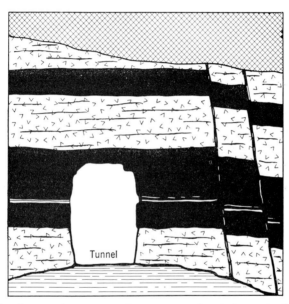

FIGURE 7-3. Layers of magmatic chromite in rocks of the Bushveld Complex, Lydenburg, South Africa. (After H. Schneiderhöhn, *Min. Boden i Sud. Afrika.*)

Such deposits obviously must be localized within the igneous rock that solidified from the magma and could not lie outside the parent body, although, as will be shown, such an occurrence is possible in late magmatic deposits. The deposits formed by crystallization differentiation may be irregular in shape or position, or they may be in stratiform layers as are the nearly horizontally emplaced igneous masses shown in Fig. 7-2. It should not be assumed that all chromite deposits have been formed in this manner, because some geologists have found evidence that there are at least some late magmatic chromite deposits. Several South African geologists believe that the extensive chromite bands in the Bushveld Igneous Complex of the Transvaal have been formed in the manner described above. The chromite layers there have remarkable continuity, over scores of miles, and surprising uniformity in thickness. Their gentle inclination, as shown in Fig. 7–3, is presumed to be due to later tilting.

Few early magmatic deposits other than chromite are known. Some platinum deposits in the Merensky Reef of South Africa are thin stratiform layers of great extent and are considered to have been formed by gravitative settlement of early-formed crystals. Many titanomagnetite and magnetite deposits were formerly thought to have originated in this manner, but later investigations disclose that the magnetite crystallized later than the rock minerals and, therefore, could not have been formed by the gravitative settlement of early-formed crystals.

Few new deposits have been added to this early magmatic group, and many have been removed.

LATE MAGMATIC PROCESSES AND DEPOSITS

Late magmatic deposits consist of igneous ore minerals that have crystallized from a residual magma toward the close of the magmatic period. They are the consolidated fractions of the magma left after crystallization of early-formed rock minerals and in this respect differ from concentrations of early-formed ore minerals described under the previous heading. Consequently, the ore minerals are later than the rock silicates and cut across them, embay them, and react with them, yielding reaction rims of alteration products around the margins of the silicate minerals. These reaction rims occurred before the final consolidation of the igneous body and are not to be confused with somewhat similar effects produced in later rock alteration by hydrothermal solutions.

Late magmatic deposits are always associated with basic igneous rocks. They have resulted from variations of crystallization differentiation, gravitative accumulation of a heavy residual liquid, and liquid separation of sulfide droplets (called liquid immiscibility), like drops of oil out of water, or other modes of differentiation.

This late magmatic group now includes most magmatic deposits of iron and titanium ores, and also some chrome ores, that formerly were considered to have been formed in an early magmatic stage. The group is gaining representatives at the expense of the early magmatic group. Included in the late magmatic group are many deposits of magnetic iron ore, ilmenite—an ore mineral of titanium desired for light strong metal and for white paint—chrome, nickel-copper sulfides, platinum, and pegmatites with their metallic and nonmetallic indus-

Late Processes and Deposits

trial minerals. Since these deposits result from the crystallization of a residual magma fraction, which is almost the last part of the magma to solidify, the ore fluid would have mobility until it becomes frozen.

Thus, the residual ore fluid in the interstices of previously formed rock grains may be forcibly squirted out like water out of a squeezed sponge (Fig. 7–4, L2a), or the ore fluid may quietly drain out to settle on an underlying solid layer (Fig. 7–4, L4). This settled portion may solidify at this point, or while still fluid it may be subjected to pressure that will cause it to be injected as an intrusion (Fig. 7–4, L3a) into the outer solidified portion of the parent igneous body or even into the surrounding rocks. Its mobility permits concentration and also intrusion. If the residual ore fluid does not drain out of the interstices between the silicate crystals or is not squeezed out it will then solidify as disseminated grains around the previously formed silicate crystals (Fig. 7–4, L2), and the result will be no concentration and no economic deposit.

The essential difference between early magmatic and late magmatic deposits is that the early deposits must lie within the igneous body at the place of settling and, as the constituents accumulate as solids, there is no mobility after accumulation. However, in the late deposits accumulation is through mobility, and the deposit may lie snugly and conformably enclosed within its host rock or cut across its internal structures or those of other rocks in disrupting fashion.

RESIDUAL ACCUMULATION. In Chapter 4 the progress of differentiation of magmas by crystallization was reviewed from the initial stage through to final consolidation. It was pointed out that in most magmas the more basic minerals crystallize first and that the residual magma becomes progressively richer in silica, alkalies, and water, with the silicic minerals the last to crystallize. In certain types of basic magmas, however, the residual magma becomes enriched in iron and titanium. It has been widely observed in many types of basic rocks that magnetite was the last, or almost the last, mineral to crystallize and lies molded around or fills the interstices between early-formed grains of calcic feldspar or the dark rock minerals. A restudy of many titanium-bearing magnetite deposits has also disclosed that in such deposits the magnetite or titanomagnetite crystallized last and surrounds, cuts across, and even embays the earlier-formed rock minerals. This factual evidence is indisputable.

Given a somewhat basic magma, perhaps with a higher than normal content of iron, undergoing differentiation at depth, the early-formed

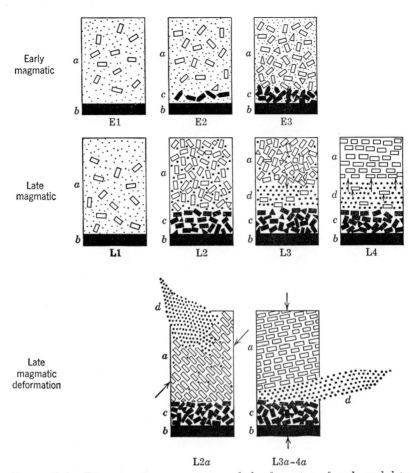

FIGURE 7–4. Diagrammatic representation of the formation of early and late magmatic deposits. *Early magmatic:* E1: early stage of crystallization of magma *a* after formation of chill zone *b*. E2: more advanced stage with settling of early-formed chromite crystals *c* on chill zone *b*. E3: final stage with early magmatic segregation of chromite *c*. *Late magmatic:* L1: same as E1; L2: formation of layer of sunken, early-formed, basic rock crystals *c* resting on chill zone *b* with mush of later-crystallizing silicates above, whose interstices are occupied by residual magma enriched in iron oxides. If consolidation should occur at this stage the result would be late-formed disseminated iron oxides but not an ore body. L3: down-draining, mobile, residual magma enriched in iron oxides settling on solidified layer *c* and floating up lighter silicate crystals, giving rise to final stage L4, and the formation of a concordant ore body in which a few silicate crystals are still trapped. L2*a*: L2, before consolidation, subjected to pressure by which the mobile iron-enriched, residual magma is filter-pressed out from the interstices and injected into other parts of the crystallized magma or into the invaded rocks to form a late magmatic injection. L3*a*–4*a*: the still mobile, iron-enriched, gravitative settled layer *d* of L3 subjected to pressure and squeezed out, or tilted and decanted, to form a late magmatic injection that may be concordant or nearly so.

crystals will be calcic feldspar and dark rock minerals. At first there will be a few crystals in a large volume of liquid (Fig. 7-4, L1) and they will tend mostly to sink, since their specific gravity will be greater than that of the liquid (Fig. 7-4, L2). As crystallization proceeds there will be a mush of crystals and liquid, in which the crystals will be growing larger and new ones will form. The liquid portion, meanwhile, will be growing richer and richer in excess iron, and its density may come to exceed that of the crystals so that they will tend to float upward and ride on the heavier liquid (Fig. 7-4, L3, 4). This process would tend to concentrate the liquid separately from the crystals and may be a factor in accumulation.

There may, however, be no such separation of liquid from crystals at this time. Crystal formation may proceed until the crystals touch each other and form an interlocking meshwork with the intercrystal spaces filled with liquid, which by this time would be still more enriched in iron. If uninterrupted crystallization proceeds at this stage, the result will be a granular basic igneous rock of which the interstitial material will be grains of magnetic iron ore (Fig. 7-4, L2). A great many basic igneous rocks do disclose just such interstitial magnetite, which was the last substance to crystallize. The basic sills of Connecticut and of the Palisades of the Hudson, for example, contain from 2 to 10 percent of such interstitial iron ores.

However, during the stage of a mesh of crystals with interstitial residual iron-rich magma, two other possibilities may happen: (1) the iron-rich residual liquid may drain downward to collect below, and (2) the liquid may be pressed out of the mesh of crystals and injected elsewhere. Each of these will now be considered.

In the first possibility (Fig. 7-4, L2, 3, 4), during early crystallization the heavy minerals olivine and pyroxene would settle through the liquid to form a solid bottom layer. In the later stage of a meshwork of touching crystals with interstitial iron-rich residual liquid, the iron and titanium oxides would become more and more enriched in the liquid as rock-mineral crystallization proceeds. It might reach the point of being composed chiefly of such metals. The liquid would also be a heavy one. It would, therefore, tend to trickle downward through the interstices of the silicates and collect on a lower solid layer of early-formed crystals (Fig. 7-4, L3). To make room for itself, however, it would have to float up the lighter silicates, which gradually would be pushed upward toward the top of the magma chamber. This upward displacement by the heavy settling magma high in iron ores would tend to jostle the floating crystals, breaking apart some that were touching, bending them, and crushing their edges. This may be the explanation for some of the phenomena of marginally broken and bent crystals of plagioclase feldspars in some anorthosite rocks, which usually has been ascribed to filter pressing (the second possibility to be considered further on). In this manner accumulation of a heavy residual magma would occur *within* the body of the intrusive and be *concordant* with the structure of the intrusive (Fig. 7-4, L4). The proportion of silicate constituents to metallic oxides in this residual magma would determine, upon final solidification, the grade of the resulting iron-ore deposit. If the silicate constituents were high, the ore would be low-grade; if the proportion of metallic oxides were high, the ore would be high-grade. It is probable also that some enrichment in the metallic oxides would take place during the solidification of the residual magma itself by the elimination through crystallization of some of the late silicate constituents. At Lake Sanford, New York, where titaniferous magnetite bodies are being mined in anorthosite and gabbro, one ore body contains about 35 percent oxides of iron and titanium, and another body contains about 62 percent; a third thin body contains up to 75 percent oxides.

The process outlined above would explain some of the troublesome features that have bothered geologists recently. Studies of titaniferous magnetite bodies have disclosed not only that the metallic oxides are later than the silicates but that some bodies lie within the mass of the intrusive and are concordant with the grain alignment of the enclosing rock. These have been an enigma. Other discordant bodies that cross the grain structure have been readily explained by filter pressing, to be discussed later, but no adequate explanation has hitherto been advanced for the concordant ore bodies enclosed within the intrusive. Herein may lie the explanation. Concordance would be expected because the enriched residual liquid would lie upon a floor of settled crystals of earlier crystallization, and crystals above the residual liquid would rest concordantly upon the liquid. Moreover, the residual liquid would accumulate, not at the bottom of the magma chamber, but well above the bottom where a considerable thickness of early-formed olivine and pyroxene crystals had already settled. Also, the midstage crop of lighter crystals would float above the heavy residual liquid. Thus, the metallic oxides would be expected to lie in the midsection of the intrusive and concordant with the grain structure both above and below the ore layer.

Late Processes and Deposits

Let us see if there is any field support for the above hypothesis. The Palisades of the Hudson offer unique opportunity to study small-scale magma differentiation and crystal settling, for here is a horizontal sill of basic igneous rock almost 1,000 feet thick and readily accessible from top to bottom. It has been carefully studied recently by F. Walker. It is true that it contains no deposit of iron ores, but nevertheless it illustrates in part the principles of their accumulation. Walker showed that, above a bottom frozen layer representing the quickly chilled portion of the original magma, there is an olivine-rich zone, above which lies a zone richer in basic pyroxenes. His vertical sections and analyses show that iron ore is scarce in the olivine zone and the olivine-diabase zone above it, which occupy the lower two-thirds of the section, and is most abundant in the upper third. He states, "In the rapidly cooled contact modifications small well formed crystals [of iron ore] of early crystallization are prominent; in the coarser central portions, the mineral [iron ore] is . . . of somewhat later crystallization. In the . . . [higher zone], it occurs as irregular skeletal masses . . . of still later formation." And again he states, ". . . in the coarsest upper portions [iron ore] is sufficiently abundant to be a major constituent." Mr. Walker also observed, "This tendency for the volatiles in diabases to hold back iron oxide and expel it as iron ore or chlorophaeite, just before the last stages of crystallization . . . in Scottish rocks." Many similar examples have been observed elsewhere. This example of the Palisades illustrates two features in relation to the above hypothesis: (1) that a zone of early-formed crystals (olivine), scarce in iron ore, accumulated at the bottom of the sill; (2) that the greatest quantity of iron ore occurs in the upper part of the sill, but still within it, and that the iron ore is of later crystallization than most of the rock silicates.

Again, in the titaniferous iron ores of the Adirondack region, Osborne recognized two types of ore occurrence, one concordant and another discordant with the primary structure of the host rock. The discordant type was accounted for by filter pressing, but the concordant bodies proved extremely puzzling. Those in anorthosite were thought to be due to filter pressing because the plagioclase feldspars showed crushed edges and bent lamellae indicating movement, but concordant bodies in gabbro remained an enigma because no crushing of the rock minerals was evident. The facts disclosed would fit in with the hypothesis formulated above, since gravitative liquid accumulation would bring about concordance and would account for the crushing and bending of the early-formed plagioclase feldspar and the lack of crushing in the pyroxene of the gabbro, since the pyroxene in part would crystallize out of the residual liquid after accumulation.

As a third and last field illustration let us refer to the remarkable Bushveld Igneous Complex of South Africa. Here is a gigantic intru-

FIGURE 7-5. Layers of titaniferous magnetite in rocks of Bushveld Igneous Complex. Magnet Heights, South Africa. (Courtesy of Edward Sampson.)

sion of basic igneous rocks exhibiting such a striking primary grain layering that it is called "pseudostratification" because it is almost as regular as stratification in sedimentary rocks. Several thousand feet above its base are stratiform layers of titanomagnetite (also of chrome ore) concordant with the grain structure of the enclosing rock (Fig. 7-5). These layers of iron ore have excited geologists because of their continuity over tens of miles with remarkable uniformity of position, thickness, and content. How they were formed has been a

Late Processes and Deposits

puzzle. At first they were thought to have been formed by the settling of early crystals of magnetite, but the magnetite being later than the rock silicates precludes this explanation. The later age of the iron ore is shown conclusively under the microscope, since it surrounds, penetrates, cuts across, and embays earlier feldspar crystals. Consequently, the formation of the iron ore must be a late magmatic feature, but how to account for the position and occurrence of the iron layer has intrigued geologists. Hydrothermal replacement has no basis in fact; late magmatic filter pressing seems eliminated, since the concordance and lack of evidence of crushed crystals (even though some feldspar crystals are bent) oppose this view; likewise, a proposed idea of separate intrusions of iron-ore fluid is not supported by the contact and internal structures.

The possibility of late gravitative liquid accumulation, as proposed above, must be entertained and does fit in with the field facts. The iron ore rests on an even floor which gives all appearance of having been solid when the magnetite came to rest so sharply upon it. Plagioclase feldspar crystals lie above it with their longer axes aligned parallel to the floor (Fig. 7-4, L4). The bottom part of the magnetite layer is almost pure ore with only a few feldspar plates; the upper part contains more and more feldspar crystals until at the top it is composed of feldspar crystals and no visible iron ore. The feldspar plates are aligned parallel to the iron-ore layer as they would if they had been floated up and pushed up by down-trickling residual iron-rich magma which, seeking rest on a floor, would have to shoulder the feldspar crystals upward. Moreover, the feldspar crystals exhibit bent lamellae, indicating jostling and some slight squeezing. The bottom part of the iron-ore layer would purify itself by excluding the residual silicate material by crystallization, which crystals floating upward would give an upper part higher in silicate crystals and lower in iron-ore minerals. This process of formation would also account for the continuity and uniformity of thickness exhibited by the iron-ore layers.

The facts just mentioned would also be adequately explained had the iron-ore minerals accumulated as early-formed settling crystals, but, as pointed out, this explanation cannot be, as the ore minerals crystallized later than the rock minerals. There are, of course, some additional complications to this somewhat oversimplified picture, such as the formation of the floor on which the iron ores rest, the problem of composite intrusions of magma before differentiation, and the reasons why iron oxides should accumulate in the residual magma, but they are beyond the scope of this chapter.

RESIDUAL LIQUID INJECTION. An iron-rich residual liquid accumulated in the manner described above may be subjected to pressure exerted from below, from the sides, or from above by the weight of overlying rocks. Some crustal disturbance and adjustment is generally to be expected where magma bodies have been formed and intrusion occurs. Fluid under pressure will tend to move to the place of less pressure, which may be provided by shear zones, fractures, contacts, or other places of weakness. The squeezing out of a residual magma from its place of repose, which may be in the interstices of a mesh of crystals or an accumulated floor layer, has been aptly termed *filter pressing*. As applied to the formation of mineral deposits one may presume that crystallization has proceeded to the stage that the residual liquid occupies only a minor part of the magma chamber, at which time it may then be highly enriched in the iron oxides. Filter pressing at this stage may operate in two ways: (1) it may squeeze out mobile interstitial liquid, filtering liquid from crystals, or (2) it may squirt out a body of already accumulated liquid.

In the first method (Fig. 7-4, L2a) the liquid would be forced to move to some other site. It might be injected with vigor into another part of the crystallized parent igneous body where relief of pressure could take place, such as along the primary structure where it would occupy what might appear to be concordant position. More likely it would be injected across the primary structure, giving discordant relations. Or it might be injected along the contact between the parent body and the intruded rocks to form a contact body. Again, it might be injected into the outer invaded rocks along a fissure to form a dike, such as the dike of cumberlandite (magnetite, silicates, apatite) in Rhode Island, or as a sill between stratified rocks, or in other forms. Given favorable conditions, the resulting body after solidification may be large enough and high enough in metallic oxides to constitute an ore deposit.

The squeezing out of residual liquid from crystal interstices could take place only by mashing the crystals together, and even then some of the liquid must remain. It would follow that the rock crystals should show some of the effects of such mashing in the form of broken corners and edges and strained, bent, and cracked crystals. This is just what is found in the crystals of feldspar in many anorthosites where filter pressing is thought to have taken place. Osborne described such effects of mashing in the anorthosite rocks associated with titaniferous magnetites in the Adirondack region which he considered had resulted from filter pressing.

Late Processes and Deposits

Similar injections of iron oxides might occur if before solidification a body of accumulated residual magma high in iron oxides were subjected to pressure (Fig. 7-4, L3a–4a). Since the residual liquid here does not lie in crystal interstices, no accompanying mashing of rock crystals would be expected. The liquid might be squirted along the primary structures of the parent rock or across them or into adjacent rocks. If a high degree of purification had gone on by ejection through crystallization of much of the silicate constituents in the residual liquid before injection, mineral bodies of high purity might result.

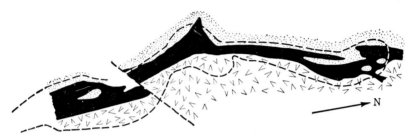

FIGURE 7-6. Map of part of Rektor ore body, east of the Kirunavaara deposit, Sweden. Mass of magmatic magnetite with apatite intrusive into quartz-porphyry (stippled) and Rektor altered volcanics (*v*'s) and containing inclusions of both. Black is ore; dashed lines represent open cut. (After Per Geijer, *Swedish Geol. Survey, 1950.*)

Injected ore bodies formed under the conditions described above would generally transect the primary structures of the parent or invaded rocks. They would exhibit the intrusive relations of normal intrusive rocks, and the ore minerals would surround, cut across, corrode, and react with earlier-formed rock minerals. These reactions are strictly magmatic, however; they take place before final consolidation. If the injected iron-rich fluids are rich in volatiles some pneumatolytic or hydrothermal reactions might occur.

Many deposits of magnetite and ilmenite, which were formerly called early segregations, are now known to be late magmatic injections of this type. This has been shown rather conclusively for many of the titanomagnetite deposits of the Adirondack region of New York State. The huge magnetite deposit of Kiruna, Sweden, with its hundreds of millions of tons of high-grade iron ore, is definitely an injection and encloses angular fragments of the adjacent country rocks (Fig. 7-6). It should be considered as a late magmatic differentiation product that elsewhere underwent accumulation of high purity and was then injected into its present site. Some of the South African chrome ore is later in formation than the enclosing rock silicates and at Dwars River Bridge even encloses a piece of anorthosite, which is supposedly its parent rock (Fig. 7-2). The new great deposits of ilmenite (titanium ore) in northern Quebec, exceeding 300 million tons of ore, intrude the enclosing anorthosite, cross the grain structure, and include blocks and fragments of coarse anorthosite. This likewise appears to

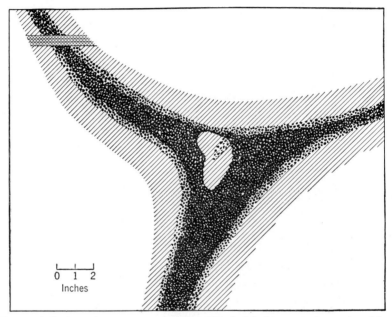

FIGURE 7-7. Branching massive chromite "dike" (black) exposed on vertical face of serpentine. La Coulee, New Caledonia. (After J. C. Maxwell, *Econ. Geol.*)

be a late magmatic injection into the parent host rock. The chromite "dike" at La Coulee, New Caledonia (Fig. 7-7) looks suspiciously like an injection.

The unique platinum pipes of iron-rich peridotite and chromite of South Africa, in carrot-shape form, cut vertically across the pseudo-stratification of the Bushveld Igneous Complex rocks for hundreds of feet (Fig. 7-8). Their origin has always been a puzzle. Their cylindrical form consists of an outer basic igneous rock (pyroxenite) grading in toward a more basic rock (olivine dunite), in the center of which is the rich ore consisting of platinum with chromite and ultrabasic igneous rock. Many dikes of one of the most basic of rocks (hortonolite-dunite) cut across the central part of the core—dikes of materials that are generally thought to have crystallized early. The pipes certainly appear to be injections across primary rock pseudostratification. The basic to ultrabasic constituents composing them obviously could not have been early magmatic crystal settling and must be a late magmatic phase. Moreover, the centers of the pipes,

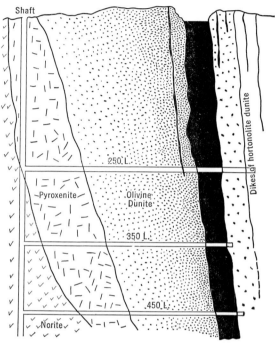

FIGURE 7-8. Mooihoek platinum pipe, South Africa, east half. Center (black) is chrome with platinum in hortonolite dunite, surrounded by olivine dunite, then by pyroxenite, and all enclosed in norite. (Based upon P. A. Wagner, modified by sketches made by author at mine.)

presumably the parts that cooled and crystallized last, are composed of the most basic constituents. It is difficult to escape the conclusion that these pipes, and the accompanying dikes, are late magmatic injections.

IMMISCIBLE LIQUID SEPARATION. There are many copper-nickel sulfide deposits associated with basic igneous rocks to which a magmatic origin has been ascribed. With growing knowledge, a hydrothermal origin has been advanced for some of these.

An early concept envisaged early crystallization and settling of the sulfides into irregularities on the floor, giving rise to ore deposits formed *in situ*. Then came the realization that the sulfides were later in age than the rock minerals, and this destroyed the concept of early crystal settling. J. H. L. Vogt in Norway investigated the separation of nickel-copper sulfides in basic igneous rocks. He concluded that they are soluble up to a few percent in basic rocks and that during cooling they may in part separate out as molten droplets and as such sink through the magma to accumulate at the bottom of the chamber as liquid sulfide segregations. In this respect it would resemble the accumulation of molten copper (matte) that collects at the bottom of a smelting furnace by draining down through the overlying molten slag. He pointed out also that, since a sulfide melt has a lower freezing point than a rock magma, the sulfide-rich accumulation would remain liquid long after the rock constituents had solidified. Its solidification would, therefore, be the last phase of magma consolidation.

The sulfide accumulation need not be a pure sulfide melt. Rather it must be thought of as an enrichment in sulfides of the lowest part of the magma, which upon consolidation gives rise to a basic igneous rock with sulfide content up to 10 or 20 percent. This, however, might constitute a valuable ore deposit. Before consolidation the sulfide-rich melt might enrich itself further by floating up crystallizing rock constituents, and under exceptional circumstances an almost pure sulfide melt might result.

Under quiescent conditions the sulfide-enriched portion might freeze at the place of accumulation, giving rise to a late magmatic nickel-copper sulfide deposit formed *in situ* at the bottom of the intrusive igneous body. Platinum is a common constituent. Such bodies would always lie along the lower margins of the accompanying basic intrusive, notably where there are depressions in the floor. Their size, naturally, is proportionate to the size of the mother intrusive. Their mineralogy is simple and monotonously uniform; Vogt noted a relatively uniform ratio of nickel to copper in the Norwegian deposits. Some examples of such deposits are: the nickel-copper deposits of the Insizwa type in South Africa (Fig. 7–9); nickel-sulfide deposits of the Bushveld, South Africa, and of Norway. Of course, all basic magmas do not contain sufficient amounts of these metals to give rise to deposits; they are exceptional features.

Conditions may not be quiescent while the sulfide-rich accumulation is still molten. Should disturbance occur at this time and pressure be exerted the mobile liquid might be squirted toward places of less pressure and enter sheared, fissured, or brecciated areas, and there consolidate as injected deposits. They might intrude fractured por-

Late Processes and Deposits

tions of the already consolidated parent intrusive or intrude the adjacent rocks and enclose broken fragments of the host and foreign rocks. They would exhibit intrusive relations and might form dikes. They could penetrate, corrode, and even replace earlier rock minerals. If the residual liquid contained considerable volatiles or water, the resulting deposits might display transitions into hydrothermal deposits. Examples of this type are to be found in South Africa and Norway, and perhaps the Frood deposit of the International Nickel Company mines at Sudbury, Ontario, may come to be considered as belonging

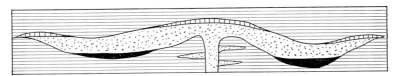

FIGURE 7–9. Generalized diagram of Insizwa type of intrusion. Black is basal zone carrrying nickel-copper sulfides. (After D. L. Scholtz, *Geol. Soc. S. Africa*, 1936.)

to this class. Many pyrite deposits have been called magmatic injections, and if there are such, of which the author is extremely dubious, they would belong to this class.

PEGMATITES. So far, in considering late magmatic deposits, we have been concerned only with those formed in association with basic igneous rocks. However, we saw earlier that when silicic magmas undergo differentiation there is a progressive enrichment in the residual magma of silicic and alkalic constituents. The late residual fluid of such silicic magmas would consist chiefly of the low-melting ingredients to crystallize last in a granitic magma, namely, quartz, alkalic feldspar, and mica, along with concentrations of the rare constituents, mineralizers, and much water. It was also pointed out that, if some of this residual liquid were tapped off and crystallized elsewhere, the resulting product would be a pegmatite. Pegmatites commonly occupy fissures (Fig. 4–3), and hence are called pegmatite dikes; many, however, are irregular in form (Fig. 7–10).

The "mineralizers" and water in the residual pegmatitic fluid play a double function: they are solvents and transporters of many rare metallic and nonmetallic substances that become concentrated in the residual magma, and their presence facilitates coarse crystallization of the individual minerals of pegmatites. Consequently, pegmatites are generally coarse-grained, and individual crystals may be several inches or even a few feet across; rarely, they exceed 10 feet or more, and single crystals may weigh several tons. In addition to water, the volatile substances consist of compounds of boron, fluorine, chlorine, phosphorus, sulfur, and other, rarer elements. By decreasing the viscosity of the magma and lowering the freezing point, they permit the large crystal growth.

The composition of the pegmatites depends in part on the original composition of the magma from which they spring, but also in larger

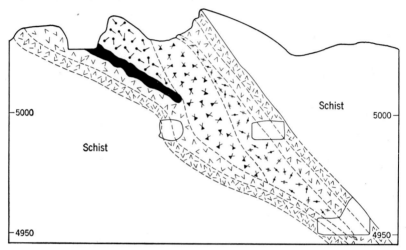

FIGURE 7-10. Cross section of a uranium-bearing (black) pegmatite showing zonal arrangement of pegmatite minerals. Bob Ingersoll Dike No. 2, Keystone, South Dakota. (After L. R. Page, *Econ. Geol.*)

part upon the stage of crystallization at the time they are withdrawn from the magma chamber. Most of them are offshoots of granitic magmas, and hence they have the general mineral composition of granitic rocks. If the residual fluid of a granitic magma is tapped off well before the completion of crystallization, it will consist largely of the ingredients of the minerals that are later to crystallize. Water of course will be present, and the residual fluid will be somewhat enriched in mineralizers. This will be an aqueo-igneous stage—a transition between a strictly igneous stage and a hydrothermal stage, leaning more to the igneous—and is referred to as the pegmatitic stage. Such relatively early withdrawals will yield simple *pegmatite dikes*, which are merely varieties of igneous rocks. They consist chiefly of quartz, feldspar, and mica with only very minor quantities of the uncommon

Late Processes and Deposits

metallic and nonmetallic minerals. They may, however, be mined for their mica or feldspar content.

If the withdrawal of residual liquid takes place during a more advanced stage of magma crystallization much of the feldspar will already have crystallized, but particularly there will be a greater concentration of mineralizers, and rare and uncommon constituents. The resulting pegmatites are commonly characterized by druses or irregular openings lined with crystals and may contain minerals of tungsten, tin, uranium, beryllium, titanium, phosphorus, chlorine, fluorine, lithium, molybdenum, and other elements, and perhaps a few common minerals of ore deposits, although sulfides are rare. These pegmatites have often been referred to as pegmatite veins rather than dikes. Indeed, some of them are valuable mineral deposits and are mined for many kinds of gemstones and for ores of tungsten, tin, columbium, tantalum, titanium, beryllium, and lithium, as well as for many nonmetallic minerals of industry. They are treasure-houses for the mineralogist.

Pegmatites most commonly occur as clusters or swarms of dikes or lenses around the outer margins of intrusive granitic rocks or in the adjacent rocks. Their length may range from a few feet to a thousand

Some Common Economic Products That Result from Magmatic Concentration

Substance	Chief Minerals	Examples of Localities
Beryllium	Beryl, aquamarine	Brazil, India, Argentina
Chromium	Chromite	South Africa, Turkey, Rhodesia, Russia
Chrysoberyl	Chrysoberyl	Ceylon
Columbium	Columbite	Nigeria
Copper	Chalcopyrite	Southwestern Africa
Diamond	Diamond	South Africa
Garnet	Pyrope, others	South Africa, India
Iron	Magnetite, some hematite	Kiruna, Sweden
Iron-titanium	Titanomagnetite	Adirondacks, Wyoming, South Africa
Nickel-copper	Chalcopyrite, pentlandite, polydymite, sperrylite	South Africa, Norway, Ontario(?)
Peridot	Peridot	Levant, Egypt, Burma
Platinum	Native platinum	South Africa, Russia
Platinum metals	Osmium, iridium, palladium, others	South Africa
Quartz gemstones	Many	World wide
Sapphire	Sapphire	India—pegmatites
Tantalum	Tantalite	Brazil, Belgian Congo
Tin	Cassiterite	Malaya
Titanium	Ilmenite	Quebec, Norway, Adirondack
	Rutile	Scattered pegmatites
Topaz	Topaz	World wide
Tungsten	Wolframite	Scattered pegmatites

feet or more, and their width from a fraction of an inch to many scores of feet.

Not all pegmatites are simple crystallization of late magmatic fluids. Some display many complications of growth and evidences of alteration by later fluids that attack or replace the primary minerals. Their own residual liquids left over when the original minerals crystallize may be out of chemical adjustment with the earlier-formed crystals and may change them to new minerals as, for example, original potassium feldspar may be changed over to soda feldspar. In addition, new pulsations of magmatic fluid may arise along the same channels and replace the older minerals by newer ones. Rarely, the whole pegmatite may be of secondary replacement origin.

SELECTED REFERENCES ON MAGMATIC CONCENTRATION

Ore Deposits of the Western United States—Lindgren Volume. Am. Inst. Min. Met. Eng., New York, 1933. Chap. 11, Part I, "Magmatic Segregations," by J. T. Singewald, Jr. General discussion of United States deposits.

"Magmas and Igneous Ore Deposits." J. H. L. Vogt. *Econ. Geol.* 21:207–233, 309–332, 469–497, 1926. Fundamental discussion of magmatic processes.

"Certain Magmatic Titaniferous Ores and Their Origin." F. F. Osborne. *Econ. Geol.* 23:724–761, 895–922, 1928. Filter pressing applied to titaniferous magnetites.

"Anorthosite Area of the Laramie Mountains, Wyoming." K. S. Fowler. *Am. Jour. Sci.* 5-19:305–315, 373, 403, 1930. Segregation of titaniferous magnetites.

"The Iron Ores of the Kiruna Type." Per Geijer. *Swedish Geol. Surv.* No. 367, Stockholm, 1931. Occurrence and origin.

"Magmatic Nickel Deposits of the Bushveld Igneous Complex." P. A. Wagner. *South Africa Geol. Surv. Mem.* 21, 1924.

"The Bushveld Igneous Complex, Transvaal." A. L. Hall. *South Africa Geol. Surv. Mem.* 28, 1932. Detailed occurrence and origin of this great magmatic body and its included magnetite, chromite, nickel, and platinum.

"Magmas and Ores." Alan M. Bateman. *Econ. Geol.* 37:1–15, 1942. Theoretical discussion of various origins of magmatic ore deposits.

"The Chromite Deposits of the Bushveld Igneous Complex." W. Kupferbürger and B. V. Lombaard. *South Africa Dept. Mines, Geol. Ser. B* No. 10, 48 pp., 1937. Detailed geology of all chrome deposits of South Africa.

Internal Structure of Granitic Pegmatites. E. N. Cameron, R. H. Jahns, A. H. McNair, and L. R. Page. Econ. Geol. Pub. Co. Mono. 2, 1949.

"Chromite Deposits of the Eastern Part of the Stillwater Complex, Montana." J. W. Peoples and A. L. Howland. *U. S. Geol. Surv. Bull.* 922–N, 1940. Montana low-grade deposits.

"Chromite, Lomagundi and Mayie, Southern Rhodesia." F. E. Keep. *Southern Rhodesia Geol. Surv. Bull.* 16, 1930. Descriptions of the Great Dyke occurrences.

Selected References

"Varieties of Chromite Deposits." Edward Sampson. *Econ. Geol.* 26:833–839, 1931, and 24:632–641, 1929. A discussion of origin and relation to differentiation.

"Geology and Chromite Resources of Camagüey, Cuba." D. E. Flint, J. Francisco de Albear, and P. W. Guild. *U. S. Geol. Surv. Bull.* 954–B, 1948. Geology, occurrence, and origin of refractory deposits.

"Chrome Ores of the Western Bushveld." C. J. F. van der Walt. *Geol. Soc. South Africa* 44:79–112, 1942. Discussion of ores and chemical relations.

"Beryl-Tantalite Pegmatites of Northeastern Brazil." W. D. Johnston, Jr. *Geol. Soc. Amer. Bull.* 56:1015–1069, 1945. The largest beryl district of the world; in pegmatites.

"Titaniferous Magnetite Deposits of Lake Sanford, New York." R. C. Stephenson. *A.I.M.E. Min. Tech.* Vol. 9, No. 1, pp. 1–25, 1945. Petrography, mineralogy, and geology; a late magmatic residual liquid injection origin is advocated.

Economic Mineral Deposits. 2nd Edit. Alan M. Bateman. John Wiley & Sons, New York, 1950. Chap. 5·1, Magmatic concentration; Chap. 12, platinum; Chap. 14, iron, nickel, chromium; Chap. 15, beryllium, tantalum, columbium, titanium; Chap. 21, lithium; Chap. 23, corundum; Chap. 24, gemstones.

CHAPTER EIGHT

Contact-Metasomatic Processes

Perhaps the most impressive and terrifying natural phenomena are the great explosive volcanic outbursts that are seen and recorded from time to time. The gigantic forces of nature that are emitted from within the earth with impelling speed and appalling vigor are the effect of gases and vapors that have escaped from an underlying magma to the surface. They testify also to the vast volume of gases that are released; witness the colossal gas jet of the 1906 Vesuvius eruption mentioned in Chapter 5. Gaseous emanations may continue long after explosive volcanic activity in the form of fumaroles and solfataras, some of which may deposit sublimates such as sulfur at the surface.

These gaseous constituents originally were in solution in the magma, and as long as the magma remained liquid and the pressure unchanged they remained in solution. With release of pressure, they escape. Most magmas, however, do not reach the surface but are shut in deep within the earth's crust. Consequently, their gaseous constituents rarely reach the surface. However, if some relief of pressure occurs around the magma chamber, gases and vapors will come out of solution and stream through the liquid magma, like the sparkling gas out of an opened soda bottle, toward the place of pressure relief. A gas or vapor bubble streaming through a magma serves as a collector for other gases of small vapor pressure. They will be distilled into the bubble regardless of pressure, and thus they collect and transport other substances out of the magma.

The gaseous constituents may be released at any time during the cooling and crystallization of the magma, provided relief of pressure occurs. With progressing crystallization, however, there would tend to be an increasing concentration of the volatiles into the diminishing residual liquid. Thus, during the later stages of crystallization a greater concentration of volatiles would be available, and they would have greater opportunity to dissolve metals and other substances.

Escape from the magma during this stage would, therefore, provide the optimum conditions for mineralization by gaseous or, as it is often called, *pneumatolytic* action.

The escape of high-temperature volatiles, during the consolidation of the magma, into the adjacent rocks produces in them profound effects near the contacts. These effects have been divided by Barrell into two types: (1) the effects of heat, without any appreciable accessions from the magma, giving rise to *contact metamorphism;* and (2) the effects of heat combined with additive materials brought out from the magma chamber, called *contact metasomatism.* Contact metamorphism does not give rise to mineral deposits except in a few rare cases, but contact metasomatism does give rise to valuable and distinctive deposits.

Contact metamorphism manifests itself by some changes upon the margins of the igneous intrusive body itself, but chiefly by its pronounced effects upon the invaded rocks, particularly if they are carbonate rocks such as limestone or dolomite. The contact effects are generally with deep-seated intrusive rocks of intermediate to silicic composition but are uncommon with basic igneous intrusives. The effects consist of a baking or hardening of the invaded rocks surrounding the intrusive body, and commonly their thorough transformation. Under the conditions of such high temperatures and pressures the invaded rocks become unstable, and their constituents recombine to form new minerals that are stable under the changed conditions. In an impure limestone rock consisting of carbonates of calcium, magnesium, and iron, along with some impurities such as clay and quartz, recombinations of these ingredients take place. The calcium of the calcite and the silica of quartz form calcium silicate (wollastonite); dolomite, quartz, and water form calcium magnesium silicate (tremolite) or, with the addition of some iron, form the mineral actinolite; calcite, clay, and quartz form garnet. Similarly, other changes occur, giving rise to an extended suite of unusual and characteristic high-temperature minerals. If the invaded rocks are relatively pure limestone or dolomite they are changed over to marble; sandstone is changed to quartzite, and shale to hornfels. In all of these alterations the original rocks undergo little change in chemical composition except for the loss of some volatile constituents, such as carbon dioxide, and minor additions of magmatic volatiles. It is chiefly just recrystallization. The result of these changes is the formation around the intrusive of a contact-metamorphic *aureole* that varies in shape and size according to the shape and size of the intrusive and the character and structure of the invaded rocks.

Contact metasomatism differs from contact metamorphism in that great quantities of materials have been introduced from the magma into the invaded rocks, in addition to the metamorphic changes mentioned above. The introduced constituents have replaced the minerals of the invaded rocks in whole or in part. The resulting mineralogy is thus more varied and complex than that produced by heat metamorphism alone. The alteration is not simply metamorphic but is metasomatic, under conditions of high temperature and pressure; hence the distinction between the two. If the magmatic gases and vapors happen to be highly charged with the constituents that go to make up mineral deposits, *contact-metasomatic deposits* result, particularly in the favorable environment of carbonate rocks. Such deposits in the past have often been called contact-metamorphic deposits (and also pyrometasomatic deposits), but the usage of contact-metasomatic deposits is preferred here since it connotes the much-described relationship with intrusives and their metasomatic or replacement origin, and also the fact that they are characterized by a distinctive assemblage of high-temperature minerals.

This process of mineral formation was early recognized by von Cotta in 1865 at Banat, Hungary, and later established by Vogt in 1894 as the ore-forming process at Christiania (Oslo), Norway. Lindgren first recognized the type in America at Seven Devils, Idaho, and since then several American geologists were led to the realization of the importance of this type of mineral-forming process and helped unravel the intricacies of its mode of formation.

THE PROCESS AND EFFECTS

The high-temperature contact-metasomatic effects of deep-seated intrusives upon invaded rocks is brought about by actual transfer of heat and constituents by gaseous emanations. Transfer of heat by conduction is probably very minor. The emanations may change entire beds of limestone to complex silicate rocks generally high in garnet, called *tactite* or *skarn*. Huge additions and subtractions of material are involved. Where the emanations carry ore constituents metallic or nonmetallic mineral deposits may be formed spasmodically within the contact aureole, but of course all magmas do not yield mineral deposits.

For the formation of contact-metasomatic mineral deposits certain requirements are essential: (1) certain types of magma are necessary; (2) the magma must contain the ingredients of mineral deposits; (3)

Process and Effects

it must be intruded at depths not too shallow; and (4) it must contact reactive rocks.

TEMPERATURES. The temperature at the immediate contact must have been that of the intruding magma, which in siliceous magmas ranges from 500° to 1,100° C. Outward from the magma contact the cooling effect of the invaded rocks would bring about a gradual decline in temperature. Also, there would be a gradual lowering of the temperature during the slow cooling of the intrusive. Consequently, there may have been a range of several hundred degrees in temperature during the formation of the contact minerals.

Certain minerals, called *geologic thermometers*, give some idea of the temperature that prevailed during their formation. For example, high quartz has been identified which shows that it was formed above 573° C; low quartz has been also noted. Wollastonite indicates a temperature below 1,125° C, and andalusite less than 1,000° C. Some andradite garnet indicates a temperature below 800° C. Thus, contact metasomatism probably takes place at temperatures between 400° to 800°, or possibly even higher.

RECOMBINATION AND ACCESSIONS. Both recrystallization and recombination of constituents take place in the alteration zone. By recombination, minerals that for convenience may be designated *AB* and *CD* may be changed to *AC* and *BD*. With additions of materials from the magma these minerals may become *ACX* and *BDY*, where *X* and *Y* represent accessions. In this manner many complex minerals are built up. The magmatic accessions consist chiefly of metals, sulfur, silica, boron, chlorine, fluorine, potassium, magnesium, and some sodium. These substances, it has been noted earlier, are characteristic of fumaroles.

The volume changes of mass that have occurred in the contact zone are almost beyond belief. Lindgren in his study of the contact-metasomatic deposits at Morenci, Arizona, showed that for every cubic meter (a cube only a little more than 3 feet to a side) of the altered limestone, 1,015 pounds of calcium oxide and 2,625 pounds of carbon dioxide have been carried away, and 2,935 pounds of silica (SiO_2) and 2,600 pounds of iron oxide have been added. These astonishing figures of 1.8 tons of material removed and 2.8 tons added per cubic meter give some idea of the enormous wholesale transfer of materials during contact metasomatism.

The transfer of material apparently may commence shortly after intrusion and continue until well after consolidation of the outer part of the intrusive. The outer part of the intrusive itself commonly shows some alteration, indicating solidification before the contact-metasomatic action. In general, the early stage consists of recrystallization and recombination as a result of heat effects, with or without magmatic accessions, which mostly occur later. Magnetite and hematite form with the silicates and later than they, and generally the formation of sulfides of the metals follows the iron oxides.

It is generally considered that the transfer of the substances takes place by gaseous emanations, but some evidence indicates that high-temperature liquid emanations have operated in the later phases of contact metasomatism.

RELATION TO INTRUSIVES. Contact metasomatism is restricted exclusively to intrusive magmas. Extrusive bodies and small dikes may produce a little baking, hardening, or other minor effects at the contact but never mineral deposits.

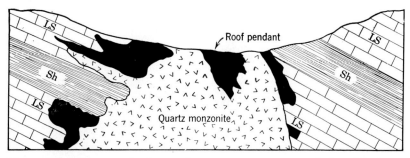

FIGURE 8–1. Diagram of relation of contact-metasomatic ore bodies (black) to quartz monzonite intrusive and to structure of limestone. Roof pendants of limestone projecting down into monzonite are particularly favorable loci for ore.

The magmas that yield mineral deposits are mostly silicic ones giving rise to rocks of intermediate composition, such as granodiorite, quartz, diorite, quartz monzonite, or monzonite. The most silicic rocks, such as granite, rarely yield contact-metasomatic deposits, and this is true also of basic rocks. The reason is probably that the silicic group have a higher water content, whereas the basic are relatively dry.

The deposits are mostly associated with the large intrusives such as batholiths, stocks, and intrusive bodies of similar size, and not with the smaller intrusives such as dikes and sills. Intrusives whose roofs are irregular in form, and particularly those in which deep pendants of overlying sediments project downward into the intrusive, are especially favorable for the development of contact metasomatism. This is notably the case at Mackay, Idaho (Fig. 8–1), where large roof pendants have been extensively garnetized and mineralized.

The depth of intrusion is also an important factor because contact-metasomatic deposits are found only with deep-seated granular igneous rocks. Deposits are not found with those rocks that cool rapidly at or near the surface and have a stony or glassy texture, probably because the emanations are rapidly lost to the atmosphere. Most granular intrusives have probably crystallized at depths greater than one mile. At Philipsburg, Montana, the depth was probably 6,500 feet, and at Marysville, Montana, Barrell concluded that the intrusive had reached within 4,000 feet of the surface. Generally similar depths have been inferred for intrusives associated with contact-metasomatic deposits in Utah and New Mexico.

RELATION TO INVADED ROCKS. Both composition and structure of the invaded rocks affect the character, localization, and extent of contact metasomatism.

Carbonate rocks are the most susceptible to attack because calcite and dolomite are so readily broken up under the conditions that prevail. Pure limestone and dolomite readily recrystallize and recombine with the magmatic accessions. Impure carbonate rocks are affected even more, as such impurities as silica, alumina, and iron are ingredients already available to enter into new combinations. The entire rock adjacent to the intrusive may be converted to a mass of garnet rock with contained ore bodies.

Other sedimentary rocks may undergo much alteration but seldom are they, particularly the clay rocks, hosts for large ore bodies. Invaded igneous rocks are the least affected, probably because their composition is less strikingly different from the intrusive igneous rock. Thus, the rocks most congenial to the formation of contact-metasomatic mineral deposits are sedimentary rocks and particularly limestone and dolomite.

The structure of the sedimentary rock is important in determining the position and extent of contact metasomatism. Sedimentary beds that dip inward toward the intrusive (Fig. 8–1) provide good channelways for the upward and outward escape of magmatic emanations, in contrast to slower migration across the bedding. Fractured sediments also provide greater opportunity for ingress of emanations. Large faults that trend upward and outward from the intrusive serve as through channelways that conduct the emanations farther from the intrusive and localize the metamorphism and mineralization in tabular form.

THE RESULTING MINERAL DEPOSITS

The mineral deposits resulting from contact metasomatism are generally small as compared with some of the great "porphyry copper" or sedimentary deposits. They are mostly disconnected bodies that occur within the contact zone like scattered plums in a plum pudding. They are difficult to find, as they exhibit few "signboards" pointing to their presence, and costly exploration and development are necessary to discover and outline them. They are vexatious deposits to exploit because of their relatively small size, their capricious distribution within the contact aureole, and their abrupt terminations. Most of the deposits are within a few hundred feet or less from the igneous contact, and many lie on the outside of the contact aureole. Roof pendants and large inclusions in the intrusive are especially favorable loci for deposits. Large faults may localize deposits along them and permit deposits to be formed as much as 2,000 feet from the contact.

The deposits are notably irregular in outline and may have almost any shape. Generally they are more or less equidimensional, but ramifying projections along bedding planes and fractures may enhance the irregularity of outline.

The outstanding feature of contact-metasomatic deposits is the unusual and distinctive high-temperature nonmetallic mineral assemblage. The following is a list of the more common ones:

andradite garnet	hedenbergite	vesuvianite	tourmaline
grossularite garnet	hastingsite	ilvaite	axinite
wollastonite	epidote	diopside	scapolite
tremolite	zoisite	fluorite	ludwigite
actinolite	forsterite	chlorite	chondrodite
albite	anorthite	micas	topaz
quartz			

Vesuvianite, fluorite, and the last six in the list contain mineralizers.

The metallic minerals include oxides of iron, titanium, and aluminum, the common sulfides of base metals, compounds of arsenic, antimony, tellurium, and tungsten, and precious metals.

Table 7 shows some of the chief types of deposits formed by contact metasomatism, along with the chief minerals and a few examples of deposits.

Selected References

TABLE 7
Types of Mineral Deposits Formed by Contact Metasomatism, with the Chief Constituent Minerals and Examples

Deposit	Chief Metallic Minerals	Some Examples of Deposits
Iron	Magnetite, hematite	Cornwall, Pa.; Iron Springs, Utah; Fierro, N. Mex.; Banat, Hungary
Copper	Chalcopyrite, bornite, pyrite, pyrrhotite, sphalerite, molybdenite, iron oxides	Some deposits at Morenci and Bisbee, Ariz., and Bingham, Utah; Cananea and Matehuala, Mexico; Suan, Korea
Zinc	Sphalerite with magnetite, sulfides of iron and lead	Hanover, N. Mex.; Kamioka, Japan
Lead	Galena, magnetite, and sulfides of iron, copper, and zinc	Magdalena, N. Mex.; Inyo County, Calif.
Tin	Cassiterite, wolframite, magnetite, scheelite, pyrrhotite	Pitkäranta, Finland; Saxony, Germany; Dartmoor, England; Malaya
Tungsten	Scheelite with molybdenite and minor sulfides	Mill City, Nev.; Bishop, Inyo County, Calif.; King Island, Australia
Molybdenum	Molybdenite, pyrite	Yetholm, Australia; Azegour, Morocco; Buckingham, Quebec
Graphite	Graphite	Adirondacks, N. Y.; Clay County, Ala.; Buckingham, Quebec; Ceylon; Madagascar; South Australia
Gold	Gold, arsenopyrite, magnetite, and sulfides of iron and copper	Cable, Mont.; Hedley, British Columbia; Suan, Korea
Manganese	Manganese, iron oxides, silicates	Långban, Sweden
Emery	Magnetite, corundum, ilmenite, spinel	Virginia; Peekskill, N. Y.; Turkey; Greece
Garnet	Garnet and silicates	Widespread
Corundum	Corundum with magnetite, garnet, and other silicates	Peekskill, N. Y.; Chester, Mass.

SELECTED REFERENCES ON CONTACT METASOMATISM

Ore Deposits of the Western United States—Lindgren Volume. Am. Inst. Min. Met. Eng., New York, 1933. Chap. 11, Part III, "Pyrometasomatic Deposits," by Adolph Knopf, pp. 537–556. Brief review of western occurrences illustrating Lindgren's classification.

"Clifton-Morenci District, Arizona." W. Lindgren. *U. S. Geol. Surv. Prof. Paper* 43, 1905. Outstanding work on contact metasomatism, particularly the transfer of materials from magma to invaded rocks.

"Occurrence of Ore on Limestone Side of Garnet Zones." J. B. Umpleby. *Calif. Univ.* (*Berkeley*) *Pub.* 10:25–37, 1916. Review of contact-metasomatic deposits showing that ore bodies lie on the outside of the garnet zone.

"Ore Deposits at the Contacts of Intrusive Rocks and Limestones." J. F. Kemp. *Econ. Geol.* 2:1–13, 1907.

Economic Mineral Deposits. 2nd Edit. Alan M. Bateman. John Wiley & Sons, New York, 1950. Chap. 5·3, contact metasomatism.

"Geology and Ore Deposits of the Mackay Region, Idaho." J. B. Umpleby. *U. S. Geol. Surv. Prof. Paper* 97, 1917. Detailed account of contact-metasomatic deposits and estimates of magmatic accessions.

"Contact Metamorphism at Bingham, Utah." W. Lindgren. *Geol. Soc. Amer. Bull.* 35:507–534, 1924. Metasomatic changes and estimates of materials added and subtracted.

"Contact Metamorphic Tungsten Deposits of the United States." F. L. Hess and E. S. Larsen. *U. S. Geol. Surv. Bull.* 725–D, 1922.

"Iron Ore Deposits at Cornwall, Pa." W. O. Hickok. *Econ. Geol.* 28:193–255, 1933. Contact-metasomatic magnetite deposits.

"Contact Pyrometasomatic Aureoles." H. Schmitt. *A.I.M.E. Tech. Pub.* 2357, 1948. Localization of tactite zones by structure.

CHAPTER NINE

Hydrothermal Processes

PRINCIPLES

Thus far, the mineral-forming processes that have taken place within the magma chamber have been considered first, then those that have been formed just outside the magma chamber, both resulting during magma consolidation. It is now appropriate to follow magmatic processes a step further and consider the deposits that result as near after effects of magma consolidation. They are called *hydrothermal processes* and *hydrothermal deposits*.

The discussions in Chapters 4 and 5 have revealed that magma differentiation and magma consolidation yield an end product of magmatic fluids in which there may be concentrated the metals that may have been present in the original magma. These hydrothermal solutions carry out in solution the constituents of mineral deposits from the consolidating intrusive to the site of deposition and are considered to be the major factor in the formation of epigenetic mineral deposits. They are liquids to start with, or become liquids, and behave as chemical solutions operating under elevated temperature and pressure. They are hottest near the igneous source and gradually lose heat with increased distance from the intrusive source. Thus, when deposition of their mineral content takes place they give rise to high-temperature hydrothermal deposits nearest the igneous source, to intermediate-temperature deposits at some distance outward, and to low-temperature deposits farthest outward. Lindgren has designated these three groups as *hypothermal, mesothermal,* and *epithermal* deposits, according to the temperatures, pressures, and geologic relations under which they were formed, as indicated by the contained minerals.

It is only natural that the source of mineralizing solutions has been from early times a subject of enquiry. Before the time of Werner (see

Chap. 3) the ore solutions were considered to be diffuse ascending waters of uncertain source. Werner advocated that they were descending waters derived from a primeval ocean. The concept that mineralizing solutions are of magmatic origin gained ground rapidly in the middle of the last century and has persisted to the present despite intervening controversies regarding lateral secretion, surface waters, and ascending hot meteoric waters. It has of course been realized that ascending hot magmatic waters may mingle with cold ground waters, and also that magmatic volatiles may discharge into ground water, giving rise to mineralizing solutions. The evidence that deep mines are dry indicates that, at the great depths at which most mineral deposits are formed, the hydrothermal solutions had little or no contact with meteoric waters.

The concept of the formation of many types of mineral deposits from water solutions evolved slowly as supporting evidence accumulated. Familiar chemical reactions were called upon to account for natural deposition of ore minerals in the rocks. For the evidence that hot solutions can dissolve, transport, and deposit the constituents of hydrothermal mineral deposits, the reader is referred to Chapter 5.

In their journey through the rocks the hydrothermal solutions may lose their mineral content by deposition in the various kinds of openings in rocks, to form *cavity-filling deposits,* or by metasomatic replacement of the rocks, to form *replacement deposits.* The filling of cavities by precipitation from solution may at the same time be accompanied by replacement of the walls of the cavities. Thus, there may be a gradation between these two types of mineral deposits. In general, replacement dominates under the conditions of high temperatures and pressures near the intrusives where hypothermal deposits are formed, and cavity filling dominates under the conditions of low temperatures and pressures where epithermal deposits are formed; both are characteristic of the intermediate or mesothermal deposits.

The hydrothermal group of mineral deposits are the most numerous and diverse of all the types. Most of the literature on mineral deposits has hinged around them because they are encountered everywhere and they present many problems of origin. They constitute the vast array of mineral deposits that supply the major part of the metals and minerals that form the backbone of modern industrial development and represent security in peace or war. From such deposits are won most of the gold and silver, copper, lead, zinc, tin, tungsten, mercury, antimony, molybdenum, most of the minor metals, and many nonmetallic minerals. Consequently, these deposits have been investigated, mined, and written about far more than any other type. They have given rise to many of the great mining districts of the world, and the lore of mining has largely sprung up around them.

Despite the diversity of a large group of deposits of which no two are alike, certain fundamental principles underlie the formation of each, and the reasons why they occur where they do are now fairly well understood. The nature of the hydrothermal solutions must be interpreted by inference and by analogy with certain types of hot springs. Their action is visible only in the form of mineral deposits or the alteration of the enclosing wall rocks, both of which are chemical effects. As the term hydrothermal implies, they are hot waters that probably range in temperature from 500° C down to 50° C. The mineral substances are presumed for the most part to be carried in chemical solution, but colloidal solutions are also thought to have been operative. Their chemical character is discussed in Chapter 5.

Essential features for the formation of hydrothermal mineral deposits are: (1) available mineralizing solutions; (2) available openings in rocks through which the solutions may travel; (3) available sites for the deposition from solution of the mineral content; (4) chemical reactions that result in deposition; and (5) sufficient concentration of deposited mineral matter to constitute economic deposits.

Some of these features will now be considered.

ROCK OPENINGS

If solutions are to travel from their source to the site of mineral deposition, often over long distances, connected openings in the rocks are prime essentials. When one considers that some hydrothermal deposits consist of hundreds of millions of tons of extraneous materials that have been brought in to their present sites, it means that through channelways of travel must have been available. Furthermore, for cavity-filled deposits there must have been existing cavities to fill. Equally obviously, replacement deposits could not form unless the solutions could reach the rock that underwent replacement. Consequently, interconnected openings in rocks are essential for the formation of hydrothermal deposits. Likewise, they are necessary for the existence of bodies of ground water, oil, and gas.

The various openings in rocks that may serve as receptacles for ores or that permit the movement of solutions or ions through the rock may be grouped as follows:

Original Openings in Rocks

1. Pore space
2. Crystal structure
3. Vesicles or "blow holes"
4. Lava drain channels
5. Cooling cracks
6. Igneous breccia cavities
7. Bedding planes

Openings Induced in Rocks

1. Fissures
2. Shear-zone cavities
3. Cavities due to folding and warping
   a. Saddle reefs
   b. Pitches and flats
   c. Fold crackling and slumping
4. Volcanic pipes
5. Tectonic breccias
6. Collapse breccias
7. Solution caves
8. Rock alteration openings

Most of the cavities listed above may under rare and special circumstances become filled to form varied types of hydrothermal mineral deposits. Some serve only as conduits for mineralizing solutions. Some permit ingress of materials for replacement; others, such as pore spaces, serve as receptacles or conduits for water, oil, or gas. Some play a more important role in replacement than in cavity filling. The formation of hydrothermal deposits is dependent upon them.

Pore spaces are the small openings between grains, capable of absorbing a fluid or gas. They make rocks permeable and serve as containers for ores, water, gas, and oil. The oil and gas pools of the world, as well as subsurface water supplies, are largely contained in rock pores.

Porosity of a rock is the volume of pore space expressed in percentage of the volume of rock. It ranges from almost zero in some igneous rocks to 53 percent in loose soils. If the rock grains are spheres the amount of porosity is the same regardless of the size of the spheres, although obviously there are more pores with smaller spheres. A rock composed of angular grains has greater porosity than one of spherical grains, and aggregates of fine angular grains have considerably greater porosity than those of coarse materials. For example, fine clayey sand has 53 percent porosity, but coarser material has only about 33 percent. The average porosity of 25 granites is 0.369 percent; of 29 oil sands 19.1 percent; of 14 clays 28.4 percent.

Permeability of a rock exists by virtue of its porosity, but a rock may be porous and not permeable. Permeability is dependent upon the size of pores, the amount of pore space, and particularly upon interconnection of pores. An electric light bulb has much porosity but no permeability. The smaller the pores, the greater the surface exposed, the greater the friction, and the smaller the flow of liquid through them. If fine pores become wet the large exposed surface exerts a very tight hold on the contained fluid, and water-wet shales and clays become essentially impermeable. Coarse-pored rocks, on the other hand, even if of low porosity, are quite permeable provided the pores are connected.

Crystal structure openings that lie between the atoms of a crystal may permit the diffusion through such openings of ions of smaller radius than the atoms that make up the crystal. Mica is a particularly favorable example in this respect, as its atoms are bound together less tightly in one direction than in others. This diffusion through crystal structures aids replacement and permits additions of materials within crystals.

Bedding planes are well-known features of nearly all sedimentary rocks, which permit ingress of hydrothermal solutions and replacement of adjacent walls by ore.

Vesicles or "blow holes" are openings produced by expanding vapors in the upper parts of many basaltic lava flows. They are tubular in shape and commonly are spaced less than 2 inches apart. If the vesicles are filled, the rock is called an amygdaloid. If they are closely crowded together they form a cellular rock like a sponge, called scoria. Vesicles filled by copper have given rise to large ore deposits.

Volcanic flow drains form in lava flows when the outside of a lava flow has solidified and liquid lava in the interior drains out, leaving a pipe or tunnel.

Cooling cracks are formed as a result of contraction in cooling igneous rocks. They may be regularly spaced joints that divide the rock into columns, cubic or rectangular blocks, or irregular cracklings.

Igneous breccia cavities are openings between angular fragments and blocks of igneous rock enclosed within a matrix of finer igneous materials. There are volcanic breccias, formed by explosive volcanic activity, and intrusion breccias. Both may be quite permeable.

Fissures are continuous tabular openings formed by rupturing of rocks and are generally of considerable length and depth (Fig. 9–3). Fissures may or may not be faults. Thus, faults are fissures but not all fissures are faults. Such openings constitute long and continuous channelways for solutions, and when occupied by introduced mineral matter they form *fissure veins,* one of the commonest types of mineral deposits.

Shear-zone cavities result when a rock has been subjected to shearing along a zone of rupture, giving rise to innumerable closely spaced, more or less parallel, discontinuous surfaces. The thin sheetlike openings, mostly of infinitesimal size, make excellent channelways for solutions, as evidenced by the copious water flows where they are cut in tunnels and mines. Because of the minute openings, only minor open-space deposition can occur, but the large specific surface available makes shear zones favorable localizers of replacement deposits.

Fold openings formed by flexing and folding of sedimentary strata give rise to: (1) *saddle reef* openings at the crests of close folds composed of alternate hard and soft beds (Fig. 9–12); (2) *pitches* which are highly inclined; and *flats* (Fig. 9–13), which are openings formed by the parting of beds under gentle slumping; (3) longitudinal cracks along the crests and troughs of folds (Fig. 9–14).

Volcanic pipes result when explosive volcanic activity bores pipelike openings through rocks. The fragments and dust blown out may fall back or be washed back into the pipe, forming an angular breccia with openings between the rock fragments (Fig. 9–17). Pipes are thus excellent confined conduits for hydrothermal solutions, which may give rise to cavity-filling or replacement deposits.

Breccias may be formed by the fragmentation of any brittle rock; folding, faulting, intrusion, or other tectonic forces form *tectonic breccias;* or collapse of rock overlying an opening gives rise to *collapse breccias;* sedimentation forms *sedimentary breccias*. The openings between the fragments provide space for circulation of solutions and for mineral deposition.

Solution openings, such as caves and enlarged fissures in soluble rocks, supply channelways and open spaces for cavity filling.

Rock alteration openings result from the alteration of wall rocks by hydrothermal solutions. Tests show that they are more permeable than unaltered rocks and permit ingress of mineralizing solutions.

The part played in mineralization by the various openings enumerated above is discussed further in a later part of this chapter.

MOVEMENT OF SOLUTIONS THROUGH ROCKS

The movement of solutions through rocks is favored where long continuous openings such as fissures are available or where there are smaller interconnected openings as in shear zones, vesicular lava beds, or permeable sediments. To supply the scores of millions of tons of ore in some deposits, huge quantities of solutions must have been necessary. Hence, large channelways must have been available, and they must have been fairly well confined or else their mineral matter would have become too dispersed. A widespread volcanic breccia, for example, might be quite permeable throughout, but, because of this widespread permeability, mineralizing solutions would be so dispersed that the minerals deposited from them would be too diffuse to constitute ore. The large channelways serve as the main freight lines, and small fractures, cleavages, or pore spaces constitute distribution lines for solutions to reach the front lines of mineral deposition. Here diffusion acting over short distances may deliver the mineral matter to the actual point of deposition. Thus, it seems that rock openings are essential to transmit the mineral content of solutions to sites of deposition.

The size of rock grains is important, not only in the flow of solutions through rocks but particularly in aiding chemical reactions between rock matter and solutions. The surface area of a unit mass of a finely divided substance is called the specific surface. With large surface area and small particle size, specific surface is large, pores are small, and permeability is low. However, these features are opportune for mineral deposition because a large specific surface permits greater contact between rock and solution and, therefore, greater opportunity for reaction between them.

EFFECT OF HOST ROCKS

For some unknown reason, reactive wall rocks not in equilibrium with adjacent solutions exert a profound effect upon hydrothermal mineral deposition, particularly in replacement. This is shown by field evidence the world over. Certain carbonate beds are congenial to ore deposition, whereas others are not. Metallizing solutions seem almost to have been empowered with peculiar discrimination in passing by certain limestone beds and replacing others at Kennecott, Alaska; Leadville, Colorado; Bisbee, Arizona; and other places. The preference of silver-lead-zinc ore for dolomitic limestone over pure limestone in northern Mexico has been pointed out by Hayward and Triplett. However, at Santa Eulalia, Mexico, the reverse is true, and massive sulfide ore in limestone terminates abruptly at dolomite. The selective mineralization of "greenstone" in preference to adjacent "porphyry" and granite has been noted in the gold camps of northern Ontario. At Mt. Bischoff in Australia tin ore widens in slate but is pinched in quartzite, and at Rio Tinto, Spain, copper ore widens in porphyry and is pinched in slate (Fig. 14-4). These examples and many others that could be cited illustrate that ore deposition, particularly replacement, is influenced by the character of the host rock.

FACTORS AFFECTING DEPOSITION

The deposition of minerals from hydrothermal solutions is mainly the result of chemical changes, reactions between the solutions and wall rocks or vein materials, and changes in temperature and pressure. Various causes of mineral deposition are given in Chapter 2.

CHEMICAL CHANGES. It is inevitable that hydrothermal solutions in their long journey through the rocks must undergo some chemical changes by reaction with the wall rocks.

Reactions with silicate rocks would make them alkaline or more alkaline. The alkalinity or acidity alone (pH) may determine when deposition will occur. In replacement, the substitution of new minerals for older ones can, of course, take place only by reaction between solution and solid. In general, solids more soluble than those in solution go into solution, and the less soluble ones are deposited. Strongly reactive wall rocks, such as limestone, that are out of equilibrium with the solutions bring about a rapid chemical change accompanied by deposition. Colloidal solutions may encounter ions of opposite electrical charge which bring about deposition from solution. There are also other complicated chemical reactions, beyond the scope of this discussion, that might tend to bring about deposition of minerals from solution.

TEMPERATURE AND PRESSURE. We have seen that hydrothermal solutions are initiated under conditions of elevated temperature and pressure. In general, a drop in temperature and pressure decreases solubility and promotes deposition.

In their outward journey the solutions lose heat to the rocks through which they pass, the rate of loss being dependent upon the capacity of the wall rocks to conduct heat away, and this in turn depends upon the amount of solution moving past and upon chemical reactions that give out heat. In the initial stages of circulation with cool wall rocks, the temperature drop will be relatively rapid, but continued flow of the solutions will heat the wall rocks to the temperature of the solutions, at which time the loss of heat will slow down. The nature of the rock openings also affects heat loss. Rapid flow through a straight-walled fissure would entail less loss of heat than flow through the intricate openings of a breccia with large specific surface, and the initial drop in temperature would be rapid. Such a drop in temperature would favor deposition. Once heated up, however, the breccia would not remove much heat, and some resolution might occur. As

Wall Rock Alteration

the solutions travel farther from their source more loss of heat would take place and more deposition would occur. Eventually, as the solutions become warm instead of hot, the last of the substances and the most soluble carried in solution would be deposited.

The solutions in their upward journey through zones of lower pressure would also undergo a lowering in pressure, which promotes precipitation. If pressure release should occur suddenly by passage from a constricted opening to an open porous rock, an action similar to the outflow of steam escaping from a pipe into open air, a sudden dumping of mineral content might result, thus localizing mineral deposition.

WALL ROCK ALTERATION

Extended field observations have shown that the wall rocks of hydrothermal mineral deposits have undergone some alteration. The walls

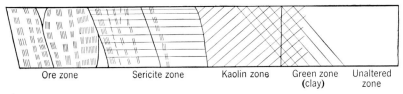

FIGURE 9–1. Cross section showing wall rock alteration outward from fault vein at Butte, Montana. (Modified from R. Sales and C. Meyer, *Quart. Jour. Colo. School of Mines*, 1950.)

parallel to a fissure vein commonly disclose a zone of uniform width of alteration parallel to the fissure, whose width varies according to the width of the vein. If veins are closely spaced the alteration halo of one vein may merge with that of another so that a large volume of rock may be altered, as in the central zone at Butte, Montana. With the "porphyry coppers" the entire ore deposit and a halo beyond the ore is profoundly altered. The alteration is more pronounced with medium- and high-temperature deposits than with low-temperature deposits.

The nature of the hydrothermal rock alteration also varies somewhat with the temperature of formation of the ores and with the kind of rock. Dark rocks are changed to a light gray; light-colored rocks are also changed in composition. The original rock minerals are altered to a white mica (sericite) and in places to silica. Outside this more intense sericite alteration there is generally a zone of less intense alteration containing clay minerals (Fig. 9–1). This zone merges into unaltered rocks.

The presence of such a halo of rock alteration indicates hydrothermal action, which generally means the existence of hydrothermal mineral deposits that may be hidden. Thus, an alteration halo can be used as a practical guide in ore-finding.

LOCALIZATION OF HYDROTHERMAL MINERALIZATION

With waning mineral resources, the urge to discover more in order to sustain national economy is ever present. Consequently, any clues that can be deciphered as to the whereabouts of mineral deposits hold great practical as well as scientific importance.

For this purpose we can turn to the information gained from the study of existing deposits as to why they occur where they do, and in a scientific manner apply the information so gained to the search for new deposits. Mineral deposits are extremely sparse occurrences within very wide areas. Concerted events must have taken place to cause them to come into existence in given localities. Such events are not accidents of nature but are due to specific geologic causes. Some of these causes seem clear; others are still an enigma that geologic studies are attempting to solve. For the most part there is no single cause for the localization of ore in a specific locality, but rather there is a coincidence of causes, which we need to unravel. Among these are the location of intrusives, the chemical and physical character of the host rock, structural features, kinds of rock openings, changes in size of openings, and depth of formation.

INTRUSIVES. Since most hydrothermal solutions spring from igneous intrusives, the position of the parent intrusive determines in a broad way the localization of the ore. In many places this is shown without much doubt by a zonal arrangement of minerals about an intrusive, as in Bingham, Utah, where copper ores occur in the intrusive, with contact-metasomatic ores adjacent to the intrusive, copper-zinc ores near by, and silver-lead ores more remote. In other places there is no evidence of any intrusive that supplied the hydrothermal solutions, as in the zinc-lead region of the Tri-State district or at Kennecott, Alaska. There may, of course, be undisclosed intrusives at depth. So firmly has the concept of localization of ores by intrusives become fixed in the minds of geologists that many are prone to assume without evidence that the ores sprang from known intrusives near by. Thus, dikes are often assumed to be the source of specific ores, and much needless exploratory work has been done on such assumptions, whereas both dikes and ores more probably arose from an underlying intrusive body. Cupolas on the tops of intrusives (see Chap. 14) in a few places seem to localize ores near by, and at Bisbee, Arizona, irregular outliers of an underlying intrusive have clusters of ore bodies about them.

CHARACTER OF HOST ROCK. Hydrothermal deposits may be formed in any kind of rock, but certain ones are definitely more congenial to ore formation than others, as was previously mentioned. In cavity-filling deposits, the opening, more than the character of the containing rock, localizes the ore, although both the physical and chemical nature of the host rock may play a part in determining the exact place of deposition as well as determining the location and shape of the cavity. For example, brittle rocks shatter more readily than nonbrittle rocks and thereby localize fractures or breccias; carbonate rocks permit solution openings. In replacement deposits the chemical character of the host rock plays an important role. However, regardless of how chemically favorable a host rock may be, ore deposition cannot occur unless rock openings are available to afford sites for cavity filling or to permit entry of solutions for replacement. Permeability is necessary, and this may be supplied by interconnected openings of fissibility, brecciation, cleavage planes of minerals, joints, minor fractures, or other features. From this it may be seen that the influence of the host rock in localizing hydrothermal mineral deposits may be physical, chemical, or both.

STRUCTURAL FEATURES. Various structural features are effective in localizing mineral deposits. Some geologists perhaps go too far in assuming that structural features are the chief or only controls of localization of deposits. However, they have often been shown to be the dominant controls.

Fissures perhaps dominate among structural features in localizing mineral deposits. They serve a dual purpose: they are themselves loci of mineral deposits, and they act as channelways to conduct ore fluids to rocks susceptible of replacement. The intersection of fissures with favorable rocks is often utilized in searching for replacement deposits. Both are necessary. There may be a limestone congenial to replacement but no replacement in it can occur unless the solutions can travel to it, and, conversely, the conduit may be available but favorable rocks are lacking.

Multiple fissures and *shear zones* are important localizers of mineral deposits in a manner similar to that of fissures. *Fissure intersections* are particularly favorable sites. Close pitching *folds* and *drag folds*, such as those of the great Homestake gold mine, have localized many valuable deposits. *Breccias* are unusually favorable ore loci for both cavity-filling and replacement deposits.

MINERAL SEQUENCE

In hydrothermal deposits sequences of mineral deposition, and commonly repetition and overlapping, are customary features. In cavity-

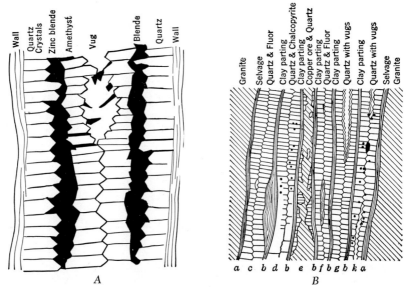

FIGURE 9–2. A, diagram of quartz crystals that have grown outward from vein walls toward open center, forming comb structure, and a vug in center containing latest crystals. B, crustified tin vein in which reopening and redeposition of crusts have occurred. Redruth, Cornwall, England. (After J. A. Phillips, *Ore Deposits*.)

filling deposits the ore is built up in successive layers, or crust by crust, called *crustification*, a younger crust being deposited on an older one (Fig. 9–2). In places, eight or ten minerals may occur or recur in one mineral sequence. The last minerals to form are perched on older crystals in unfilled central portions of the deposit, called *vugs*. The cause of such mineral sequences in cavity filling is generally considered to be the decreasing solubility of the minerals in solution under decreasing temperature and pressure, the least soluble minerals being deposited first and the most soluble last. However, relative solubility and alkalinity or acidity (pH) and other factors also enter in.

Study of mineral sequences (paragenesis) throws light on the chemical character, continuity in time, and changing character of the mineralizing solutions. It also gives a hint of possible changes in character of ore in depth inasmuch as the earlier minerals of a sequence might be expected in greater quantity in deeper, hotter zones.

CAVITY FILLING

The filling of open cavities in rocks was for a long while thought to be the only mode of formation of mineral deposits. Later, when it was realized that the great massive replacement deposits were not the filling of large openings, the number of deposits ascribed to cavity filling became smaller. However, it is probably still true that there are more deposits formed by the filling of cavities than by any other ore-forming process, and of these deposits the fissure vein is the most common.

Cavity filling consists of the deposition from solution of minerals in rock openings. The first minerals to be deposited line the walls of the cavity and grow inward toward the opening where the solution exists. Commonly they have inward-pointing crystal faces. In a fissure, both walls are successively covered by later crusts until the fissure vein is filled and a crustified vein results (Fig. 9–2B). In breccias successive crusts may surround the rock fragments, with the crystals pointing toward the interstices. In some places the crusts of opposite walls do not meet, and central open spaces, or vugs, remain, in which beautiful natural crystals may be perched. These are eagerly sought by mineral collectors because they often prove to be treasure-houses of the lovely and rare crystals that adorn museums.

Cavity filling involves two separate processes: (1) the formation of the opening to start with, and (2) the deposition of the minerals. The two processes are generally separated by a long time interval, but in a few instances, as in some fissure veins, both may proceed simultaneously.

The deposits that result from cavity filling may be grouped as follows:

1. Fissure veins.
2. Shear zone deposits.
3. Stockworks.
4. Saddle reefs.
5. Ladder veins.
6. Fold cracks.
7. Breccia deposits.
8. Solution cavity deposits.
9. Pore-space fillings.
10. Vesicular fillings.

Each of these types of deposits will be discussed separately in the order given above.

1. FISSURE VEINS

Fissure veins are the earliest described type of bedrock deposit, and the lore of mining has grown around them. In simple terms, a fissure vein is a fissure in the rocks filled with mineral matter. It forms a tabular ore body of which two dimensions are much greater than the third. Fissure veins are the most important and most widespread of the cavity fillings and yield a great variety of metals and minerals.

The formation of a fissure vein involves (1) the formation of the fissure itself, and (2) the filling of the fissure with mineral matter. The two may be separated by a long interval of time, although in some cases the formation of the fissure and its filling may be essentially contemporaneous.

Fissures are formed by tension, compression, or torsion forces operating within the earth's crust by means of which rocks are split or fissured. Their formation may involve faulting, or the movement of one wall past the other. Most fissure veins are faults, but those formed by tension forces commonly have no fault movement. Where faulting has occurred the walls may be smoothed, polished, and slicked, and are then called slickensides. Fissures may also be formed or enlarged at the time of mineralization by the intrusive force of the mineralizing solutions which acts as a wedge from below and spreads the rocks apart along some crack or line of weakness. Commonly, fissure veins suffer reopening and additional mineral deposition, which may be repeated more than once. The reopening generally occurs along one wall because the vein is weakest at this place.

There are several varieties of fissure veins, such as simple single cracks, a sheeted zone of closely spaced cracks, and linked, chambered, and lenticular, as may be seen from Figure 9–3. Each of these may have crustified or massive filling.

Most fissure veins are only a few feet in width, although some attain a few tens of feet. In length they mostly range from a few hundred to a few thousand feet, but some extend a few miles. Their depth is measured in hundreds or thousands of feet. Depths of over a mile have been reached on fissure veins in Ontario, California, India, and elsewhere. Few veins are vertical; most of them are highly inclined. The inclination is referred to as *dip*, which is measured as the angle between the vein and the horizontal. The horizontal course of a vein is called the *strike*. Most veins curve gently along both strike and dip, and they exhibit irregularity in width with pinches and swells. Such irregularities in veins whose walls have moved may serve to hold the walls separated and provide openings, as in Figure 9–4.

Fissure Veins   113

Fissures seldom occur alone but tend to be in groups, and if the fissures of a group are of the same age and have approximately parallel strike and dip they constitute a *fissure system*. There may be one or

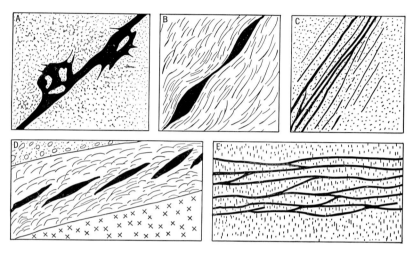

FIGURE 9–3. Diagrams of varieties of fissure veins. *A*, chambered vein (after Becker); *B*, dilation veins in schist; *C*, sheeted vein, Cripple Creek, Colorado; *D*, en echelon veins in schist; *E*, linked vein.

several systems. Different systems intersect each other, and later ones generally dislocate earlier ones. At Butte, Montana, there are seven separate fissure systems, each of which dislocates and displaces

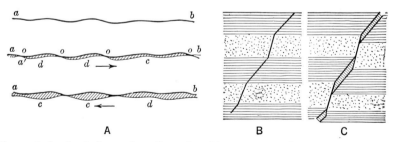

FIGURE 9–4. *A*, pinches and swells produced by movement along irregular fissures, with walls supported at opposite bulges (de la Beche); *B* and *C*, openings produced by movement along fissure that bends slightly in different rocks.

earlier ones (Fig. 9–5). The fact that fissures commonly occur in systems always holds the hope that an unknown vein may be discovered parallel to a known one.

Two intersecting systems may differ in age, as in Figure 9–6E, or they may be of the same age, as in Figure 9–6D. Those differing in age generally displace each other, carry different ores, and carry sepa-

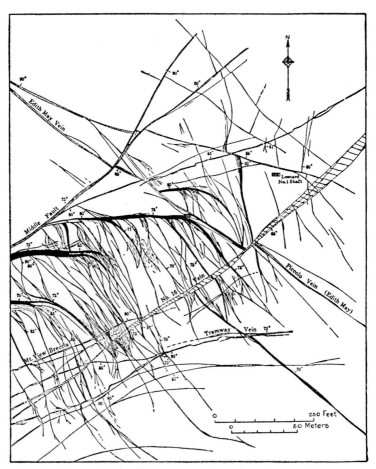

FIGURE 9–5. Part of 1,200-level Leonard mine, Butte, Montana, showing fissure systems with E-W, NW, NE veins and faults; also "horsetail" structure of some E-W veins. (Reno Sales, A.I.M.E.)

rate legal rights, whereas intersecting systems of the same age carry similar ores and do not displace each other. Five different fissure systems occur in the Freiberg district of Germany, each carrying different ores. The legal features referred to have given rise to a great deal of mining litigation. Intersecting fissures of the same age form junctions, and the first locator of a mining claim on a vein is entitled

Fissure Veins 115

to follow it down the dip (within the end lines of his claim) regardless of whether its dip carries it below the surface of another property. He is also entitled to underground junctions of the original vein. A faulted intersection, however, does not carry junction rights. Consequently, the interpretation of junctions or fault intersections is of practical and legal interest.

Fissures and veins tend to respond physically in passing from one rock formation to another because of differences in behavior between

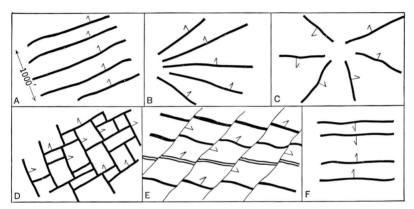

FIGURE 9–6. Classes of fissure systems. A, parallel system; B, fan-shaped; C, radial; D, intersecting cognate; E, intersecting systems; F, conjugated.

brittle and nonbrittle rocks when under stress and also because of compositional differences. A strong fissure in one rock may split into a number of stringers in passing into another rock, or it may pinch, be deflected, or die out, as may be seen in Figures 9–7 and 9–8.

The termination of a fissure vein is an always present fear for a miner, as this means that the "pay runs out." Fissures cannot extend unlimited distances, however, and all must terminate. Chiefly, fissures terminate: (1) against other fissures or faults (Fig. 9–6E); (2) upon entering an uncongenial rock formation (Fig. 9–7A, D); (3) by gradually pinching out (Fig. 9–7C, F); (4) by being cut off by later intrusives. Since a fissure cannot enter a rock formation younger in age than itself, it is obviously essential to know the relative age of the fissure and of the rocks that lie in the path of its strike and dip.

An intriguing feature of a fissure vein is the depth to which it may be expected to extend. The owner naturally hopes that it may go down for a mile or more, and any information that bears on this point is eagerly sought. There are some hints, however, that help the geolo-

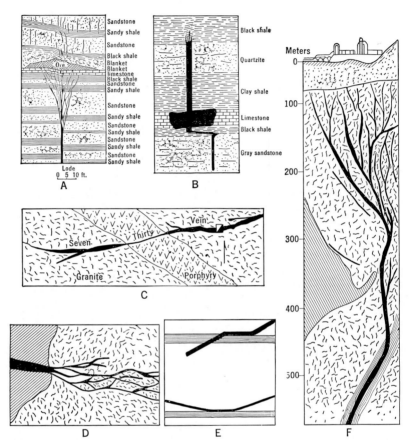

FIGURE 9–7. Relation of fissures to physical character of rocks. *A*, strong vein in brittle rocks that breaks into stringers and disappears upon entering nonbrittle rocks. (F. L. Ransome, *U. S. Geol. Survey.*) *B*, strong fissure at Ouray, Colorado, deflected in lower black shale and dying out in upper black shale. (J. D. Irving, *U. S. Geol. Survey.*) *C*, vein at Georgetown, Colorado, constricted in passage through porphyry. (J. E. Spurr and G. H. Garrey, *U. S. Geol. Survey.*) *D*, division of strong vein into stringers upon entering schist. Gottlob Morgengang vein, Freiberg, Germany. (R. Beck, *Nature of Ore Deposits.*) *E*, change of course (upper) and reflection (lower) of a vein upon encountering an incompetent rock. *F*, strong fissure in schist diverging upward into minor fissures in brittle andesite, Mazarron, Spain. (After R. Pilz, *Zeit. Prakt. Geol.*)

gist to make better than a wild guess as to the expected depth. Most fissure veins are formed deep below the earth's surface. They are seen on the surface because the overlying rock has been eroded away to expose the vein. The intersection of the vein with the surface, called the *outcrop*, may be near the top, middle, or lower part of the vein, depending upon the coincidence of depth of erosion. If the surface cuts across the upper part of the vein then a considerable depth of vein may lie below the surface, but if the lower part of the vein is

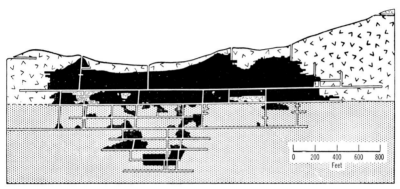

FIGURE 9–8. Decline in ore content from upper (stippled) wall rock to lower (dotted) wall rock. Black is ore. Tomboy mine, Telluride, Colorado. (H. E. McKinstry, *Mining Geology*, copyright 1948 by Prentice-Hall.)

intersected only the roots of the vein may be expected below (Fig. 9–9). In some places the geologic formations may permit an estimate of the amount of erosion. Another type of estimate can sometimes be made in gold veins by studying the amount of placer gold that has collected from the wearing away of the gold veins. Still another hint can be gained by the degree and intensity of supergene sulfide enrichment (see Chapter 11) that may have taken place. Where rich secondary ores overlie lean primary ores, much erosion is indicated. For example, a copper vein with primary ore averaging 1 percent copper overlaid by 500 feet of secondary ore with 5 percent copper, must have undergone a minimum of 2,000 feet of oxidation to supply the secondary ore, provided the grade and width of ore persisted unchanged above.

In fissure veins the values are rarely evenly distributed throughout. They tend to be concentrated toward either wall or in the center. Horizontally, some parts may be rich or lean or barren, and similarly, vertically. Large concentrations are called *ore shoots* (Fig. 9–10).

These vagaries of distribution of values constitute one of the hazards of mining fissure veins.

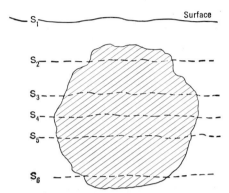

FIGURE 9-9. Relation of length and depth of fissure to outcrop length. Shaded area is a longitudinal section of a simple fissure vein that pinches out in all directions. Dashed lines represent successive surfaces at different stages of erosion. With surface at S_1 the vein is blind, i.e., does not outcrop. With surface at S_2 considerable depth of vein may be expected. With surface at S_6 only the roots of the vein lie below, and therefore only shallow depth may be expected.

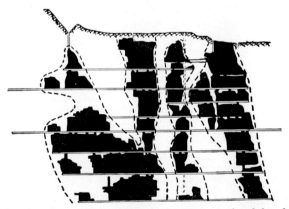

FIGURE 9-10. Ore shoots shown in vertical section, outlined by dashed lines. Black is mined-out areas. Horizontal lines are mine levels spaced 100 feet apart. Blue Bell mine, Arizona. (After V. C. Heikes, *U. S. Geol. Survey*.)

Some of the world's most famed ore deposits, both ancient and recent, are fissure veins. Included among them are some of the deepest and richest mines of the world, as well as some of the oldest worked

Shear Zone Deposits

deposits. They have yielded vast treasures of gold a[nd have] contributed largely to the world supply of copp[er,] tungsten, mercury, fluorspar. They have been som[e con-]tributors of antimony, cobalt, and uranium. A b[rief list of] fissure vein deposits is given in Table 8.

TABLE 8

Some Important Fissure Vein Deposits

Ores	Examples
Gold	Cripple Creek, Camp Bird, San Juan, Colorado; Mother Lode, Grass Valley, California; El Oro, Veta Madre, Mexico; Porcupine, Kirkland Lake, Ontario; Kalgoorlie, Australia
Silver	Sunshine, Idaho; Tonopah, Nevada; Fresnillo, Pachuca, Guanajuato, Mexico; Potosi, Huanchaca, Bolivia; Cobalt, Ontario; Tintic, Utah
Silver-lead	San Juan, Colorado; Clausthal, Freiberg, Germany; Przibram, Austria
Copper	Parts of Butte, Montana, and Cerro de Pasco, Peru; Walker mine, California
Lead	Clausthal, Freiberg, Germany; Przibram, Austria; Linares, Spain
Zinc	Butte, Montana; Sardinia
Tin	Llallagua, Huanani, Oruro, Bolivia; Cornwall, England; Erzgebirge, Germany
Antimony	Hunan, China; Mayenne, France
Cobalt	Cobalt, Ontario; Annaberg, Germany
Mercury	New Idria, California
Molybdenum	Temiskaming County, Quebec
Radium, uranium	Great Bear Lake, Canada; Joachimsthal, Czechoslovakia; Katanga, Belgian Congo
Tungsten	Boulder County, Colorado; Kiangsi, China; Chicote, Bolivia
Fluorspar	Illinois-Kentucky region
Barite	Missouri; Harz, Germany
Gems	Colombia

2. SHEAR ZONE DEPOSITS

A shear zone is a tabular-shaped zone of closely spaced rock shearing. The thin, sheetlike, connected openings serve as excellent channelways for mineralizing solutions, and some deposition of minerals takes place within the seams and crevices. The amount of open space is of course too small to accommodate bulk minerals, but gold with pyrite forms workable ore at Otago, New Zealand. Shear zones are generally wide and long and because of their large specific surface are favorable sites for replacement and give rise to many large and valuable deposits.

3. STOCKWORKS

Stockworks are a mass of rock interlaced by a network of small, closely spaced ore-bearing veinlets. The individual veinlets rarely exceed an inch or so in width or a few feet in length, and they are spaced a few inches to a few feet apart. The interveinlet areas may in part be impregnated by ore minerals. The veinlets consist of open-space fillings that may exhibit crustification and druses. The whole crackled mass of rock is mined as a unit and may attain dimensions

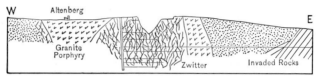

FIGURE 9–11. Stockworks of tin ore, Altenberg, Germany. (After Dalmer, *Zeit. Prakt. Geol.*)

of several thousand feet (Fig. 9–11). The shape of the deposits is irregular, and the walls merge into unprofitable country rock.

The stockwork veinlets are formed by crackling attendant upon cooling and contraction of the upper and marginal parts of intrusive igneous rocks and by the formation of irregular fissures produced by torsional or tensional forces. One unusual example at the great Veta Madre vein in Guanajuato, Mexico, was formed by the upper wall of the inclined vein riding over a hump in the lower wall, which produced a great crackled area in the upper wall rock the veinlets of which became filled with gold-silver ore.

Stockworks yield ores of tin, gold, silver, copper, lead, zinc, molybdenum, cobalt, and mercury. Much of the bedrock tin ore of the past has come from large stockworks at Altenberg and Zinnwald, Germany; Mt. Bischoff, Tasmania; Cornwall, England; and New South Wales. Disintegrated stockworks have yielded stream tin in the Netherlands Indies and Malaya. Silver ores are mined from a stockwork at Fresnillo, Mexico; gold-silver ores at Quartz Hill, Colorado; and low-grade gold ores at the Alaska Juneau mine in Alaska.

4. SADDLE REEFS

If one sharply arches a stack of writing paper, openings form between the sheets at the crest of the arch. Receptacles of similar shape, later filled by ore, are formed when alternating beds of hard and soft rocks are sharply folded. They resemble the cross section of a saddle, hence their name.

Ladder Veins

The saddle reefs of Bendigo, Australia (Fig. 9–12), are the type examples and have yielded over $300,000,000 in gold to a depth of 4,600 feet. The saddles are repeated in bed after bed down the axis of the fold. Some 15 lines of folds are known, and 5 have proved highly productive. The individual saddles are not more than 5 to 20 feet across, and the depth of the legs is mostly less than 100 feet. One saddle at Bendigo, however, has been followed horizontally for 9,000 feet. The ore is localized by the fold cavities and is mostly cavity filling, although replacement occurs also.

5. LADDER VEINS

When dikes of igneous rocks cool they contract and form joints that extend across the dike at right angles to the walls. If the dikes are vertical the joint cracks are almost horizontal, and their vertical spacing resembles the rungs of a ladder. If the cracks are filled with ore and are closely spaced the whole dike may be mined as a single ore body. The Palisades of the Hudson River represent similar joints that are formed vertically across a horizontal intrusive body. The individual veinlets are restricted to the dike, and their transverse direction is small.

The mode of formation by cooling contraction is not applicable to all deposits called ladder veins because some of the fissures extend beyond the walls of the dikes, and the total width of the cracks is greater than could occur by contraction. Such ladder veins have probably been formed by stresses exerted on the enclosing country rock in opposite directions on either side of the dike, causing cracking in a brittle dike.

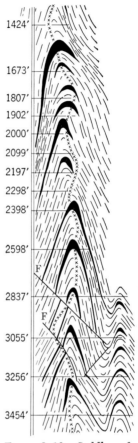

FIGURE 9–12. Saddle reefs in vertical section at Great Extended Hustlers shaft, Bendigo, Australia, with basic dike along "center country." (After T. W. Edgeworth David, *Geology of Commonwealth of Australia*, Vol. II.)

Ladder veins are not numerous or important but do constitute several gold deposits in Victoria, Australia, and in California; molybdenum ore in New South Wales; and copper ores in Telemarken, Norway.

6. FOLD CRACKS

As may be seen from Figures 9–13 and 9–14, cracks occur where sedimentary beds have been gently slumped or along the crests of arch

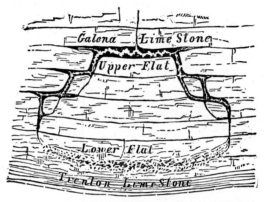

FIGURE 9–13. Pitches and flats containing lead-zinc ore, Mississippi Valley. (After T. C. Chamberlin, *Geology of Wisconsin*.)

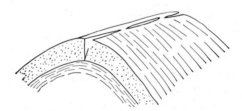

FIGURE 9–14. Tension cracks along anticline.

folds. This gives rise to *pitches* and *flats,* of which there are numerous lead-zinc examples in the upper Mississippi Valley, or to brecciation, or to tension cracks in uparched and downarched folds. The mineralization, chiefly of lead, zinc, and copper, is all cavity filling.

7. BRECCIA FILLINGS

Breccias are characterized by a haphazard arrangement of rock fragments with numerous intervening openings that permit ready entry of mineralizing solutions and afford excellent sites for mineral deposition (Fig. 9–15). Beautiful ores may result because of delicate crusts

FIGURE 9–15. Face of ore at depth. Number 2 vein, Lake Shore mine, Kirkland Lake, Ontario, showing breccia type of gold quartz ore. (Courtesy of W. T. Robson.)

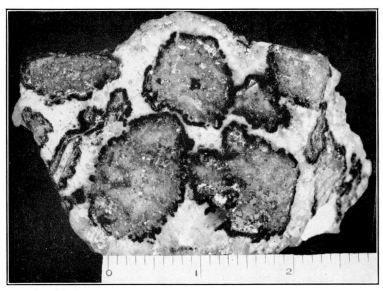

FIGURE 9–16. Rock breccia fragments coated by crusts of sphalerite (black) and quartz (white) forming cockade ore. Morococha, Peru. (Courtesy of F. M. Chace.)

of different-colored minerals around each fragment (Fig. 9–16), and often finely formed little crystals may be seen projecting into open spaces between fragments. Breccias are formed in several different ways and yield different kinds of mineral deposits.

Volcanic pipe breccia deposits result from explosive volcanic activity by which pipelike holes are drilled upward through rocks. The rock fragments and dust that are blown out fall back or are washed back into the pipe; some are materials arrested in the pipe on the way to be blown out. The pipes are vertical or highly inclined and are roughly circular or oval. The fragmented materials, confined by solid walls, make ideal channelways for mineralizing solutions and provide much open space for mineral deposition. The pipes yield gold, silver, copper, lead, and zinc. The Bassick breccia pipe in Colorado is 30 by 100 feet across and has been mined for gold to a depth of 1,400 feet. Figure 9–17 shows another example. The great Braden copper mine in Chile is considered to be a volcanic breccia pipe several thousand feet across, and the famed Cresson breccia pipe of Cripple Creek, Colorado, has yielded fabulous wealth in gold; one vug alone yielded $1,200,000.

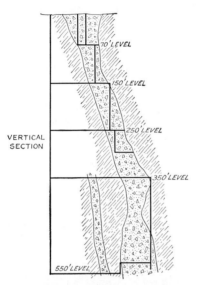

FIGURE 9–17. Gold-bearing volcanic breccia pipe in schist, Bull Domingo mine, near Lake City, Colorado. (W. H. Emmons, *U. S. Geol. Survey.*)

Collapse breccias result from the collapse of the roof of a large underground solution opening. The collapse may extend upward several hundred feet and be bordered by crackled and fissured rock. The resulting breccia is characterized by large angular fragments with little interfragment material and, therefore, with much open space between fragments available for mineral deposition. Breccias of this type have also been called such diagnostic names as "cave collapse breccias," "crackle," and "founder" breccias. The Cactus pipe in Utah (Fig. 9–18) has been mined to a depth of 900 feet, and the Emma chimney in Utah has yielded silver-lead-zinc ore to a depth of 1,000 feet.

Pore-Space Fillings

Tectonic breccias are produced by folding, faulting, intrusion, or other tectonic forces and have been referred to variously as crush, rubble, crackle, and shatter breccias (Fig. 9–19). They likewise yield much open space available for mineral deposition. Mineralizing solutions have availed themselves of the opportunity to deposit zinc ores to form large breccia deposits of this metal in Tennessee (Mascot, Jefferson City), Virginia, and the Tri-State zinc district (Kansas-Oklahoma-Missouri).

8. SOLUTION CAVITY DEPOSITS

The caves that are so widely known in limestone regions offer unusual opportunity for mineral deposition, although most caves are not filled by ore because they are shallow-seated features and are mostly above the reach of mineralizing solutions. Most caves are solution enlargements of pre-existing openings, such as joints, fissures, bedding planes, or fissure intersections, forming caves, galleries, and gash veins.

Small caves may be almost completely filled by mineral matter, but large caves generally contain only peripheral crusts of ore minerals, among which may be unusually large and beautiful crystals. Caves are common containers of lead-zinc ores in the Mississippi Valley region, and of oxidized ores of copper, lead, lead-silver, zinc, vanadium, fluorspar, and other metals and minerals in various parts of the world. The deposits are mostly small, although they may be rather rich.

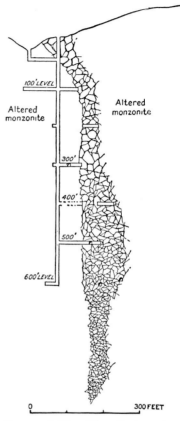

FIGURE 9–18. The Cactus pipe, Utah, composed of breccia ore. (After B. S. Butler and A. Locke, *Econ. Geol.*)

9. PORE-SPACE FILLINGS

Rock pores may contain ores, in addition to oil, gas, and water. Copper ores occupying pore spaces in sandstones are known in southwestern United States, in Permian red beds of the Urals, in the Don Basin and Djeskasjan, Russia, and in Germany and England. Similarly, lead ores occur in Rhenish Prussia and silver ores at Silver Reef, Utah. Quicksilver ores impregnate sandstones in California and Arkansas. Uranium and vanadium ores occupying sandstone pores are actively mined in Colorado, Utah, and Arizona.

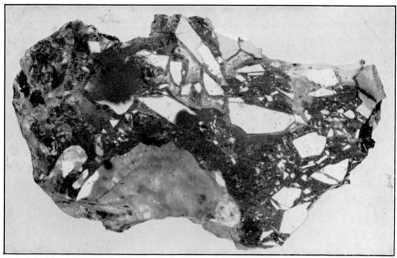

FIGURE 9–19. Tectonic breccia of rhyolite (light) containing silver-lead-zinc ore. Collins mine, Arizona. (Courtesy of H. Carlisle.)

10. VESICULAR FILLINGS

In basaltic lava flows, which remain liquid at lower temperatures than rhyolitic lavas, escaping gases that expand toward the freezing top form small tubelike openings that are termed vesicles. These are commonly filled by white nonmetallic minerals that give the black rock a startling "bird's-eye" appearance, called amygdaloid. Less commonly the vesicles are filled by native copper, the most notable example being at Keweenaw Peninsula, Michigan, where there are 400 basaltic lava flows of which many have amygdaloid tops. A number of these are commercially metallized with copper, and in a belt 100 miles long 26 highly productive mines since 1845 have yielded over $350,000,000 in dividends from ores that average about 0.8 percent copper. One lode has been mined continuously for a depth of 10,000 feet down the dip. Similar but less important vesicular fillings are known in Alaska, the Canadian Arctic, Brazil, Norway, and Siberia.

SELECTED REFERENCES ON CAVITY FILLING

Genesis of Ore Deposits. Franz Pošepný. Am. Inst. Min. Eng., New York, 1902. Pp. 54–188. A classic account of various types of cavity-filling deposits.

"Structural Relations of Ore Deposits." S. F. Emmons. In *Ore Deposits.* Am. Inst. Min. Eng., New York, 1913. Pp. 26–64. A general discussion of rock openings filled by ore.

Types of Ore Deposits. Edited by H. F. Bain. Min. and Sci. Press, San Francisco, 1911. Pp. 77–130 by H. F. Bain and E. R. Buckley. Openings of lead-zinc ores. Pp. 324–353, by R. A. F. Penrose, Jr. Causes of ore shoots.

"Structural Control of Ore Deposition." C. D. Hulin. *Econ. Geol.* 24:15–49, 1929. Kinds of openings that control ore deposition.

"Vein Systems of Anastre Basin, Colorado." W. S. Burbank. *Colo. Sci. Soc. Proc.* 13, No. 5, 1933. An interesting group of vein systems.

Ore Deposits as Related to Structural Features. Edited by W. H. Newhouse. Princeton Univ. Press, Princeton, N.J., 1942. Brief résumés of many hydrothermal deposits that exhibit localization of ore by structural features.

Structural Geology of Canadian Ore Deposits—A Symposium. Geol. Div. Can. Inst. Min. and Met., Montreal, 1948. This volume contains many descriptions of Canadian deposits that exhibit structural controls.

"Hydrothermal Differentiation." H. Neumann. *Econ. Geol.* 43:77–83, 1948. Formation of hydrothermal solutions by differentiation.

"Factors in the Localization of Mineralized Districts." C. D. Hulin. *A.I.M.E. Tech. Pub.* 1762, 1945. Igneous intrusion, structural controls, and mineralization.

Economic Mineral Deposits. 2nd Edit. Alan M. Bateman. John Wiley & Sons, New York, 1950. Hydrothermal processes, 94–136; gold deposits, 425–443, 448–453; silver, 457–474; copper, 491–498; lead-zinc, 532–545; tin, 548–552; minor metals, 599–621.

REPLACEMENT

Metasomatic replacement, or simply replacement, as it is generally called, is the most important process in the emplacement of hydrothermal mineral deposits. It dominates in the high- and medium-temperature deposits, and is important in the low-temperature deposits. It is also the controlling process in supergene sulfide enrichment as in contact metasomatism.

Replacement is a process of essentially simultaneous solution and deposition by which new minerals are substituted for earlier-formed minerals. By means of replacement wood may be transformed to silica (petrification), a single mineral may take the place of another, retaining its form and size (pseudomorphism), or a large body of ore may take the place of an equal volume of rock. The replacing minerals are carried in solution, and the replaced minerals are carried away in solution.

It used to be thought that large bodies of ore in limestone had been formed by the filling of huge solution caves, but, with the growing knowledge of replacement and particularly with information as to how it could be recognized, it was realized that such bodies had been formed by replacement. Likewise, many bodies that formerly were thought to have been the filling of fissures and of rock pores are now known to have been formed by replacement.

THE PROCESS

When mineralizing solutions encounter minerals that are unstable in their presence, substitution may result and replacement take place. The interchange is practically simultaneous, and the resulting mineral body occupies the same volume and may retain the identical structure of the original body. It is as though from a brick wall each brick were removed one by one, and a silver brick substituted in its place. The end result would be a wall of exactly the same size and shape, even to the minutiae of brick pattern, except that it would be composed of silver instead of clay. This is how replacement proceeds except that the parts interchanged are infinitesimally small—of molecular or atomic size. Consequently, the shape, size, and texture may be faithfully preserved even below the visible magnifications of the microscope. Since the change is one of volume for equal volume it cannot be a simple molecular interchange, such as takes place in a chemical laboratory, because if a heavier molecule were substituted for a lighter one shrinkage would result. Thus ordinary balanced chemical equations do not express what actually happens in volume-for-volume replacement; they only indicate the trend and end products of the exchange. The process is not yet fully understood.

PROCEDURE OF SUBSTITUTION. The interchange must be by particles of molecular or ionic size. The growing mineral is in sharp contact with the vanishing one, and between them there must be a thin film of solution that supplies by diffusion the replacing substance and removes the replaced materials. The instant that space is made available by solution, some of the replacing mineral will separate out from the film. Thus the replacing mineral will continuously advance against the host and grow at its expense. A constantly advancing front will be presented against the host as long as there is a supply of new material.

Replacement

Where solution is supplied to a center, such as a pore space, growth may proceed outward in all directions beyond the pore wall and form shapeless grains or crystals bounded on all sides by crystal faces. We saw earlier that in cavity filling the growth of crystals is outward from a wall toward an open space; hence one end of the crystal is attached and exhibits no crystal faces. Therefore, the doubly terminated crystals are diagnostic of replacement.

The question arises as to how the new material arrives at the point of deposition. The problem is more striking in an unfractured cube of pyrite that is undergoing replacement by chalcocite from the outside under surface temperatures. After the cube has been replaced, say to one-quarter of its depth, then the replacing copper must penetrate a dense layer of chalcocite in order to arrive at the interior front of replacement, and the dissolved iron of the pyrite must escape outward through the same dense layer. Obviously the necessary quantity of solution cannot flow bodily through the dense chalcocite layer. Diffusion is probably the answer. Ions must move from the point of supply to the point of deposition. But diffusion is considered to be exceedingly slow and hence effective over short distances only. Therefore, it probably cannot be called upon as a means of transporting huge volumes of replacing substances over long distances. However, it is an effective means of supplying and removing the products and by-products of replacement over the short distances at the actual front of replacement. Fissures and other types of openings described in the early part of this chapter serve as the main freight lines to conduct the hydrothermal solutions, and minor openings serve as distributors to the front of replacement.

It is obvious that the more openings available, even at the actual front, the more readily will replacement proceed. Thus porous and crackled rocks afford optimum physical conditions for replacement.

The refuse of replacement is probably removed through conduits similar to those that permitted ingress, and is swept away eventually to be dispersed within the mass of the ground water.

STAGES AND GROWTH OF REPLACEMENT. Microscopic examination of replacement ores reveals that they are commonly built up in stages and that early replacement minerals are themselves replaced by later ones. The first-formed metallic minerals, of which pyrite is a common one, may be replaced by later sulfides. The pyrite is attacked along the margins and particularly along minute fractures. Replacement

130 Hydrothermal Processes [Ch. 9

of the walls of the fractures extends outward until little islands of pyrite may be left between intersecting veinlets (Fig. 9-20). Still later metallic minerals may in turn replace those of the second generation. Perhaps eight or ten such stages may result. Among the com-

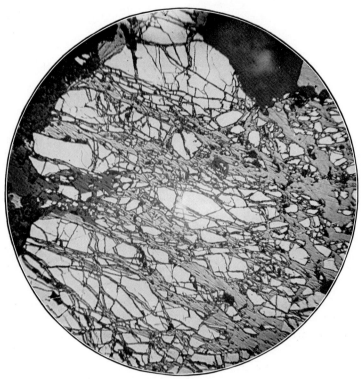

FIGURE 9-20. Replacement of pyrite (white) by zinc sulfide (gray) along cracks, leaving residual grains of pyrite included in zinc sulfide. Magnified 90 times. (L. C. Graton, *Econ. Geol.*)

mon minerals pyrite is generally followed by copper minerals, and in turn by zinc, lead, and silver minerals.

The growth of replacement deposits may advance in one or more of three ways: (1) From fissures the walls are first replaced, and the replacement advances outward with a bold face of massive ore sharply abutting unreplaced country rock (Fig. 9-21A). This produces massive sulfide bodies such as those of Kennecott, Alaska; Bisbee, Arizona; or Leadville, Colorado. (2) The growth may take place with a bold front but, preceding it, like skirmishers in front of an army, is a

Replacement   131

fringe of disseminated replacement where partial replacement is going on at many small centers (Fig. 9–21B). This gives rise to massive ore bodies fringed by lower-grade ore which gradually merges into barren country rock. (3) The third method is by multiple-center growth, as in Figure 9–21C. These centers may grow and coalesce to form almost massive ore, or they may remain as separate centers giving rise to disseminated replacement deposits of which the great "porphyry coppers" are examples.

AGENCIES OF REPLACEMENT.   Replacement takes place through the action of gases, vapors, hot-water solutions, and cold-water solutions; water dominates in all of them. The hot-water solutions that give rise

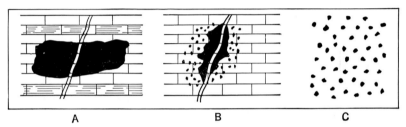

FIGURE 9–21. Types of replacement: A, Bold-face advance; B, outer fringe of disseminated replacement; C, multiple center.

to hydrothermal deposits are considered to be mostly alkaline, although in some cases acid waters operate.

Replacement may occur under almost any condition of temperature and pressure of solutions above 0° C, and its character varies somewhat according to the temperature and pressure. The formation of replacement deposits by cold meteoric waters is confined mostly to relatively soluble rocks, such as limestone which has been replaced to yield deposits of iron ore, manganese, phosphate, and other materials. Likewise, cold surface waters are notably effective in the replacement that operates in the zone of oxidation and supergene enrichment of metallic mineral deposits (see Chapter 11). Such waters are effective in the formation of great deposits of carbonates and silicates of copper and zinc, and of huge deposits of secondary copper sulfides. In general, the ore minerals formed by replacement from cold solutions are of simple composition.

With warmer solutions the ranges and intensity of replacement increase, and more kinds of rocks are affected. The metallic minerals formed are mostly simple sulfides and sulfosalts, and the gangue minerals are chiefly carbonates, quartz, and simple silicates. At intermediate temperatures extensive wholesale replacement of rocks takes place, and huge mineral deposits may be formed. A wider range of

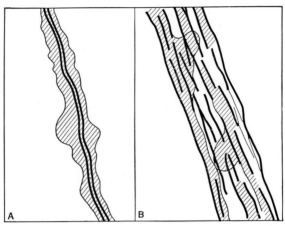

FIGURE 9–22. Replacement lodes developed (A) along a single fissure and (B) along a shear zone.

minerals occurs in such deposits, and earlier introduced minerals are replaced by later ones. At high temperatures hardly any rock may escape replacement, even to the most refractory ones. Silicates containing boron, chlorine, and fluorine are common. Sulfides and iron oxides develop in coarse textures, and the minerals may be complex ones.

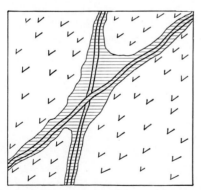

FIGURE 9–23. Replacement ore localized by fissure intersections.

The character of the host rock also affects replacement. Although every rock is susceptible to replacement, naturally the readily soluble carbonate rocks are the most widespread hosts for replacement mineral deposits. In granular igneous rocks replacement is commonly selective in that certain minerals, such as the dark minerals or feldspar, may be selectively replaced by ore minerals. This is notable in the disseminated "porphyry copper" deposits.

Replacement 133

LOCALIZATION OF REPLACEMENT. It was shown in the first part of this chapter that various structural, physical, and chemical features of rocks serve to orient and localize hydrothermal deposition at some particular site. These features are extremely important in connection with the localization of replacement deposits.

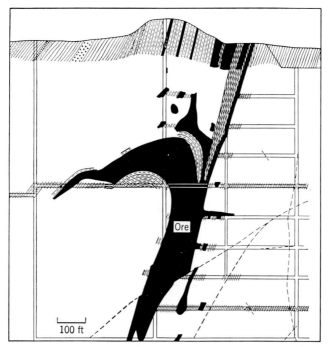

FIGURE 9-24. Large replacement deposit of the great Broken Hill lead-zinc deposit in Australia, localized by folds. Black is ore. (After W. H. Newhouse, *A.I.M.E.*)

The chemical character of the host rock alone may be the controlling factor in localizing ore, but structural features generally operate in conjunction. Of the various structural features previously mentioned, fissures, sheeted zones, and fissure intersections are prime localizers of ore, forming replacement lodes (Fig. 9-22) and intersection expansions (Fig. 9-23). Pitching folds and drag folds (Fig. 9-24) have been important in localizing ore at many places, notably at the great Homestake Gold mine.

Sedimentary features, particularly bedding, bedding planes, and contacts are especially effective in localizing replacement bodies (Fig. 9–25).

CRITERIA OF REPLACEMENT. It is one matter to speak of a replacement deposit but a different matter to recognize one when seen. Many features involved in their origin, however, leave recognizable signs that are diagnostic of the process, if properly interpreted. Some can be seen only in field relations, some in hand specimens, some only

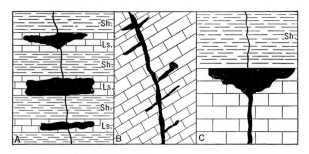

FIGURE 9–25. Relation of replacement bodies to sedimentary features (black is ore): A, to intercalated limestone beds; B, to bedding planes; C, to overlying impervious bed.

under the microscope. Some of the criteria afford positive identification, some negative. Since an extraneous mineral can gain entry into a solid rock only by filling a pre-existing opening or by making its own space through replacement, then the absence of those features characteristic of cavity filling provides dependable negative criteria for replacement.

One outstanding feature of replacement deposits, namely, unsupported nuclei, played an important role in first leading to the elucidation of replacement. These nuclei are unsupported residuals of country rock that escaped replacement when the rock surrounding them was converted to ore (Fig. 9–26). If the ore body had been formed by filling an open cavity, obviously such residuals would have rested on the floor of the cavity. If the residuals contain bedding planes in alignment with those of the wall rock (Fig. 9–26A) the evidence of replacement is even stronger.

A second criterion of replacement is the preservation of pre-existing rock structures in the ore, as shown in Figure 9–26B. Other preserved features are fossils, cross-bedding, phenocrysts of igneous rocks, schistosity of metamorphic rocks, folds, and faults.

Replacement 135

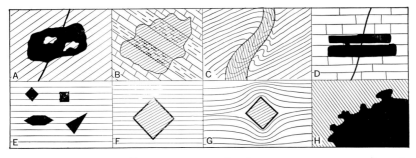

FIGURE 9-26. Features illustrating criteria of replacement. *A*, unsupported residuals; *B*, preserved strata; *C*, preserved folds; *D*, ore abutting bedding; *E*, doubly terminated crystals; *F*, pyrite cube truncating bedding, in contrast to *G*, where cube has grown by pushing aside bedding; *H*, irregular outlines.

A third criterion is the transection by an irregular ore body of rock structures that continue on either side of the ore, particularly if such structures also continue into the ore (Fig. 9-26*A*, *B*, *C*).

Tightly enclosed doubly terminated crystals that are alien to the original rock are also diagnostic of replacement, particularly if they transect several individual rock grains, as in Figure 9-26*E*. Minerals that exactly replace others of quite different composition, retaining the shape and volume, are called pseudomorphs; a cube with the form and striations of pyrite may consist of chalcocite. Such pseudomorphs are a criteria of replacement.

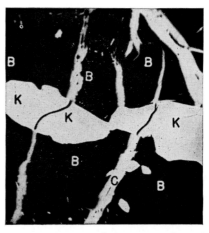

FIGURE 9-27. Replacement veinlets of chalcocite copper ore that widen by replacement in mineral B (bornite) but not in mineral K. Magnified 260 times. (P. Krieger, *Econ. Geol.*)

Wavy veinlets the walls of which do not match and which transect several differently oriented crystals and widen in some and narrow in others, as in Figure 9-27, denote replacement.

The outlines of an ore body, with characteristic protuberances and embayments into the country rock, as in Figure 9-26, also indicate an origin by replacement.

RESULTING MINERAL DEPOSITS

From the foregoing discussion it will be seen that replacement can result in many different forms of deposits, which may be divided into replacement veins, massive bodies, and disseminated bodies.

REPLACEMENT VEINS. Replacement veins or replacement lodes are those that have been localized along single fissures or sheeted zones, with replacement of the walls parallel to the fissuring (Fig. 9-22A). They thus resemble fissure veins, and many that are called fissure veins are really replacement lodes since their emplacement has been at the expense of the fissure walls rather than by the filling of open space. Such lodes are generally much wider and of more irregular width and outline than fissure veins, and the walls are seldom sharply defined. The gold veins of Kirkland Lake, Ontario, the copper veins of Kennecott, Alaska, and the lead-silver veins of the Cœur d'Alene in Idaho are examples.

MASSIVE DEPOSITS. There are many massive ore bodies whose size may be merely that of a house to huge bodies whose larger dimensions may be measured in thousands of feet. They are characterized by extremely irregular and wavy outlines and large variations in dimensions. The origin of the large bodies was earlier ascribed to magmatic injections or the filling up of pre-existing huge caves. However, when the criteria of replacement are applied to them they disclose features inconsistent with other modes of origin but diagnostic of replacement. Generally the deposits consist mostly or entirely of introduced minerals, and included rock matter constitutes only a small or negligible part. Some pyritic replacement bodies consist of stupendous masses of almost pure yellow sulfides, such as those at Noranda, Quebec, Cerro de Pasco, Peru, Hudson Bay mine at Flin Flon, Manitoba, or Rio Tinto, Spain. More than 500 million tons of massive pyrite were originally introduced into the rocks at Rio Tinto.

Such bodies result from the almost complete replacement of the host rock. Although most common in limestone rocks, they occur widely in various kinds of rocks.

DISSEMINATED REPLACEMENT DEPOSITS. In disseminated replacement deposits, in contrast to the massive type, the introduced ore minerals constitute only a few percent of the rock mass. They are peppered through the host rock in the form of specks, grains, or blebs, generally accompanied by small veinlets. The amount of introduced gangue minerals is also small. The total content of metallic minerals ranges from 2 to 10 percent of the mass, and the ores therefore are mostly low grade. The boundaries are vague, passing imperceptibly into less mineralized rock and gradually into waste rock, the ore limits being determined by the economically workable grade of the ore.

Disseminated deposits may reach huge sizes, which permit large-scale mining operations in many places by means of open pits, and this permits the utilization of low-grade ores such as the large "porphyry coppers." Since the workable parts of the deposits are generally fringed by envelopes of lower-grade material, it follows that a slight reduction in the grade of rock that can be profitably exploited permits the pushing outward of the ore boundaries and thereby brings about a large increase in size and tonnage of the workable portions of the deposits. Increased efficiency of extraction, lowered costs, or higher metal prices, which will allow the grade of workable copper ore to be lowered, for example by 0.25 percent, may increase the ore reserves by tens of millions of tons. This has happened over and over again in most of the "porphyry copper" deposits, so that today the grade of ore that can be profitably extracted is sometimes only half of that originally considered to be the economic cutoff.

Some idea of the extent of the process of disseminated replacement may be realized by glancing at the immensity of the operations and the developed ore reserves of some of the large deposits: the Chile Copper mine at Chuquicamata, Chile, is reported to have ore reserves in excess of 850 million tons which to date have averaged about 1.64 percent copper; the great Utah Copper mine of the Kennecott Copper Corporation at Bingham, Utah, many years ago reported ore reserves of 640 million tons averaging 1.07 percent copper, although the grade is now around 0.8 percent copper, and has handled over 100,000 tons of ore and 100,000 tons of waste per day; the Northern Rhodesia Copper Belt has indicated ore reserves of over 550 million tons, and the Climax Molybdenum mine in Colorado has ore reserves of over 200 million tons averaging 0.7 percent molybdenum. Extensive disseminated lead ores occur in southeastern Missouri.

FORM AND SIZE. As has been pointed out, the form of replacement deposits is determined largely by the structural and sedimentary features that localize them. Some representative forms may be seen in

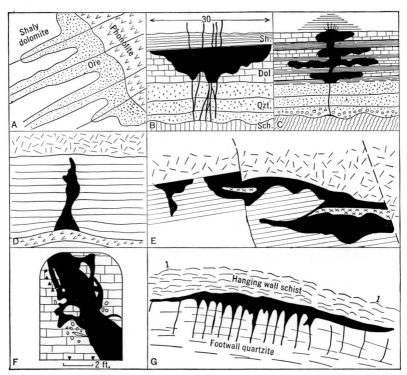

FIGURE 9-28. Forms of replacement deposits. A, along fissures and where phonolite dammed solutions, Mineral Farm, Black Hills, South Dakota; B, ore restricted below shale and abutting dolomite bedding, Union mine, South Dakota; C, Portland mine, South Dakota; D and E, cross and longitudinal sections of Iron Hill, Leadville, Colorado. (After J. D. Irving, *Econ. Geol.*) F, sketch of replacement vein, Jumbo mine, Kennecott, Alaska (A. M. Bateman); G, relation of ore to fissures in quartzite and to overlying schist, Ferris Haggarty mine, Encampment, Wyoming. (After A. Spencer, *U. S. Geol. Survey.*)

Figures 9–28 and 9–29. The size has a wide range from mere cracks containing high-grade ore to that of the dimensions of one of the limestone bodies at Leadville, Colorado, which was 3,500 feet long, 1,600 feet wide and 200 feet thick. Replacement lodes may reach several

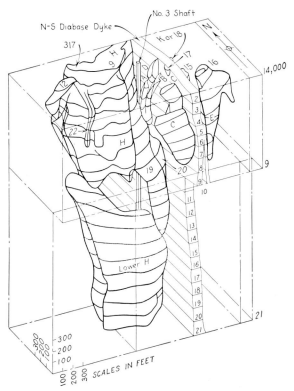

FIGURE 9–29. Massive replacement deposits of sulfide ore in isometric projection, Horne mine, Noranda, Quebec. Numbers of dike refer to mine levels (21 = 2,100 feet depth). (After Peter Price, *Bull. Can. Inst. Min. and Met.*)

thousand feet in length and depth and up to 200 feet in width. The greatest of the disseminated deposits, that of Chuquicamata, Chile, has a length of 10,500 feet and a maximum width of 3,600 feet.

ORES FORMED AND EXAMPLES. Outside of some sedimentary and nonmetallic deposits, replacement processes have given rise to some of the world's largest and most important mineral deposits. Some examples of the chief deposits of different metals and minerals are given in Table 9.

TABLE 9

SOME EXAMPLES OF REPLACEMENT DEPOSITS

Ores	Type and Examples of Some Important Deposits
Iron	Magnetite: Dover, New Jersey; Lyon Mountain, New York. Hematite: Iron Mountain, Missouri. Limonite: Oriskany ores, Virginia
Copper	Disseminated: Utah Copper, Utah; Nevada Consolidated, Nevada; Chino, New Mexico; Ray, Ajo, Miami-Inspiration, and Morenci, Arizona; Northern Rhodesia Copper Belt; Chuquicamata, Braden, and Potrerillos, Chile. Lode: Kennecott, Alaska; part of Bingham, Utah; Magma, Morenci, and Bisbee, Arizona; Britannia, British Columbia; Cerro de Pasco, Peru. Massive: Bisbee and United Verde, Arizona; Rio Tinto, Spain; Noranda district, Quebec; Flin Flon, Manitoba; Granby, British Columbia; Boliden, Sweden
Lead	Massive: Leadville, Colorado; Bingham, Utah; Sullivan, British Columbia; Santa Eulalia, Mexico; Broken Hill, Australia. Lode: Cœur d'Alene, Idaho; Park City and Tintic, Utah. Disseminated: southeastern Missouri
Zinc	Massive: Leadville, Colorado; Bingham, Utah; Sullivan, British Columbia; Flin Flon, Manitoba; Silesia. Lode: Broken Hill, Rhodesia; Park City, Utah; Franklin Furnace, New Jersey; Trepca, Serbia
Gold	Massive: Noranda, Quebec. Lode: Homestake, South Dakota; Kirkland Lake, Ontario. Disseminated: Juneau, Alaska; Witwatersrand, South Africa
Silver	Massive: Leadville, Colorado; Bingham, Utah. Lode: Park City and Tintic, Utah; Cœur d'Alene district, Idaho; Cerro de Pasco, Peru; Santa Eulalia, Mexico
Tin	Transvaal; Australia
Mercury	Lode: Almaden, Spain. Disseminated: New Almaden, Calif.
Molybdenum	Disseminated: Climax, Colorado; Utah Copper, Utah
Manganese	Massive: Leadville, Colorado; Potgietersrust, South Africa
Barite	Lode: Missouri
Fluorite	Lode: Illinois-Kentucky field
Magnesite	Massive: Manchuria; Washington; California
Kyanite	Disseminated: North Carolina; Virginia

SELECTED REFERENCES ON REPLACEMENT

"The Nature of Replacement." W. Lindgren. *Econ. Geol.* 7:521-535, 1912. A fundamental discussion of the process of replacement.

"Metasomatism." W. Lindgren. *Geol. Soc. Amer. Bull.* 36:247-261, 1925. Development of ideas; processes involved and problems raised.

"Replacement Ore Deposits and Criteria for Their Recognition." J. D. Irving. *Econ. Geol.* 6:527-561, 619-669, 1911. Chiefly forms and criteria.

"Influence of Replaced Rock on Replacement Minerals of Ore Deposits." B. S. Butler. *Econ. Geol.* 27:1-24, 1932. How the host rock affects the minerals deposited by replacement.

"The Mechanics of Metasomatism." G. W. Bain. *Econ. Geol.* 31:505-526, 1936. Relation of metasomatism to fine pores.

Selected References

Ore Deposits as Related to Structural Features. Edited by W. H. Newhouse. Princeton Univ. Press, Princeton, N. J., 1942.

Structural Geology of Canadian Ore Deposits—A Symposium. Geol. Div. Can. Inst. Min. and Met., Montreal, 1948.

Economic Mineral Deposits. 2nd Edit. Alan M. Bateman. John Wiley & Sons, New York, 1950. Replacement, 137–157; precious metals, 441–447, 459–462; nonferrous metals, 480–491, 498–524, 536–545; others, 595–597.

"Structure of Part of the Northern Black Hills and the Homestake Mine, Lead, South Dakota." J. A. Noble, J. O. Harder, and A. L. Slaughter. *Geol. Soc. Amer. Bull.* 61:321–351, 1949. Relation of gold ore to rock structure; well illustrated.

Tintic District. Paul Billingsley. *16th Int. Geol. Cong. Guidebook* 17:101–124. An outstanding example of geological field work and structural control of ores.

The Porphyry Coppers. A. B. Parsons. Am. Inst. Min. Met. Eng., New York, 1933. A story.

Copper Resources of the World. 2 vols. 16th Int. Geol. Cong., Washington, D. C., 1935.

CHAPTER TEN

Sedimentary Processes and Cycles

Sedimentation involves the derivation of the materials of sediments, their transportation, deposition, and consolidation. It is this process that gives rise to the common sedimentary rocks—conglomerates, sandstones, shales, and limestones. It also gives rise to many valuable mineral deposits of iron, manganese, copper, phosphates, sulfur, clays, bleaching earths, magnesite, salt, gypsum, soda, potash, borax, nitrates, coal, petroleum, and others. Most of these substances may be regarded merely as exceptional varieties of sedimentary rocks that happen to be valued because of their chemical or physical properties. The sedimentary mineral deposits are composed of both inorganic and organic substances, and their source, like that of any sedimentary rock, is other rocks or deposits that have undergone disintegration, the ultimate source of course being generally the igneous rocks and magmas.

The formation of sedimentary deposits involves, first, an adequate source of materials; second, the gathering of the materials by solution, suspension, or other means; third, the transportation of the minerals to the site of deposition if that is necessary; fourth, the deposition of the materials; and fifth, their compaction. Subsequent changes may of course take place, and some mineral deposits are dependent upon such changes.

The source materials, solution, transportation, and deposition are generally similar for the various types of deposits, but pronounced variations give rise to radically different end products. There are many different processes involved some of which are simple and some complex. Those involved in evaporation are quite different from those that give rise to beds of coal or iron ore. Consequently, the sedimentary cycle of each product will be taken up separately, after consideration of some general principles.

SOME PRINCIPLES INVOLVED IN SEDIMENTATION

In contrast with the mineral-forming processes that operate at depth, considered in preceding chapters, sedimentation processes are essentially surface ones. They involve rocks, weathering, soil formation, surface waters, atmospheric constituents, the heat of the sun, and animal and plant life.

SOURCE OF MATERIALS. The materials that enter into sedimentary mineral deposits have been derived for the most part from the weathering of rocks, or to a minor extent from the weathering of former mineral deposits. The atmosphere supplies some constituents, and other constituents have passed through an intermediate organic stage or have been held in the oceans.

Under the relentless attack of the atmosphere and water most rocks succumb to mechanical and chemical disintegration and decay. Some of the constituents remain behind to form soils, others are dissolved in the surface waters, and some are carried as fine suspended particles in the same waters. Different rocks, of course, yield different ingredients. Feldspathic and aluminous rocks yield the clay minerals, some of which are removed in suspension, and also sodium, potassium, calcium, and magnesium, which may be carried away in solution. Some igneous rocks yield the phosphorus that goes to make phosphate beds, and boron for borates. Iron is derived chiefly from basic igneous rocks but also from sediments and metamorphic rocks. Clarke and Washington have shown that in the earth's crust the average content of iron is 5.05 percent, and Eckel has calculated that the portion of the earth's crust beneath the United States to a depth of 1,000 feet contains over 275,000 billion tons of iron, of which only about 0.01 percent has been concentrated into commercial deposits in a ratio of about 80,000 to 1. Clearly one need not look beyond the ordinary rocks for an adequate source of the iron in deposits. Clarke and Washington also estimate that manganese makes up 0.09 percent of the earth's crust, there being about 56 times as much iron as manganese. Using this proportion and applying it to Eckel's estimate for iron, we can calculate that there should be 5,000 billion tons of manganese beneath the United States to a depth of 1,000 feet—surely an adequate source of supply. Other constituents of sedimentary deposits are also obtained, chiefly in solution, from the weathering of surficial rocks or of deposits of minerals contained in the rocks.

SOLUTION AND TRANSPORTATION. The chief solvents of the constituents of sedimentary deposits are carbonated water, humic and organic acids, and sulfate solutions. The clay minerals are not transported in solution but in suspension.

Carbonated waters are effective solvents of iron, manganese, phosphorus, calcium, sodium, potassium, and others, and organic matter aids the solution of some of these. The iron is carried as ferrous carbonate, and the other substances also as carbonates.

Humic and organic acids derived from decomposing vegetation are effective solvents of iron. The hydroxy acids dissolve large quantities of iron, but the weak organic acids dissolve remarkable quantities and are the most effective of all natural solvents. Experiments by Moore and Maynard led them to conclude that the iron of sedimentary deposits was probably taken into solution and transported as ferric oxide hydrosol stabilized by organic colloids, although small amounts may have been carried as salts of organic acids or adsorbed by organic colloids. The idea of colloidal solution and transportation of iron is becoming more firmly established. The knowledge of colloidal solution and transportation of manganese is as yet inadequate, but it is probable, as with iron, that organic compounds are important solvents and that manganese may be removed in the colloidal state.

Sulfate solutions are effective solvents of iron, manganese, copper, and other substances, but they are not sufficiently abundant to effect large-scale solution and transportation. Sulfate solutions form from the weathering of pyrite, which is a widely distributed mineral.

The substances dissolved during weathering will remain in solution as long as the solution does not undergo any appreciable physical or chemical change. Some or all of the iron or manganese, however, may be lost during transportation if the solutions traverse limestone or are subjected to other agencies of deposition. If the metals escape these hazards they may be transported to bogs, lakes, playas, or the sea, where quantity concentration can take place. Most of the substances of the sedimentary deposits, except for coal, are transported by streams and subsurface waters. For the most part they reach the sea, but some are arrested en route or find a site in inland bodies of water or interior land basins. The sea becomes the great receptacle of the salts that are poured into it annually by most of the rivers of the world.

DEPOSITION. The great bulk of the materials of sedimentary mineral deposits are deposited on the floors of seas, and a minor amount in lakes, swamps, and on land surfaces. They are deposited mechanically,

Principles Involved

chemically, or biochemically. The manner of deposition depends upon the nature of the solvent and on the place of deposition, whether in a sea or swamp, and the products of deposition will depend upon these conditions.

From *bicarbonate solutions* substances such as iron and manganese may be deposited by: (1) losing carbon dioxide from the solution; (2) oxidation; (3) plants; (4) bacteria; (5) replacement of sea-bottom shells; (6) reaction with sea-bottom clay and colloidal silica. Calcium phosphate is precipitated from solution in the presence of calcium carbonate. Calcium carbonate is precipitated chiefly by small forms of life in bodies of water, but also by evaporation.

From *organic solutions* iron and manganese are precipitated by: (1) oxidation of ferrous and manganous carbonate to the higher oxides of iron and manganese; (2) bacteria; (3) plants; (4) hydrolysis; (5) reaction with alkalies; (6) sea-water electrolytes acting on ferric oxide hydrosols.

From *sulfate solutions* these metals are thrown down by reacting with calcium carbonate, by oxidation or hydrolysis, or by reactions with silicates. There is a famed river in Spain, called the Rio Tinto, into which pours a great volume of iron sulfate solution derived from the oxidation of pyrite of the huge pyritic replacement deposits. Near the deposits the water is green but by oxidation lower downstream changes to brownish yellow. Iron oxide is precipitated as soon as it reaches the salt water of the sea.

Biochemical deposition plays an important role in deposition, particularly in the precipitation of iron and manganese. Bacteria swarm almost everywhere, and three distinct groups of iron bacteria are recognized, as well as another group of manganese-precipitating bacteria. The iron bacteria deposit ferric hydroxide from both organic and inorganic solutions, and experiments have shown that the soil bacteria of natural waters throw down iron hydroxide. Phosphorus is also precipitated by bacteria, but most of it is removed from solution by vertebrates or shellfish. It is generally agreed that the sedimentary phosphates have been concentrated through organic agencies. Sulfur is also deposited by bacteria. Many forms of shell life extract calcium carbonate from sea water.

Catalytic deposition may be a common means of natural precipitation that has not been sufficiently considered. C. Zapffe found that the municipal well water supply of Brainerd, Minnesota, was choking the mains by deposition of manganese and iron oxides that contained abundant manganese-precipitating bacteria. He could remove the iron by aeration but not the manganese. After many experiments he found that the manganese mineral pyrolusite acted as a catalytic agent that precipitated the manganese out of the water supply. This same type of catalytic action may go on in nature.

Colloidal deposition is readily brought about by ordinary sea-water electrolytes, which cause instantaneous precipitation of iron oxide from colloidal iron solutions.

Deposition by evaporation, giving rise to evaporites, is a widespread depositional process in lands of much sunshine and little rain. Some of the past geologic periods have been characterized by world-wide aridity, when great deposits of common salt, gypsum, potash, and other salts have been laid down. The evaporites may be deposited from sea water, saline lakes, or ground water.

Mechanical deposition of suspended matter carried by streams takes place by settling when the velocity of the water is lessened below its ability to carry the suspended matter. Thus when a stream enters a lake the suspended matter settles to the bottom. This occurs on a grander scale when streams reach the oceans and ordinary settling is accelerated by the action of salt water. Also, hillside streams may enter valleys where the water sinks into the ground, leaving their transported solids behind. Mechanical settling of woody matter takes place in swamps where coal beds are born.

CONDITIONS AND PLACES OF DEPOSITION. The conditions and places of deposition may in large part determine the mineral composition of the resulting deposits and their size, purity, and distribution. Restricted basins of deposition yield small deposits, but large marine basins of deposition afford opportunity for the formation of sedimentary deposits that are measured in thousands of square miles. Also, iron deposited under swamp conditions forms as iron carbonate, but under marine conditions as the more desired iron oxide. The chief places of deposition are bogs, lakes, swamps, peneplain depressions, and the seas.

In *bogs and lakes,* which are of course restricted in size, the deposits formed are small and local. Glaciated areas are favorable localities. Iron is deposited as bog ore, chiefly as the hydrous oxide. Manganese may accompany bog iron ore or be deposited by itself as impure manganese wad.

Swampy basins are the sites of coal deposition and to a minor extent of iron and manganese ores. The iron is deposited chiefly through loss of carbon dioxide and, in the presence of plants, in the form of ferrous carbonate (siderite). Sedimentary beds of siderite are common in coal measures. If it is deposited along with accumulating vegetation it forms *black band iron ore,* which is associated with coal beds and looks like coal. If it is deposited in the coal measures but not along with coal it forms *clay ironstone,* an impure siderite. Manganese is similarly deposited, and such sedimentary iron and manganese may later be enriched through weathering to higher-grade deposits.

Interior drainage depressions are the sites of deposition of saline minerals through evaporation. Salts may also be deposited at or near the surface as the result of evaporation of ground water in arid regions.

Marine deposition accounts for the vast majority of all sedimentary deposits. The seas are the great repositories of the salts washed in from the lands and of the solid materials dumped near the shores or across shallow mediterranean basins. Beds of sedimentary rocks are laid down, the coarser materials nearer shore, the finer materials farther out. Other sedimentary beds are laid on top of them, and as accumulation proceeds the sea floor tends to sink and receive still more beds. Shallow-water areas such as marine lagoons or long narrow continental seas are particularly favorable sites of deposition. Here extensive beds of iron or manganese ores or phosphates may accumulate. In offshore areas beds of commercial carbonate rock or clays may be laid down. The deeper ocean floors receive only minor depositions such as iron silicate minerals. Arms of the sea may become almost cut off in arid regions where evaporation may bring about deposition of the salts contained in the sea water.

THE CYCLE OF IRON

The iron dissolved by surface waters during weathering moves largely in streams ever onward toward the sea. Some of it may not reach the sea because it may be trapped en route. For example, if the iron-bearing solutions contact limestone on the way the iron may be deposited from solution, or organic matter may do the same, or the solutions may lose their carbon dioxide content or come to rest in an inland basin. Such depositions probably will not give rise to commercial deposits, or if they do they are likely to be small as in bog iron ores or impure coal measure beds. For extensive oxide precipitation the iron in solution must reach a sea.

The iron solutions that reach the shallow seas give rise to the largest iron-ore deposits in the world. Apparently the optimum conditions exist where sluggish streams incapable of transporting much suspended matter enter from deeply eroded, low-lying coastal areas of basic rocks. Thus little sediment accumulates with the iron ore. Shallow waters are indicated because the associated rock beds display ripple and current markings, worm burrowings, and broken fragments of shells that have been pounded by wave action. The margins of the sea bottoms were even exposed to the sun at times because the associated beds show mud cracks that form when clayey sediments become dried. There are also impressions of raindrops made on the soft muds while the muds are exposed to the air. Oscillations of the strand line took place during the sedimentation of iron, and at times the basins of deposition were shut off from the sea. Marine life, now represented by enclosed fossils, was not dwarfed; therefore no unusual conditions of environment are indicated. The beds above and below the iron-ore beds consist of sandstones and mud rocks with some limestone. The sluggish streams, relatively high in iron and low in suspended matter, introduced the iron, which was probably carried as carbonate or as a colloid. Deposition of the iron would take place by the means discussed in the preceding subsection.

The iron was deposited either in the form of hydrous iron oxide (limonite) or as ferric oxide (hematite). Some of it was deposited in the form of oölites, which resemble fish roe, some of it replaced or coated broken shells on the sea bottom, some of it was precipitated as an iron mud, perhaps as a colloidal gel. In some places some suspended clayey matter was deposited at the same time, thereby diluting the iron with undesirable silica and alumina, as in the sedimentary iron-ore beds of Central Europe, thus giving rise to iron-ore beds high in impurities and hard to treat. In other places a little calcium carbonate was deposited at the same time, also diluting the iron ore, but, since calcium carbonate (or limestone) is added to iron ore in order to smelt it, this impurity is not undesirable. The great Clinton iron-ore beds of Alabama, which give rise to the steel industry of Birmingham, Alabama, are fortunate in this respect. Eckel pointed out that in the Clinton iron-ore beds the minor associated clayey matter is higher in alumina and iron than are normal clay rocks, thus suggesting that the iron was derived from deeply weathered igneous rocks or limestones.

The marine character of such iron-ore beds is indicated by the marine fossils they contain, by the oölites, by the associated sediments, and by the size of the basins of deposition. Their sedimentary origin, however, has not been unquestioned. The iron is considered to have been transported as iron bicarbonate or as a colloid in organic solutions and to have been deposited either as hematite or limonite. In the extensive sedimentary iron-ore beds of Newfoundland there is also quite a little siderite and the iron silicate, chamosite. It is rather puzzling why hematite is precipitated in some places and limonite in others since, because of the dehydrating effect of salt water, one would expect hematite instead of the hydrous limonite. Ordinarily, when entering a shallow aerated sea where calcium carbonate is present and where there is a high hydrogen ion concentration (pH), iron solutions will deposit hematite. With a lower pH, siderite might be expected. The little oölites are interesting because some of them disclose alternating concentric onion-skin shells of hematite and chamosite.

In some places several iron-ore beds have been deposited of which more than one has proved to be workable. A noteworthy and puzzling feature of these sedimentary iron-ore beds is the long period of continuous deposition of iron ore that took place over very extensive areas. For example, the Big Seam of Clinton iron ore at Birmingham, Alabama, according to Eckel, attains a thickness of over 30 feet, averages 10 feet, and has been tested over a width of 10 miles and a length of 50 miles. Outcrops of Clinton ore are almost continuous for a length of 700 miles, so the main basin of deposition must at least have been this long and 50 miles wide. Eckel states that in one part of this field the unbelievable quantity of 5,000 million tons of iron ore was laid down in continuous deposition, and that the Wabana basin of Newfoundland probably had a continuous deposition of 7,000 million

TABLE 10

Deposits Formed and Examples

Type	Character	Chief Localities
Marine oxides	Hematite or limonite	Clinton oölitic hematites of United States; "Minette" oölitic limonites of Lorraine and Luxemburg; Newfoundland; Krivoi Rog, Krusch, Russia; Brazil
Bog ores	Limonite	Quebec, Maine, Sweden, Russia
Carbonate	Black band ore, clay ironstone	Pennsylvania; Ohio; Middleboro, north Staffordshire, Lowmoor, Dowlais, Clyde Basin, Great Britain; Westphalia, Saarbrücken, Germany; Russia
Iron silicates	Thuringite	Thuringia, Switzerland, South Africa, Russia
	Chamosite	Cleveland Hills, England; Newfoundland
	Greenalite-glauconite	Widely distributed

tons. The reserves of the Lorraine sedimentary iron ores in France are estimated at over 5,000 million tons. It is evident that iron accumulation was on such a vast scale that purely local conditions cannot be invoked to explain it. Sedimentary iron ores are also mined in Germany, Luxemburg, Russia, South Africa, and Brazil. Some examples of the chief localities are given in Table 10.

THE CYCLE OF MANGANESE

The cycle of sedimentary manganese almost duplicates that of iron. The two metals are derived from the same source, are dissolved by similar or the same solutions, may be transported together as similar chemical compounds, and are deposited as oxides and carbonates by the same agencies, although generally separately. There may also be some admixture with minor manganese in iron-ore deposits and iron in manganese deposits.

An interesting feature in connection with the formation of sedimentary manganese ore is this separation from iron during deposition. Where deposition occurs from carbonate solutions separation takes place because manganese carbonate is more stable in solution than iron carbonate, hence is carried farther and is thus separated from the iron. Similarly, iron oxides are less soluble than manganese oxides and so are deposited from solution first. The dioxide is the most stable of the manganese oxides, is the easiest formed, and is therefore the principal manganese ore mineral. The acidity or alkalinity (pH) of the solutions is also a factor in bringing about separation of iron and manganese since iron oxides precipitate at a low pH and manganese oxides at high pH. Neutral precipitation (pH 7) gives simultaneous precipitation of carbonates. From sulfide solutions, according to Dunnington, separation of iron and manganese is affected by their different reactions with calcium carbonate, which causes deposition of ferric hydroxide but has no effect upon the manganese sulfate until it is exposed to both air and calcium carbonate simultaneously.

Deposition of manganese ore may occur both in fresh and salt water, in lakes or bogs, or in the sea. It parallels that of iron. Marine depositions, chiefly as the dioxide, have taken place mainly in shallow waters, but it is also known in deep-sea deposits. Under near-shore conditions manganese oxide hydrosol or bicarbonate solutions yield oxides or carbonates of manganese, or both. They commonly form as

Cycle of Manganese

oölites, which, along with included marine fossils, indicate a marine origin for the manganese. The deposits are associated chiefly with shales and limestones.

DEPOSITS AND EXAMPLES. Sedimentary manganese carbonate is widely distributed but is nowhere commercial. Beds of relatively pure manganese carbonate are known in Newfoundland, Arkansas, Maine, Minnesota, South Dakota, California, the Appalachian states, Wales,

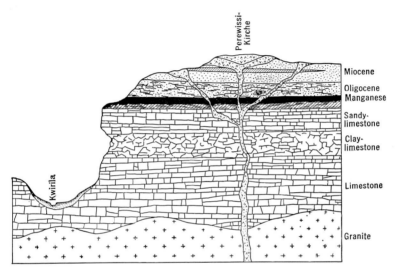

FIGURE 10–1. Section across manganese bed at Tchiaturi, Caucasus, Russia. (After de la Sauce, *Abh. z. Prakt. Geol. u. Berg.*)

Belgium, and Russia. They rarely exceed a few feet in thickness or contain more than 20 percent manganese. Their chief importance is in the light they throw on the origin of manganese and in the fact that they supply preliminary concentrations of manganese, which upon weathering may yield marketable deposits of secondary manganese oxides. Many of the large deposits of manganese ores are formed in this manner.

The great Russian sedimentary oxide deposits of Tchiaturi in Georgia (Fig. 10–1) and Nikopol in the Ukraine are the largest manganese deposits in the world and formerly supplied over half the world's needs of this essential ore. The deposits occur as horizontal beds of sandy clay from 6 to 10 feet thick containing oölites and nodules of manganese oxides carrying from 28 to 33 percent manganese, which is hand-sorted to a shipping grade of 42 to 53 percent manganese. The manganese is thought to have been precipitated directly from sea water by algae and bacteria living in the littoral zone of those Tertiary seas. Similar but smaller deposits occur also in west Siberia, middle Volga, south Urals, and Greece.

As shown in Chapter 16, the distribution of manganese is important in international affairs as it is essential as a purifier in the making of all steel as well as in making manganese steels, bronzes, and other alloys and is, therefore, in great demand in all industrialized countries. Countries deficient in manganese, like the United States, must depend upon imports to sustain their steel industries.

THE CYCLE OF PHOSPHORUS

The sedimentary cycle of phosphorus is puzzling and fascinating. Dissolved from the rocks, some of it enters the soil from which it is abstracted by plants; from them it passes to the bodies of animals and is returned via their excreta and bones to accumulate into deposits. These in turn may undergo re-solution and reach the sea, and there the phosphorus is accumulated or deposited by sea life, embodied into sediments, and returned to the land upon uplift, when a new cycle may start. Some of the phosphorus of the sea is absorbed by fish life, and the fish in turn are eaten by birds whose excreta have built up great phosphate deposits on islands of the Pacific.

Phosphates are quite soluble in carbonated water and, in the absence of calcium carbonate, will stay in solution. The phosphate in limestone resists solution in natural waters. Some phosphoric acid in solution reaches the sea, where it is extracted by organisms; some is redeposited as secondary phosphates, which may be redissolved; and some is retained in the soil. Swamp waters rich in organic matter dissolve phosphates, and some phosphorus compounds probably enter solution as colloids. Phosphorus is transported by streams probably as phosphoric acid and as calcium phosphate. Some is transported by birds and animals.

SPECIAL CONDITIONS OF DEPOSITION. Commercial beds of phosphate are formed only under marine conditions in the form of phosphorite. The beds have been formed from early to recent geologic periods and extend with remarkable regularity over thousands of square miles. They are sparingly fossiliferous and are interlaminated with marine fossiliferous beds. These features together with their oölitic character indicate a marine origin.

Mansfield considers that the accumulation of sufficient phosphates to yield sedimentary beds must have required unusual marine conditions. Other deposition of sediments must have ceased, to permit sufficient accumulation of pure phosphatic materials to form the oölites that aggregated into phosphate beds. The presence of hydrocarbons and iron sulfide in the deposits indicates deposition under reducing conditions. Mansfield thinks that the material represents a slow accumulation, shut off from the open sea, of phosphatic debris under anaerobic conditions and that during climatic oscillations there were long periods of cool temperature. These conditions favored the growth of life in the shallow waters and reduced the activities of denitrifying bacteria, which reduction in turn curtailed the deposition of calcium carbonate and favored the concentration of the phosphatic solutions from which the phosphate oölites were formed.

Mansfield considers that two features are essential to obtain phosphate accumulation: (1) a combination of favorable geographic features, and (2) the presence of some agent to fix the phosphoric acid in relatively insoluble form. He thinks this agent may be fluorine, which is present in one of the phosphate minerals. He has noted an association of phosphate deposition and of volcanism from which the needed fluorine could have been supplied. These may provide the unusual conditions necessary for phosphate deposition.

EXAMPLES OF DEPOSITS. The great phosphate deposits of the world have been formed by the process of sedimentation. Those of Algeria, Tunisia (Fig. 10–2) and Morocco together yield the largest produc-

FIGURE 10–2. Cross section of phosphate beds of the area from southern Tunisia to Algeria, contained in folded Eocene rocks. (L. Cayeux, *Res. Min. France d'Outre-Mer, IV.*)

tion in the world of this greatly desired fertilizer mineral. The Moroccan deposits occur as horizontal beds along with limestones, marls, and clays in which there are 3 to 4 beds up to 2.5 meters thick. The Algeria-Tunisia deposits are similar, and yield about 2 million tons annually and have probable reserves in excess of 10,000 million tons. The western United States deposits underlie parts of Utah, Wyoming, Idaho, and Montana, and extend into Canada. The beds are in the

154 Sedimentary Processes and Cycles [Ch. 10

Phosphoria formation of Permian age, and the chief bed is about 5 feet thick. Some 300,000 acres of phosphate beds have been surveyed and contain 6,000 million tons of phosphate; some 2 million acres remain to be surveyed. These beds give some realization of the enormous tonnages of materials that have been formed by sedimentation.

Other sedimentary phosphate deposits are found in Egypt and Russia. The large phosphate deposits of Florida, which supply most of the United States' fertilizer and chemical consumption are not strictly sedimentary beds but are pebble deposits derived from a sedimentary phosphate bed. The guano deposits of the Pacific islands also are not sedimentary.

THE CYCLE OF SULFUR

Sulfur is a common element in the earth's crust. It is a copious constituent of volcanic gases and magmatic emanations and is well known in hot springs. Thus new supplies are constantly being added. Most of it exists in the form of sulfates or sulfides. A stupendous reservoir of sulfur exists in the oceans in the form of calcium sulfate (gypsum-anhydrite), which ranks fourth in abundance among oceanic salts, and as potassium sulfate. As we shall see later, under the Cycle of Chemical Evaporites (p. 188), widespread beds of gypsum-anhydrite have been deposited up to many hundred feet in thickness, but that is sulfate, not sulfur.

The sulfur of sedimentary deposits is derived from the bacterial reduction of sulfates in solution, from sulfates in rocks, and from hydrogen sulfide given by volcanic emanations. Most of the sulfur is transported as sulfates or sulfide in solution, but some may be carried as colloidal sulfur.

Sulfur is deposited from sulfates and hydrogen sulfide in bodies of water that are low in oxygen and where reducing conditions and bacteria that exist without oxygen (anaerobic bacteria) prevail. The sulfates are reduced by the bacteria to hydrogen sulfide, which in turn oxidizes to native sulfur and water. Hydrogen sulfide (stink gas) becomes so highly concentrated in some waters free from oxygen that it actually prevents the growth of marine organisms. Sulfur bacteria are also thought to deposit sulfur from the hydrogen sulfide. We do know that sulfur is a rather common constituent of marine muds.

Special conditions of sedimentation are required, however, to give rise to commercial concentrations of sedimentary sulfur. First, there must be an adequate source of sulfur compounds, the quantity available must be large, the deposition of other sediments must be at a minimum to yield layers of pure sulfur, and microorganisms must be present. Generally the sulfur beds are accompanied by gypsum deposition, which probably took place during periods of high salinity of the waters brought about by evaporation, which in turn eliminates temporarily the sulfate-reducing bacteria and, in consequence, the deposition of sulfur.

Volcanic hydrogen sulfide has been considered to be the source of the sulfur of several sulfur deposits. Kato thought that the sulfur layers in a lake at Kozuke, Japan, came from such a source, and Sagui considers that the sulfur basins in Sicily were fed by hot springs from underlying cooling lava. Murzaiev, likewise, attributed the sulfur deposits of Russia to a volcanic source.

Many experiments have been carried on to demonstrate that hydrogen sulfide can be generated by reduction of sulfates by microorganisms but not by sterile inorganic compounds. Van Delden, as a result of his experiments, estimated that microorganisms would deposit from a 10-meter zone of a lake high in hydrogen sulfide about 45 kilograms of sulfur per square meter in 100 days.

EXAMPLES OF DEPOSITS. Among the largest of the strictly sedimentary types of sulfur deposits are those of Russia, near Knibyshev. They consist of thin gypsum beds with layers of pure sulfur, laminations of sulfur and calcite, or sulfur nodules in bituminous limestone. The sulfur is either pure or bituminous. Apparently they were formed in lagoons. Hydrogen sulfide springs are numerous near the deposits.

The sulfur beds of Sicily are another example. Here are several isolated basins up to 5 miles long and half a mile across with sulfur-bearing formations that consist of cellular limestone, bituminous shale and gypsum, clay marl, and sandstone. The sulfur is disseminated through the cellular limestone and in bands of pure sulfur. It is thought to have been deposited largely by biochemical processes. The sulfur content averages 26 percent of the rock.

Other examples of sedimentary sulfur are found in Iran, Rumania, Croatia, Galicia, and Upper Silesia.

Sedimentary sulfur beds contain vastly greater quantities of sulfur than the deposits of volcanic sulfur that flow out from volcanoes (Japan) or are deposited around them. Both these types, however, are small compared with the great salt-dome sulfur deposits of the United States Gulf Coast region. These are not strictly sedimentary deposits, as the sulfur is considered to have been formed by bacterial reduction of gypsum in the salt domes, which, however, originally were sedimentary beds.

Deposits of sulfur play an increasing role in the life of all nations, as sulfur is used so extensively in heavy chemicals and in fertilizers and insecticides, pulp and paper, explosives, paints, rubber, and many other substances encountered in daily life.

THE CYCLE OF COPPER(?)

It is claimed by some investigators that certain copper deposits have a sedimentary origin, but this has not gone without dispute by others, among them the author. It is, of course, well known that copper does have a sedimentary cycle. It is dissolved readily in surface waters during oxidation of copper deposits. Copper in solution does move to basins of fresh or salt water. It has been precipitated in sea muds as sulfides and native copper and is deposited by microorganisms. Even oysters absorb copper to the amount of 1.24 to 5.12 milligrams per oyster. A copper cycle is therefore established. What is not established is that workable copper deposits have been formed as a part of the sedimentary copper cycle.

A sedimentary origin has been advanced with vigor for the famed Kupferschiefer of Mansfeld, Germany, and has been opposed as vigorously by advocates of a hydrothermal origin. These remarkable deposits have been mined from A.D. 1150 to the present and constitute the chief copper ore reserves of Germany. The sedimentary basin has an area of 22,500 square miles. Lying above a lower conglomerate bed is a thin black cupriferous shale, 1 meter thick, which Beyschlag describes as "one of the most remarkable products of the geologic ages." It is a shallow, marine, organic mud, full of land plants washed in from adjacent coasts. The sedimentary advocates believe that into this putrefying bottom there were swept cupriferous solutions derived from the oxidation of distant copper veins. The metals were thought to have been precipitated as iron-copper sulfide gels by the organic matter or by bacteria. The chief minerals are those of copper with fewer ones of iron, lead, zinc, and a number of the less common metals such as nickel, cobalt, molybdenum and vanadium. However, microscopic evidence discloses that the ore minerals cut across shale layers, replace it, and replace each other. This, claim the advocates of sedimentation, is merely a readjustment of the metals by later solutions, which are called upon to dissolve the metals and then reprecipitate them—a double and opposite duty, which would be a rather re-

Uranium and Vanadium

markable feat. The advocates of a replacement origin find normal replacement of rock by ore minerals and replacement of earlier ore minerals by later ones, just as in other replacement deposits the world over. It is difficult under this theory, however, to account for the extremely widespread distribution of ore minerals in such a very thin seam of shale. Although the arguments in favor of a sedimentary origin are strong, those in favor of a replacement origin are stronger.

The huge disseminated copper deposits of Northern Rhodesia, the greatest copper belt in the world, also have a few advocates for a sedimentary origin. The argument that most favors such an origin is the extremely wide distribution of copper, miles in extent, in relatively thin beds of altered sandstones and shales. The ore minerals very definitely cut across rock grains and replace them, and earlier sulfides replace later ones. The relationships between ore minerals and rock minerals is almost identical with those of the other great disseminated copper deposits whose replacement origin is unquestioned. Most students of the Rhodesian deposits do not agree with a sedimentary origin.

Sandstones impregnated with copper minerals occur in several countries, of which the "Red Bed" copper deposits of the western United States are examples. These have often been called sedimentary deposits, but the evidence is clear that the ore minerals are later in age than the sandstones and have been introduced subsequently by solutions.

THE CYCLE OF URANIUM AND VANADIUM

Uranium and vanadium are commonly associated in nature, so that the two are considered together. At the outset it should be pointed out that the chief sources of present-day uranium are not sedimentary deposits but are hydrothermal veins, such as those of Great Bear Lake in Canada, Shinkolobwe in the Belgian Congo, and Joachimstal in Czechoslovakia, with minor quantities in pegmatites. Sedimentary deposits, however, will probably be the chief source of uranium in the future. Similarly, the present commercial deposits of vanadium are not always associated with uranium, and the chief producers are not sedimentary.

There is a clear-cut sedimentary cycle for uranium and vanadium, but some of the stages are imperfectly known. They are transported and deposited together and separately. Released by weathering processes, they are dissolved, transported to the sea, and deposited in black sea muds rich in organic matter. The muds have been converted into shales, raised above the sea, and again subjected to weathering, releasing their uranium and vanadium to become precipitated on land or again to reach the sea, once more to be incorporated in black organic muds.

The source of uranium and vanadium must have been in igneous rocks, pegmatite dikes, or vein deposits. Their weathering released these metals for solution in surface waters. The character of the solutions is imperfectly known, but apparently solution and transportation of uranium are enhanced by acidity.

One has to unravel the sedimentary cycle of uranium by reasoning backwards. It is known that uranium is deposited in salt or brackish-water muds that are rich in organic matter and low or lacking in calcium carbonate. Places of phosphate deposition are particularly favorable for uranium deposition. Therefore one infers that uranium must have been transported to seas of past geologic ages at the time of the accumulation of the type of black muds that give rise to petroleum. The uranium content of the earth has been estimated at 0.0002 to 0.0009 percent. It occurs in marine sedimentary rocks to the extent of 0.01 to 0.02 percent uranium oxide, or a concentration of 10 to 50 times that found in the earth's crust. This is a lesser concentration than that of sedimentary manganese. Uranium occurs in sea water today to the extent of about 0.000002 grains per liter, and that in the sediments is in acid-soluble form.

More is known of the deposition of sedimentary uranium than of its natural solution. First, the uranium is restricted to marine shales, and the content is quite low in nonmarine shales. The highest uranium content is in those marine shales that are high in iron sulfide, and in organic matter of bituminous type rather than of the carbonaceous or coaly type. Therefore, there seems to be a definite relationship between uranium deposition and organic matter, and it is thought that uranium may be adsorbed by the organic matter. Second, all marine sedimentary phosphate beds carry uranium, as well as phosphatic nodules contained in marine black shales. Consequently, the conditions that favor phosphate deposition also favor uranium deposition. All phosphorites that have been tested so far contain uranium, and so do the phosphatic nodules that occur in marine black shales. Third, the marine black shales that carry uranium are thin beds of wide distribution, indicating uniform, slow, quiescent conditions of deposition. Fourth, all the high uranium shales are earlier in geologic age than the Mesozoic era. Fifth, clay minerals may have the property of attaching uranium to them and thereby absorbing it from sea water. Sixth, the mineralogy of the uranium in the black shales is unknown; it is, however, in an acid-soluble form. Thus the physical conditions of deposition are those of marine basins where accumulation was slow and the bottom water was deficient in oxygen and of near neutral hydrogen-ion concentration. As to the manner of deposition of uranium in the black muds, it may have taken place by direct deposition as an organic uranium salt; or it may have been precipitated by organisms, as suggested by McKelvey and Nelson, in the same way that marine organisms are known to remove from the sea other metals such as vanadium, copper, and radium; or it may be adsorbed by clay minerals or by the precipitation of phosphate. The hydrogen-ion concentration (pH) also seems to be an important factor, as deposition is favored with a pH 6 to 7.5, which is nearly neutral between acid and alkaline.

The sedimentary cycle of vanadium is quite similar to that of uranium. Both metals are dissolved, transported to the sea, and deposited together in beds of thin black shales or along with and under the conditions of phosphate accumulation. The sedimentary phosphate beds also carry vanadium along with uranium. The sedimentary iron-ore beds of central Europe also carry vanadium. In fact, the slags from iron-ore smelting have been utilized for their vanadium content. Petroleum-yielding shales must also have absorbed vanadium from sea water. Petroleums of the Caribbean region are pronouncedly vanadiferous, and so are the pitches that have formed from the evaporation of petroleum in this area. The petroleum soot cleaned from ships' boilers utilizing Caribbean petroleum was collected and treated for its vanadium content during World War II. Thus the same relationship between organic matter and phosphate exists for vanadium as for uranium.

Another type of sedimentary deposit containing uranium is the alluvial or placer deposit of both recent and ancient age.

Carnotite-bearing sandstones are present-day sources of uranium, but there is some question as to whether they should be considered as strictly sedimentary deposits.

EXAMPLES OF DEPOSITS. Inasmuch as the vein deposits of uranium ore are relatively restricted in size and tonnage, any large-scale industrial use for the future must look to the large low-grade, sedimentary deposits, which are extensive. Some day these may be worked for their uranium content. These consist of bituminous shales, bedded phosphate rock, and placer deposits, in addition to the carnotite-bearing sandstones.

The bituminous black shale type carries from 0.01 to 0.02 percent uranium, or an average of about $\frac{1}{3}$ of a pound of uranium per ton of rock. The richest of these are very thin alum shales containing nodular masses of hydrocarbon, called kolm, which contain up to 0.05 percent uranium oxide. Other thin oil shales have no kolm but carry up to 0.026 percent uranium oxide.

In the United States the Antrim and Chattanooga black shales in Oklahoma and other central and eastern states carry uranium, as do somewhat similar shales in Michigan and Alaska. Some black source shales of petroleum are known to contain uranium. The phosphate beds of the northwestern states and of Florida also carry uranium.

The Morrison carnotite sandstones of the western Colorado plateau are the chief United States domestic source of uranium today and are actively mined for both uranium and vanadium. The source of these metals is a mystery. The quartz sandstones are of fresh-water origin and are notably cross bedded. They contain fossil logs, vegetable matter, and saurian bones. Hess considers that the sandstones were formed in "shallow water with mobile islands, spits and shores" on to which logs and vegetation were swept by the rivers and stranded. The amazing feature is the richness of some of the logs. Two petrified logs at San Miguel River, one 100 feet long and 4 feet in diameter, yielded 105 tons of ore containing $175,000 in radium, $27,300 in uranium, and $28,200 in vanadium. Two other logs and the intervening sandstone from the vicinity of Calamity Gulch yielded $350,000—the most valuable logs ever known. It would seem to indicate that here again organic matter favored uranium-vanadium deposition but that it was not a necessary factor, as the sandstone without logs yields most of the uranium. Fischer points out that the ores occur chiefly with rolls in the sandstones and that the ore boundaries in places cross the sandstone bedding, indicating that the uranium ore was introduced later than the original sedimentation and was probably deposited from circulating ground water just after the sands had accumulated. Vanadium is commercially extracted from the same sandstone formations. Similar carnotite ores, but of very low grade, occur in Ferghana, Russia.

In Russia uranium-bearing black shales of the same age as the Swedish kolm are known to occur in Estonia and the Leningrad region. Other bituminous black shales occur in Kara Tau, Ferghana, and the Caspian Sea areas. The Caspian Sea area oil-field waters and petroleum are particularly high in radium content, and it is quite possible that the extensive petroliferous black shales of this region may prove to carry small quantities of uranium.

Other bituminous black shales will unquestionably be found to carry uranium in other parts of the world.

The world's greatest gold deposits, the Witwatersrand conglomerates, also carry minute but recoverable quantities of uranium. The origin of the gold is highly controversial. It is not known definitely whether it is of hydrothermal or of placer origin, but the thin conglomerate beds that carry the gold are definitely sedimentary, although it is uncertain whether they are marine or nonmarine. The gold-bearing conglomerates also contain uraninite and some definitely placer minerals such as chromite, diamonds, and corundum. The uraninite may also be of placer origin; if so, it would be an example of a sedimentary deposit. Since some 60 million tons of gold ore are mined and milled annually and any uranium extracted would be a by-product, the Witwatersrand holds the possibility of becoming the world's major source of industrial uranium.

THE CARBONATE CYCLE

The carbonate rocks such as limestone, dolomite, and magnesite are widely utilized in the building, chemical, fertilizer, cement, steel, smelting, and other industries. They are sedimentary deposits laid down in both salt and fresh water. Limestones are the most common form, but magnesium may in part replace calcium, giving dolomitic limestones.

Calcium is one of the eight main elements of the earth's outer crust. It and magnesium are contained in most rocks, are present in soils, are invariable constituents of plants, and are present in animals.

Calcium is released during rock weathering, and, although some remains in the soil, most of it goes into solution as bicarbonate in which form it reaches streams, lakes, and the sea. The content of calcium in river water exceeds that in sea water, indicating its removal from sea water. Some calcium is carried as a sulfate.

In the deposition of calcium carbonate, carbon dioxide plays an important role, as the solubility of calcium carbonate in water depends upon the content of carbon dioxide in solution. If the carbon dioxide content is lessened, calcium carbonate is thrown down, as in the deposition of calcium carbonate in familiar cave dripstones. The amount of carbon dioxide in the sea depends on the water temperature and on the amount in the air, which is in equilibrium with that in the water. More carbon dioxide is held in solution in cold than in warm water, a fact familiar to everyone who opens a warm bottle of soda water in contrast to an iced one. Similarly, warmed sea water loses carbon dioxide and, since it is practically saturated with calcium carbonate, deposition takes place. Aquatic plants also absorb carbon dioxide, causing carbonate deposition. In fresh water, calcium carbonate deposition is almost entirely caused by the loss of carbon dioxide; some is absorbed by organisms and shell life. In sea water, inorganic deposition takes place by loss of carbon dioxide, by colloidal deposition, and by organic deposition by plants, algae, bacteria, corals, Foraminifera, and larger shell life. Entire limestone beds may be composed of Foraminifera, nummulite shells, coral, or larger shell forms (coquina). Some calcium carbonate deposition takes place from suspended matter or from comminuted shells or corals.

The cycle of magnesium differs from that of calcium. Most magnesium salts are readily soluble, and magnesium therefore remains largely in solution in sea water unless removed through evaporation; actually it is the third most abundant salt. Direct deposition of dolomite, the double carbonate of magnesium and calcium, and of magnesite, the magnesium carbonate, may rarely take place, but sedimentary magnesite is not common. Most sedimentary dolomite and dolomitic limestones are formed by reaction on the sea floor, during sedimentation, of magnesium in the sea water with precipitated calcium carbonate, giving rise to the double carbonate. Coral reefs have become dolomitized by this process. Magnesium carbonate is also precipitated by plant life.

EXAMPLES OF DEPOSITS. Several kinds of carbonate rocks are deposited by sedimentation and differ in their quality, purity, and usefulness, depending upon the conditions of their formation.

Limestones range from pure calcium carbonate rocks to impure ones with included clay, fine sand, iron compounds, and organic matter. The pure limestones are desired particularly for agricultural and chemical purposes and for making lime. The impure varieties are sought for cement rocks. All find usefulness for building stones and road metal. The rocks may be dense and compact, or cavernous or friable. Some are composed wholly of shells, or of coral, or of minute fossil organisms visible only under a microscope. Most limestones are deposited in shallow to moderately deep sea waters free from land sediments.

Marl, a friable, incoherent, impure limestone, is deposited in lakes from calcium carbonate supplied by streams or springs. It is common in old glacial lakes because glaciers in places supplied ground limestone and yielded cold water high in carbon dioxide and, therefore, in calcium carbonate. The cold melt waters lost carbon dioxide in warmer lake waters with resultant deposition of calcium carbonate. Much marl is deposited by low aquatic plants such as Chara. Marl is utilized as a fertilizer and cement rock.

Chalk, white earthy limestone, is deposited mainly in shallow waters and is a chemical precipitate of calcium carbonate and the minute shells of organisms. Its purity makes it an excellent whitener and polisher.

Dolomite contains 54.35 percent calcium carbonate and 45.65 percent magnesium carbonate, but the proportion of magnesia is generally less and the rock then is dolomitic limestone. Iron or manganese may also be present. It may form extensive sedimentary beds and is utilized for making refractories, cements, magnesium, and other products.

Magnesite, the carbonate of magnesium, is much valued as an industrial mineral. Most of the sedimentary varieties occur in association with salt and gypsum, or shales and limestones, and are formed by deposition from concentrated waters of saline lakes. Apparently deposition has been by chemical precipitation and subsequent dehydration. Magnesium transported as magnesium sulfate reacts with calcium carbonate to yield insoluble hydromagnesite, which accumulates as a relatively pure precipitate, later dehydrated. Examples of sedimentary deposits are known in Kern County, California, and in Nevada, Idaho, British Columbia, and Germany.

Varieties of sedimentary carbonate rocks of economic interest are the following:

Kind	Use
Building limestones	Building and structural
Cement limestones	Hydraulic cements
Siliceous limestones	Hydraulic limes
Silico-aluminous limestones	Natural cements
Limestone	Flux, fertilizer, chemicals
Lime rock	Quicklime
Chalk	Cements, powders, crayons, fertilizer
Marl	Cement, fertilizer
Lithographic limestone	Fine engraving
Dolomite	Cement, refractory
Magnesite	Cement, refractory, chemical
Siderite	Iron ore *

* See section on "The Cycle of Iron."

THE CLAY CYCLE

The clay cycle differs from the other cycles previously discussed in that the constituents of sedimentary clay are not transported in solution but in suspension, and their deposition is mechanical rather than chemical or organic.

Clay is one of the most widespread and earliest mineral substances utilized by man. It carries the records of ancient races inscribed upon tablets, in brick buildings, in monuments, and in pottery. Its products portray the history of man, and in its beautiful wares we trace the development of the delicate artistry of the Chinese, the utility of the Romans, or the humor of the Incas. A wealth of artistic wares culminated in the eighteenth century, but today utility holds sway in the multitudinous uses of the varied clay products.

The term "clay" is applied to earth substances, consisting chiefly of the clay minerals, which generally become plastic when wet and stonelike when fired. These properties give clay its usefulness, because it can be molded into almost any desired form, which it retains after being heated. Widespread accessibility, ease of extraction, and adaptability to so many uses has resulted in the entrance of clay products into the wide ramifications of modern industrial civilization. Clay has many uses other than ceramic, particularly in building and manufacturing.

The constituents of sedimentary clays originate through the chemical weathering of aluminous rocks and are the chief components of argillaceous sediments. The constituents may be deposited in place to form residual clays, or they may be transported as sediments and deposited in streams, swamps, lakes, estuaries, and seas, giving rise to different varieties of clays that serve different useful purposes. The constituents were early thought to be the mineral kaolinite with some impurities, and later were held to be amorphous gels. We now know that there are some two dozen individual clay minerals most of which occur in such minute grains that they are not even discernible under high-power petrographic microscopes. It has been found by means of X-ray studies, the electron microscope, and differential thermal analyses, that the clay minerals are distinct crystalline minerals (Fig. 10–3). Their names and compositions are too complex to go into here. They exhibit the peculiar properties of base exchange by which one or more elements may substitute for all or parts of others in the crystal structures. This adds to their compositional complexities but serves

Clay Cycle 165

as a valuable economic property in refining and purifying water, oils, fats, sugars, and other substances. The chief components in the clay minerals are combinations of alumina, silica, water, potassa, magnesia, calcium, sodium, lithium, iron, and other elements. The clay minerals make up an essential part of all soils.

The deposition of the clay minerals is brought about by slow settling in water, which is speeded up in saline waters. The fineness of the

FIGURE 10–3. Electron microscope picture of kaolinite, one of the clay minerals, magnified 35,700 times, showing its structure and crystalline character. (D. W. Davis *et al.*, *Am. Petrol. Inst. Prel. Rept.* 6.)

materials permits them to stay suspended for a long time; hence much reaches the seas to give rise to extensive deposits that have now become uplifted in the lands.

Marine clays settle from suspension some distance offshore. The beds commonly are finely laminated, uniform in composition, and of great extent and thickness. Marine clays are widely distributed in Paleozoic and Mesozoic formations.

Estuarine clays, since they are laid down in shallow ocean arms, are restricted in extent and commonly contain many sandy laminations that increase toward the source of supply. Examples are found in the Chesapeake Bay region.

Lake clays are also restricted in area and contain sandy laminations. They are common in glaciated regions and are utilized chiefly for brick and tile clay. Many of them are varved, i.e., made up of thin alternating bands of coarser and finer material, each pair representing a year's growth.

Swamp clays or *fire clays* may underlie coal seams. Upright tree trunks are found in them. They are considered to have originated from suspended matter carried by low-gradient streams into coal swamps whose outer fringe of vegetation filters out the coarse material, leaving only the finest material to enter and settle in the center of the swamp. The organic acids present are considered to have purified the clay sediment. Because of their high purity and composition fire clays are eagerly sought to make high-grade refractories.

Stream clays are deposited in protected parts of flood plains during stream overflow. Thus they are pockety and grade laterally into sandy material. Different pockets vary greatly in composition. Stream clays are also deposited in isolated basins on deltas.

EXAMPLES OF DEPOSITS. Common *brick and tile* clays are widely distributed in almost every land. *Bentonite* and *fuller's earth* for bleaching and filtering are likewise widely distributed. High-grade *refractory clays* desired for making high-temperature resistant refractory materials are found in most coal regions and in many other areas.

The high-grade *kaolins* or *china clays* are not common. Where they occur ceramic centers of world renown have sprung up, and fine chinaware from them has international markets. Those of China, particularly Kiangsi, since A.D. 220 have given rise to fine wares and the employment of 1 million people. The fine kaolins of Cornwall and Devon and those of France (famed for fine Limoges and Sèvres porcelains), as well as those of Dresden, Czechoslovakia, and Bavaria are, however, not sedimentary but residual clays, to which reference will be made in Chapter 11.

THE SILICA CYCLE

Silica, or silicon dioxide, in the form of quartz, and some of the numerous other occurrences of silica, such as flint, chert, jasper, chalcedony, and diatomaceous earth, occur as sedimentary deposits of economic value. Silica is the most abundant constituent of the earth's crust.

The cycle of silica is somewhat different from that of carbonate or clay. Like carbonates it is taken into solution during weathering, and like clay most silica remains in the solid products of weathering and is transported and deposited mechanically. Quartz, the commonest form of silica, is highly resistant to weathering; therefore quartz is the chief constituent of sedimentary deposits of silica. Many sandstones are composed almost entirely of quartz, and those with 70 to 90 percent silica are common. In the weathering of rocks containing quartz, the quartz is released from accumulated minerals and mostly remains as quartz. Silica cannot decompose further upon weathering, but it may go into ionic solution or become a hydrosol and undergo transportation as such. Its solubility depends upon the pH of the solution. The higher the pH, the more silica is dissolved, and decrease of pH brings precipitation. This is just the opposite of the behavior of carbonates.

The resistant quartz is washed or carried in suspension until the body of moving water loses velocity or comes to rest, when mechanical deposition occurs. This may occur when rain wash and streams debouch into intermountain valleys, or enter lakes or desert playas, or slow down over river flood planes and deltas, or reach the sea. Most deposition of quartz sand is in the sea not far distant from shore lines. Stream, current, and wave action bring purification of the deposited sands by separating sands from pebbles, and silts and clays from the sand. Thus arise sedimentary deposits of relatively pure quartz sandstones. The commonest kinds are less well-sorted sands and gravels used for building and paving. Cleaner varieties give rise to glass sands, engine sands, and others. The well-consolidated materials give rise to such economic products as abrasive stones—grindstones, pulpstones, millstones, grinding pebbles, and sharpening stones.

The dissolved silica, made available in abundance during the weathering of silicate minerals, goes into ionic solution chiefly in the form of alkali silicates or silica hydrosol. Much of this is lost by precipitation as a gel in arid regions or by reaction with clay or other minerals. Some of it moves in true or colloidal solution in river waters and reaches lakes or seas. In fresh waters organic separation may take place by diatoms, giving rise to highly siliceous sedimentary deposits of diatomite. In sea water silica is almost negligible, being fairly completely removed through the action of such small plants and animals as diatoms, radiolarians, and sponges. Their siliceous remains become deposited to form marine sedimentary beds of silica. Some inorganic deposition gives rise to precipitates of chert, flint, and jasper.

THE COAL CYCLE

Coal is a sedimentary rock which because of its energy value happens to be of outstanding importance to man. Coal and iron ore together have been the chief factors in the development of the industrial nations of the world. Where the two have met have sprung up the great industrial and manufacturing centers of the Great Lakes region of the United States, the Midlands of England, the Ruhr. From these centers have flowed the products of coal and iron, establishing world commerce, great industrial nations, and commercially, politically, and militarily strong ones. Coal is the energy that makes the wheels go round, and the wheels are made of steel. Those countries richly endowed with coal and iron have risen above their competitors; those lacking them have become agricultural or handicraft nations. Although petroleum and gas are rapidly replacing coal as a source of primary energy, coal still produces about 70 percent of all energy sources, but in the United States its production of energy sources has dropped to about 47 percent, with 33 percent supplied by oil, 15 by gas, and 4 by water power. Coal in the form of coke is, of course, essential to steelmaking and cannot be displaced by oil.

The sedimentary cycle of coal starts in a swamp, the immediate source materials being plants, and carbon dioxide drawn from the atmosphere and surface waters. The juvenile carbon dioxide, however, sprang from igneous sources, as did the original constituents of the soil that supported the plant life. Rankama and Sahama point out that volcanic activity has been the only source of additional carbon dioxide in the cycle of carbon. That absorbed by plants and animals is largely released again; that locked up in limestone and coal in part returns again. The trees and small plants that fall into swamp water undergo partial decay, and the residue collects on the swamp bottom to form peat, the first stage of all coal. This gradually loses oxygen and hydrogen and under compaction of overlaying beds undergoes physical and chemical changes to become bituminous coal. This by metamorphism may later be changed into anthracite coal.

KINDS OF COAL AND CONSTITUTION. In nontechnical terminology four main types of coal are recognized: (1) anthracite, or hard coal, (2) bituminous, or soft coal, (3) lignite, and (4) cannel coal. Except for cannel coal, a special type, the others are divided into ranks as follows: lignite, brown coal, subbituminous, bituminous (five varieties), semi-anthracite, and anthracite. Peat lies below lignite and meta-anthracite, and graphite above anthracite.

Peat is not coal, even though it is a fuel. It is an accumulation of partly decomposed vegetable matter that represents the first stage of all coals. *Lignite* and *brown coal* are composed of woody matter embedded in macerated and decomposed vegetable matter. It is banded and jointed, and slacks or disintegrates upon drying. It has low heating value and is subject to spontaneous combustion. *Subbituminous* is an intermediate coal that resembles bituminous. It is dull, black, waxy, banded, and splits parallel to the bedding. It is a good clean coal but of low heating value. *Bituminous* coal is dense, black, brittle, banded, and breaks into prismatic blocks. Its constituent vegetable matter is not ordinarily visible to the eye. Bituminous coal ignites readily and burns with a smoky yellow flame. Its moisture is low and its heating value high. It is the most desired and most used coal in the world and serves for steam, heating, gas, and coking. *Cannel* coal is a special variety of bituminous coal that is lusterless, does not soil the fingers, is not banded, and breaks with a splintery fracture. It is made up of windblown spores and pollen. It is clean, burns with a long flame, and is sought for fireplace coal. The higher ranks of bituminous coal have the maximum heating power of all coals. *Anthracite* is a jet-black hard coal of high luster and irregular fracture. It ignites slowly, is smokeless, and has high heating value. It is restricted in distribution and is used almost exclusively for domestic heating.

Chemically, coals are made up of various proportions of carbon, hydrogen, oxygen, nitrogen, and impurities. Toward the higher ranks there is a progressive elimination of water, oxygen, and hydrogen, and an increase in carbon, which is present as fixed carbon, and in volatile matter. The volatile matter is that which burns in the form of a gas; it causes ready ignition and smoke. The fixed carbon is the steady, lasting source of heat; it produces a short, hot, smokeless flame. The fuel ratio, an important characteristic of coal, is the fixed carbon divided by the volatile matter. Ash is the residual in coal of noncombustible matter that comes from silt, clay, silica, or other substances.

Physically, banded coals are made up of partly decomposed and macerated vegetable matter, mainly vascular land plants, of which the following have been recognized (in ascending order of bacterial assailability): resins, waxes, cutin, lignose, gums, oils, fats, pigments, starches, cellulose, and protoplasm. When the vegetable matter is attacked by bacteria the most assailable substances are largely destroyed, and the least assailable, such as resins and waxes, are present in most coals.

ORIGIN OF COAL. All ordinary coals are of vegetable origin, and the banded ones are now considered to have been formed *in situ*. They originate in swamps and go through a peat stage.

The microscope reveals that the raw material of banded coals was vascular swamp vegetation not unlike that growing in present-day peat-forming swamps. Over 3,000 plant species have been identified from Carboniferous coal beds. Roots and stumps found in underclays beneath coal beds show that the vegetation grew and accumulated in place. Luxuriant vegetation flourished and consisted mainly of ferns, lycopods, and flowering plants, with conifers and other varieties also present. Ferns were treelike, rushes grew 90 feet high, and lycopods (small shrubs today) attained 100 feet in height. Bulbous and arched roots show that the trees lived in water. None of the plants are salt-water species, and the same kinds of plants are found in coals of all ranks.

The extensive distribution of individual coal seams implies swamp accumulation on broad delta and coastal plains and broad interior lowlands where shallow waters rest throughout the year. Most coals are underlain by carbonaceous shales of lake-bottom deposition. Lowlying surrounding lands are necessary, else there would be too much inflow of silt. J. V. Lewis estimates that it would take 125 to 150 years to accumulate enough material for 1 foot of bituminous coal and 175 to 200 years for 1 foot of anthracite.

The climatic conditions were mild to subtropical with moderate to heavy rainfall. Severe frosts were absent, but the climates were not without dry spells, since some of the plants have water containers in their trunks and roots. Probably the climate was like that of the Carolinas or Florida. Coals are known in places where they would not now be formed, as in Spitzbergen, Greenland, and Antarctica, indicating climatic changes since coal formation.

The change from plant debris to coal involves biochemical action producing partial decay, preserval of this material from further decay, and later processes. The type of coal depends upon the environment, the kind of plants, and particularly the duration of bacterial decay. When a tree falls on dry land it decays. The constituents are broken up into carbon dioxide and water, with which the tree started. No coal accumulates. When, however, vegetation falls into water, a similar decay sets in but more slowly. An essential to coal formation is the arrest of bacterial decay before complete destruction takes place so that there can be some residue to accumulate. This is brought about when the decay-promoting bacteria render the water toxic to themselves, and this prevents further decay of the vegetable tissues and permits their preservation and accumulation.

The biochemical changes eliminate oxygen and hydrogen and concentrate the carbon. The bacteria attack first the most decomposable constituents such as the cellulose and starches, and the resistant materials like resins and waxes, along with wood fragments, drop to the bottom of the swamps where the toxicity slows up or prevents further decay and humus accumulates.

This factor of the duration of bacterial action is most important in determining whether any coal will form or what the type of coal will be. Under normal conditions of water supply and little surface agitation of the water all but the least resistant of the vegetable matter is preserved, Heavy rains, however, will dilute the water, lessen the toxicity, and foster further decay. Floods are catastrophic to coal formation because they lower the toxicity and wash away material. Dry seasons likewise are destructive because the water level is lowered, the detritus becomes air exposed, and decay proceeds to destruction. Water level thus plays an important role. Accumulation of detritus in the swamp gives rise to peat, the initial stage of coal.

SUBSEQUENT CHANGES. After the peat stage bacteria presumably play little part, and most of the changes are chemical and physical, induced by pressure and slight increases of temperature due to deposition of overlying sediments. In peat the oxygen content has already been reduced about 10 percent. Further progressive elimination is probably caused by the combination of oxygen with carbon to form carbon monoxide and carbon dioxide, which escape.

The important chemical changes from peat to higher-rank coal are: (1) progressive elimination of water, oxygen, and bitumens; (2) conservation of hydrogen; (3) progressive increase of ulmins; (4) development of heavy hydrocarbons; and (5) increased resistance to solvents, oxidation, and heat. The chief contemporaneous physical changes are: (1) compaction, drying, and induration; (2) jointing and cleavage; (3) optical changes; (4) dehydration; (5) color change— brown to black; (6) increase of density; (7) change of luster; and (8) fracture changes from bedding to cleavage to irregular.

The change in rank is largely a result of pressure and time. The older the coals, the more likely they are to become more deeply buried, thus increasing the pressure and accelerating the metamorphism. Folding increases the rank because it brings pressure and temperature increases. This is shown in Pennsylvania, where the rank of the coal increases progressively with the intensity of the folding of the rocks, the anthracites being in the most closely folded beds.

OCCURRENCE OF COAL. Coal beds occur within what are called "coal measures," which consist of alternating beds of sandstone, shale, and clay, mostly of fresh-water origin. These indicate alternating and fluctuating conditions of sedimentation between coal and ordinary water-borne sediments, particularly as several coal seams generally occur within the coal measures. For example, in Pennsylvania there are 29 coal seams aggregating 106 feet of coal, in Alabama 55 seams; in England there is an aggregate thickness of 85 feet, and in Germany 120 feet of Carboniferous coal.

Coal seams are huge flat lenses, although some are remarkably persistent. The Pittsburgh seam, for example, underlay 15,000 square miles. The thickness of individual seams of coal ranges from a mere film up to 100 feet. The thickest bed in the United States is 84 feet; the famous Mammoth seam in Pennsylvania is 50 to 60 feet. Most coal seams range between 2 and 10 feet thick and rarely exceed 20 feet. Even these thicknesses indicate a long period of quiescent sedimentation. Many coal seams contain thin partings of shale or clay, called "bone," representing a break in accumulation with influx of sediments. Other seams divide into splits toward the margin of the basin, indicating inroads of sediment from that side during accumulation. Others display bulges and rolls of the floor or roof.

Coals occur in all geologic ages since the Devonian period (about 300 million years ago). The Carboniferous period (about 260 million years) received its name because of its world-wide distribution of coal. The Cretaceous period is the next most prolific one, and the Tertiary contains most of the lignite of the world.

In distribution, coal is world wide, but it is a rather striking fact that coal is much more common in the Northern Hemisphere than in the Southern Hemisphere. There are few countries north of the equator that are entirely lacking in coal. The countries most richly endowed in coal are the United States, Russia, Canada, England, Germany, China, and India. Fairly large deposits occur in Australia, South Africa, Manchuria, and Rhodesia. Many countries contain moderate deposits, but coal is sparsely distributed in Central and South America, Africa, Scandinavia, and the Mediterranean countries.

THE CYCLE OF PETROLEUM AND NATURAL GAS

Petroleum, although a fluid, originated from organic matter deposited in marine sediments and is largely held in sedimentary rocks. In brief, the sedimentary cycle of petroleum begins with life in the seas, presumably with the minute plants and animals, and proceeds through deposition of organic matter along with black marine muds, burial of the black shales by overlying sediments, conversion of the organic material to petroleum and gas, migration of these fluids to traps in porous rock, and preserval of the traps and included hydrocarbons free from intense folding, faulting, or strong metamorphism. There is, however, much that is still unknown about the sedimentary cycle and origin of petroleum.

Petroleum and gas are the preferred fuels of this century and in North America are rapidly displacing coal as a prime source of energy. The value of petroleum products exceeds that of any other mineral used by man, and modern industry, transportation, and warfare have become dependent upon them. It was mentioned in the section on "The Coal Cycle" that "coal . . . makes the wheels go round," but the wheels would not turn so fast if they were not oiled with petroleum lubricants. Almost 90 percent of the 2 billion barrels of petroleum consumed annually in the United States is used for gasoline, gas, and fuel oil.

Petroleum is composed of many compounds of carbon and hydrogen, with minor oxygen, nitrogen, and a little sulfur. Each of the various members of the different carbon-hydrogen series has different properties; some are liquids, some are gases, some are solid waxes, and because of this oils vary greatly in composition. Some of the carbon-hydrogen members supply high gasoline content, others little; some supply little or no lubricants, others much. Paraffin-base oils are light and yield good lubricants, whereas asphaltic-base oils are heavy, are lower in lubricants and gasoline, and may be usable only for fuel oil.

The home of petroleum and gas is in sedimentary rocks and only rarely has it moved out of this environment into other rocks. The containing rocks of commercial oil pools are sands, sandstones, limestones, conglomerates, and, rarely, fissured shale or igneous rocks. The enclosing sedimentary rocks are almost invariably marine in origin, although in rare cases fresh-water sediments are the host rocks. Oil is found in rocks of all periods, from the earliest fossiliferous rocks (late Cambrian), almost 500 million years old, to the late geologic period, the Pliocene, but the world over, beds of Tertiary age or those formed within the last 70 million years are the most prolific in petroleum.

THE ORIGIN OF OIL AND GAS. The origin of these resources is still largely a poorly understood subject, although certain phases of it are gradually becoming clarified. Older inorganic theories of volcanic origin have had so many nails driven in their coffins that there are few adherents left today. The overwhelming accumulation of evidence favors an organic origin from low forms of marine life. However, there do exist distinct differences of opinion as to just what the organic material was and, particularly, how it became converted from plant or animal life into petroleum and gas. The chief considerations, among many, that petroleum and natural gas are of organic origin are: (1) many oils are optically active, and only oils of organic derivation exhibit this characteristic; (2) nitrogenous compounds are constituents of petroleum and in nature are confined to plant and animal life; and (3) most oils contain chlorophyll porphyrin, and these are organic; (4) oils contain pigments of organic origin.

As to the kinds of organic materials that are the source of petroleum, there still exists a wide divergence of opinion, even though considerable research has been devoted to this subject. It is now generally considered that the low planktonic organisms, such as diatoms and algae, that thrive abundantly near the surface of the sea are the most probable and important source materials. The organic remains accumulate in bottom muds in depressions of shallow seas. Trask has tested thousands of bottom mud samples and has found the organic content to be constant for about 100 miles offshore and to range from 0.3 percent in deep sea oozes to 7 percent off the coast of California. He found that the average content of organic matter in 1,600 samples of recent muds was 2.5 percent, and only 1.5 percent in ancient sediments. Trask noted that nitrogenous compounds and compounds of humus and lignin make up the larger part of the organic matter and that fats and oils are present in only minor amount. He states that the composition of average plankton is 24 percent protein, 72 percent carbohydrate, and only 3 percent fat, and that in recently deposited sediments the organic matter is about 60 percent carbohydrate, 40 percent protein, and 1 percent fat. He thinks, therefore, that it is not the fats, cellulose, and simple proteins that are the mother substances of petroleum but rather the oxygen-deficient complex proteins and carbohydrates.

It is considered that the slow oxygen-free bacterial decomposition of the plant and animal remains yield the hydrocarbons. To obtain petroleum the reduction of the oxygen, nitrogen, sulfur, and phosphorus content of the organic matter is necessary, and ZoBell and others have shown that some bacteria do reduce these substances, leaving residual compounds consisting mainly of carbon and hydrogen. The greatest bacterial activity goes on in the upper few inches of the

Cycle of Petroleum

marine muds because here thrives the largest population of active bacteria. For example, one Pacific Ocean mud sample from a water depth of 3,120 feet yielded 38 million bacteria per gram of mud to a depth of 1 inch, only 940 thousand at 1 to 2 inches of depth, and merely 2,400 at 15 inches of depth of mud. Therefore with deeper burial bacterial activity largely ceases. For hydrocarbons to accumulate in the sediments there must be an absence of those bacteria that destroy and oxidize hydrocarbons, yielding carbon dioxide. Thus the presence of organic matter in sea muds, the presence of abundant bacteria, and experimental evidence that certain bacteria can convert organic materials into hydrocarbons are not definite proof but are strongly suggestive that hydrocarbons are produced in this manner.

The next step is more elusive, i.e., whether bacterial processes on the sea floor yield actual oil droplets or merely hydrocarbons that are later changed to petroleum. Since no liquid hydrocarbons have ever been found in recent sediments, the evidence favors later chemical and geologic changes after the muds have become buried by coverings of later sediments. Then pressure, somewhat higher temperature, and geologic time become involved and, with chemical actions, convert the partially decomposed organic material into petroleum. The exact nature of such chemical changes is as yet unknown. Pressure is probably insignificant, and temperature could not reach 200° C else certain constituents would be destroyed.

Radioactivity may prove to be a factor or the chief factor in the formation of petroleum. Experimental work bearing on this is under way. It was pointed out in the section on the "Cycle of Uranium and Vanadium" that recent investigations have disclosed that uranium is deposited in thin black marine muds high in organic matter, and further that a great many such bituminous black shales have been tested and found to contain uranium. Thus marine organic matter, such as might be the source of petroleum, and a radioactive mineral, uranium, have been deposited in the same beds. Further, it has been demonstrated in the laboratory that bombardment by radioactive emanations change some hydrocarbons to oily substances. It is inevitable that, under the continuous disintegration of uranium, the accompanying organic matter must have been under radioactive bombardment for long geologic periods. Add to this the further factual information that petroleum and its products contain the disintegration products of uranium. One of them, radon, occurs in some gas associated with petroleum; another, helium, is won in abundance from some oil wells; and petroleum and its associated brines are always high in uranium. Radium has been extracted from oil-field brines in Russia. Many asphalts are high in uranium. These associations are strongly suggestive that the petroleum and the disintegration products of uranium have originated at the same time.

It is also thought that some organic and inorganic substances have acted as catalysts in the transformation of organic matter into petroleum. The organic ones are thought by ZoBell to be biochemically active bacteria, and the inorganic ones may be vanadium, nickel, or other metals, and clays.

Trask thinks that the change of organic matter to petroleum may have been aided by polymerization or methylation and that the liquid hydrocarbons are capable of dissolving other organic substances such as pigments, waxes, and fatty acids.

During the conversion of organic matter into petroleum, natural gas is also formed. This is dominantly methane, and it is also of interest that methane is the gas formed in peat swamps and in the transformation of low-rank to high-rank coals.

MIGRATION AND ACCUMULATION OF PETROLEUM. The transformation of organic matter into oil does not give rise to an oil pool. Other steps are necessary. Since oil pools, where great concentrations of oil occur, are found chiefly in sandstones, it follows that the dispersed droplets of oil generated in the source muds must have migrated out of the source rocks into the sandstones, accumulated into suitable traps, and then been retained there. The chief forces that cause migration of oil are (1) compaction of the muds, (2) buoyancy, (3) capillarity, (4) gravity, (5) water currents, (6) gas pressure, (7) cementation, and (8) bacteria.

Compaction is believed to be the chief force that causes the movement of oil and gas out of source rocks into carrier beds. Source muds may contain up to 80 percent water. The weight of sediments deposited on top of the muds gradually compacts them, and the enclosed fluids are squeezed out into places of less pressure, such as the pore spaces in sands. The fluids may move up or down or laterally. Athy estimates that compaction removes 50 percent of the water when burial has reached 1,000 feet, and 85 percent at 4,000 feet. As burial of source muds progresses, there must be an almost continuous migration of the fluids into more porous beds.

Capillarity also aids migration, because, where oil-wet shales are in contact with water-wet sands, the water will move by capillary ac-

Cycle of Petroleum

tion from the coarse sand pores into the fine pores of shale and displace the oil into the adjacent sandstones.

Buoyancy also aids migration, because oil and gas, being lighter than water, tend to rise upon a water surface. This secondary migration takes place within reservoir rocks and is most effective where the pore spaces are large and where there are large volumes of fluids. This is the common arrangement in oil pools—gas and oil on top and water below. If the porous rocks are horizontal the oil will tend to rise to the tops of the permeable beds and no pronounced accumulation may occur, but if the porous beds are inclined the oil will rise up the incline and, as is shown later, accumulation may take place.

Gravity operates in the absence of water, when oil by gravity moves downward through permeable beds until arrested by impermeable beds. Generally, however, water is present.

Currents of underground water will flush oil along with them and accelerate oil migration. Such currents may be caused by compaction or artesian circulation, which may have been very effective in bringing about large-scale migration in the Rocky Mountain province.

Gas pressure is considered to be an aid to migration of oil and gas. Gas bubbles that move through oil carry oil along with them. They also lessen the viscosity of the oil and make it lighter. Gas is inserted into wells to push oil to exit wells.

Cementation of rocks, by which the grains of loose sediments become compacted and cemented together by deposition of cementing materials, reduces the pore space and forces oil to migrate.

Bacteria aid migration is three ways: (1) they release oil by dissolution of limestone rocks, also creating voids to permit migration and releasing carbon dioxide to create gas pressure that drives oil ahead; (2) according to ZoBell some bacteria have an affinity for solid rock, thereby driving off oil; (3) other bacteria produce surface-active substances that liberate oil from rock grains. All this activity aids migration, particularly within source beds.

Accumulation of oil commonly results from migration and is the collection of oil droplets into oil pools. One should not, however, think of an oil pool as a kind of underground lake into which a stick might splash. Rather it is a filling by oil and gas of the rock pores, which might be 15 to 25 percent of the volume of the rock. Oil may migrate without accumulating, or it may accumulate in noncommercial bodies, as at the top of horizontal beds. Concentrated accumulation is essential to produce commercial pools, and this in turn is dependent upon requisite reservoir rocks and traps.

RESERVOIR ROCK.   Accumulation of oil takes place only in rocks that are both porous and permeable. A rock like clay has high porosity but small permeability (see Chapter 9) and, therefore, is not a suitable reservoir rock. The most suitable ones are chiefly sands and sandstones, but also cavernous limestone rocks, rarely fissured shale or jointed igneous rocks. The cavernous limestones may yield sensational and prolific flows of oil, as in some of the Mexican and Iranian fields.

Unconsolidated oil sands of California average about 25 percent porosity, and in cemented sandstones the porosity is about 12 to 25 percent. The capacity of sand with 20 percent porosity is 1,550 barrels of oil per acre-foot, or 155,000 barrels for a sand 100 feet thick.

The greater the porosity, the greater the amount of oil that a reservoir rock can contain; and the larger the pore size, the greater the amount of oil it will yield. The percentage of the total oil recovered from a reservoir rock is surprisingly small. The Bradford sand of Pennsylvania yields only 8 percent by pumping and an additional 20 percent by flushing; Oklahoma sands yield 10 to 60 percent; unconsolidated sands of California with 25 to 40 percent porosity yield only 10 to 30 percent of the contained oil.

A confining impermeable caprock must be present to retain oil in a reservoir rock. Shale and clay are the most common caprocks, but dense limestone, dolomite, and gypsum also serve, and even well-cemented fine-grained sandstone or shaly sandstone are also effective. Water-wet impervious shales, clays, and limestones even retain gas. Good caprocks form effective seals to underlying oil and gas for scores of millions of years, but poor ones permit slow escape and loss of the mobile hydrocarbons.

GEOLOGIC TRAPS FOR OIL.   Migration in itself will not give rise to accumulation unless the upward migration is arrested. If a carrier bed is inclined, oil, gas, and water will migrate up dip until it escapes at the surface or is arrested in a trap to form an oil pool. If the carrier bed, however, is folded into an arch (anticline) the up-dip migration is arrested at the top of the arch, where it will form an anticlinal pool, the commonest kind of trap.

In the earlier period of oil exploration most of the oil discovered was in domes and anticlines (Fig. 10–4). Consequently, it was assumed that structural features were essential for the accumulation of oil. This was then called the "anticlinal theory" of oil accumulation, and a well was said to be "on structure" or "off structure." Now, how-

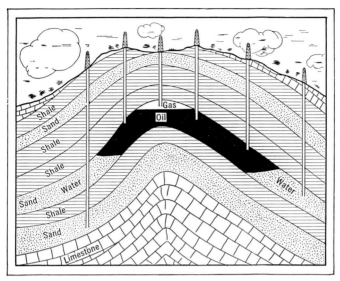

FIGURE 10-4. Oil pool located in a sand bed in an uparched (anticlinal) structure with most of the oil on the gentler sloping limb. (After D. D. Leven, *Done in Oil.*)

ever, it is realized that there are many other kinds of oil and gas traps, and these are generally divided into structural and stratigraphic traps, as follows:

Structural Traps	Stratigraphic Traps
Anticline *	Unconformities *
Dome *	Ancient shore lines *
Monocline *	Sandstone lenses *
Terraces	Shoestring sands *
Synclines	Up-dip wedging of sands *
Faults *	Up-dip porosity diminution
Fissures	Overlaps *
Salt domes *	Reflected buried hills *
Igneous intrusives	Buried coral reefs *

* The more important types of oil reservoirs.

Folds, chiefly anticlines and domes, have been the source of most of the oil yielded so far. An elongated anticline (Fig. 10-5) is the ideal oil trap because its extended flanks facilitate migration, its arch permits accumulation, and its crest or "center" generally contains domelike areas in which oil collects from all sides and is retained beneath the caprock. It is customary to draw contours on such a fold. Where the contours close is called a "closed fold," and the oil lies within the closure. Folds with gentle slopes offer larger areas for accumulation of oil. Anticlines and domes vary greatly in size. The rich Long Beach pool of California has an area of only 1,305 acres;

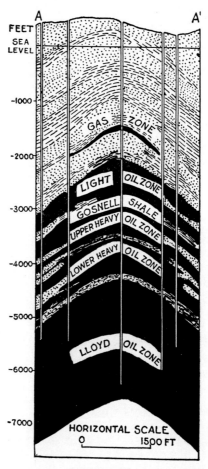

FIGURE 10–5. Ventura Avenue oil field, California, illustrating an oil pool in an elongated anticline. Cross section showing several unusually thick oil zones. (Drawn by L. C. Decius, reproduced by S. Powers, A.I.M.E.)

others may have tens or hundreds of square miles. Other types of folds are shown in Figures 10–9 and 10–10.

Faults (Fig. 10–6) have given rise to many pools, such as the Mexia-Powell field, Texas. A fault may permit a permeable carrier bed to abut an impervious bed, which arrests migration. Open faults may permit the escape of oil from lower to upper beds, or even to the surface.

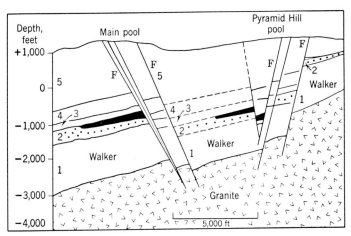

FIGURE 10–6. Oil pools trapped by faults. Section of Round Mountain field, Kern County, California. Black is oil; F, faults; 2, Vedder formation. (After Rogers, *Calif. Dept. Nat. Res.*)

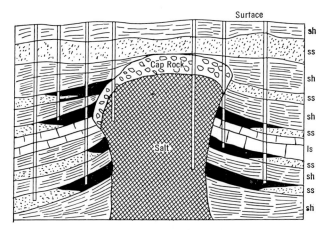

FIGURE 10–7. Diagram of a salt dome showing salt plug, caprock, and upturned strata in which oil has been trapped against the salt and penetrated by oil wells; *ss*, sandstone; *sh*, shale; *ls*, limestone.

Salt domes are a type of structure that gives rise to prolific pools in the Gulf Coast region, Mexico, Rumania, and the Middle East. They are formed by upthrust plugs of plastic salt which pierce and upturn the overlying beds, creating seals against impervious salt (Fig. 10–7).

Stratigraphic traps, as distinguished from structural traps, are those formed primarily as a result of sedimentation without necessarily involving post-sedimentation deformation. Chiefly, they are due to changes of permeability because of variations in thickness, texture, and porosity of beds formed by changing conditions of sedimentation. A common type are *unconformities,* where underlying tilted strata are

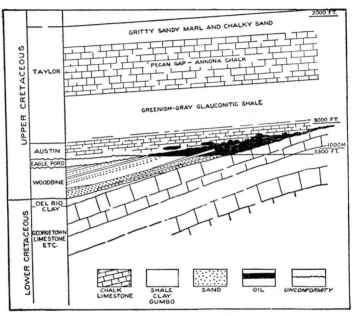

FIGURE 10–8. Section across the great East Texas field, showing oil accumulation beneath an unconformity where gently inclined strata (Austin) lie across eroded upturned older strata (Woodbine sand) in which the oil accumulated. (F. H. Lahee, *16th Int. Geol. Cong.*)

covered by overlying strata that permit accumulation beneath, as in Figure 10–8.

Buried hills that give rise to thicker strata on the flanks and thinner beds on top, as in Figure 10–9, have given rise to many pools in the Mid-Continent field.

Ancient shore lines and *overlap* of beds laid down as a sea was advancing over land have given rise to "lensing out" of porous sands into impervious beds. The greatest oil pool in the United States, the East Texas, with its 120,000 proved acres, was formed by such a trap. *Sandstone lenses* in shale and up-dip wedging of sands into shale are shown in Figure 10–10. *Shoestring sands* are long, narrow bodies of

Cycle of Petroleum 183

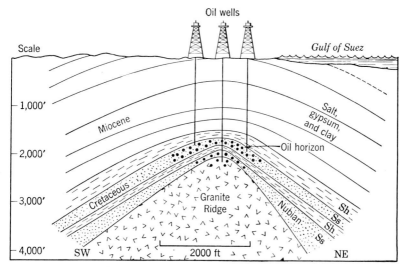

FIGURE 10–9. Buried granite ridge causing anticlinal structure of overlying beds in Hurghada oil field, Egypt. Oil occurs in the sandstones. (From S. Powers, after Hume, *A.A.P.G.*)

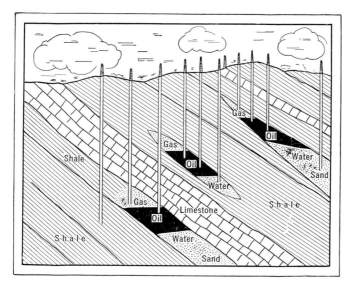

FIGURE 10–10. Diagram illustrating three types of stratigraphic traps. Left, permeable sandstone grading into impervious shale; center, a lens of permeable sandstone included in shale; right, wedge of sandstone pinching into shale. (D. D. Leven, *Done in Oil.*)

sand enclosed in shales deposited in lagoons or meandering streams. They have given rise to many small pools. Other types of traps are formed by wedging out of sands into impervious strata.

Ancient *coral reefs* or *bioherms* have been found recently to be prolific containers of oil. These are cavernous limestones that constitute the traps of Scurry County, Texas, and of Alberta.

OTHER FACTORS AFFECTING OIL ACCUMULATION. *Deformation* of strata, if gentle, gives rise to some of the types of structural traps, but if deformation is intense all oil is destroyed by metamorphism. If it is less intense only gas may be found.

Depth apparently has some effect upon oil accumulation and its retention, but it is not yet known what is the greatest depth at which oil can form. The deepest exploratory well in 1950 was 20,521 feet, and the deepest producing wells were 13,778 feet in Weeks Island, Louisiana, and 14,309 feet in Natrona County, Wyoming. The deepest oils have come in general from regions of thick Tertiary strata. In general, deeper oils are lighter and of higher quality than shallow oils.

Temperature if above 200° C would prevent the formation of petroleum. The Wasco well at a depth of 15,000 feet showed a corrected temperature of 297° F or 147° C. In California the temperature increase with depth is at the rate of 1° F for each 40 to 70 feet.

Pressure is probably not very important in the accumulation of oil, although differences in pressure act in the movement of oil from places of high pressure to those of low pressure. Of course it is of great importance in the recovery of oil, since it is pressure that causes oil and gas to flow freely from the well, often with tremendous force, giving rise at times to huge gushers with flows up to 300,000 barrels of oil per day. Shallow wells lacking pressure have to be pumped, and flowing wells ultimately have to be pumped as the gas pressure diminishes. The behavior of oil-reservoir pressure is not clearly understood. It should bear a direct relation to hydrostatic pressure, but in some wells reservoir pressure exceeds the hydrostatic, and in others it is less. At the Wasco, California, well at a depth of 5,000 feet the pressure was 2,950 pounds, or 780 pounds greater than the theoretical, but at a depth of 13,175 feet it was only 68 pounds greater.

OIL-FIELD WATERS. Most oil pools are underlain by salt waters, which are considered to be buried sea water that has undergone subsequent chemical changes. The waters of different pools have different compositions; some are not altered sea water; some are saltier than

Cycle of Petroleum

sea water. They contain less sulfates than sea water and more potassium and iodine. The composition of oil-field waters presumably changes by reaction with rock minerals, dissolving some and having part of its constituents removed by others. This compositional change is an evidence of migration through rocks.

As oil is withdrawn from a pool the salt water rises to displace it, and the encroachment of salt water in wells marks the decline of a pool.

FRACTIONATION. Bleaching clays have the power of absorbing heavier ingredients, impurities, and coloring matter from oils and are utilized widely in oil refining. This has suggested that during migration some oils pass through shales and clayey materials and undergo chemical and physical changes, or fractionation. Some very light and so-called "white oils" probably have been subjected to such changes and less apparent changes may have taken place in other oils.

CHIEF OIL FIELDS. The great oil regions of the world are the United States, the Middle East, the Caribbean lands, and Russia. The Netherlands Indies and western Canada are potentially great, and lesser oil regions lie in Rumania, Galicia, Egypt, India, Burma, Borneo, Peru, Bolivia, and Argentina. Numerous minor fields occur in many other countries.

The United States fields (Fig. 10–11) lie in six areas, namely, the Mid-Continent, which is the greatest oil-producing region of the world, California, Gulf Coast, East Central, Rocky Mountain, and Appalachian. A northward extension of the Rocky Mountain area lies along the foothill belt of the Canadian Rockies. Within each area there is a general similarity of controlling geology, oil traps, and nature of oil. The United States is underlain by about $2\frac{1}{4}$ million acres of petroliferous rocks and for long has been the greatest oil-producing country of the world.

The Middle East has undergone such astonishing development that its proved and potential oil reserves exceed those of any other region in the world. Its 225 wells each average 5,000 barrels a day compared with an average of 13 barrels a day for 430,000 wells in the United States. The many oil fields lie in Iran, Iraq, Saudi Arabia, Bahrein Island, and Kuwait, around and north of the Persian Gulf (Fig. 10–12). Oil is transported by pipeline to the Persian Gulf and to the east shore of the Mediterranean Sea. Ownership is largely British and American.

186 Sedimentary Processes and Cycles [Ch. 10

FIGURE 10–11. Distribution of oil and gas fields in the United States. (S. Powers, A.I.M.E.)

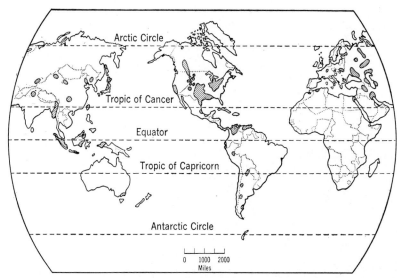

FIGURE 10–12. Outline map of world showing the distribution of the chief petroleum areas. (After C. G. Lalicker, *Principles of Petroleum Geology*, Appleton-Century-Crofts, Inc.)

Cycle of Petroleum 187

The Caribbean oil region embraces Venezuela, Trinidad, Colombia, and Mexico. Venezuela is particularly prolific and contains large potential reserves in thick reservoir strata.

Russia has long been one of the large producers of oil and is reported to have very large undeveloped potential reserves. The chief fields lie around and to the north of the Caucasus Mountains and include Baku, Brozny, Maikep-Kuban, Dagestan, and the Ural-Volga (Fig. 10–12).

Other petroleum areas are also shown by Figure 10–12.

THE CYCLE OF OIL SHALE. Although petroleum can be obtained from oil shale, the two are different products and their sedimentary cycles diverge somewhat. One is fluid petroleum; the other is compact rock which under complicated treatment may be made to yield some 30 to 45 gallons of oil per ton of rock. The organic matter of oil shales may be plant, animal, or both, but oil shales commonly originate in lakes rather than in the sea. The waxes, kerogen, and resins found in oil shales are so stable that they have not undergone change since deposition, even though some beds have been buried to depths where temperature was higher. Oil pools are not known to have originated from oil shales.

The shales are tough and in places are very thin-bedded, forming paper shales. The extensive Green River shale of Colorado, Wyoming, and Utah is a quiescent fresh-water deposit and shows very thin alternations or laminations which W. H. Bradley interpreted as indicating seasonal changes, lasting over a period of 5 to 8 million years and accumulating at the rate of 1 foot in 2,000 years. Their organic content is found to be composed of algae, but one can also see leaves, insects, fresh-water shells, and well-preserved fish. Their destructive distillation yields liquids, gases, and solids. The liquids are oil and water; the gases are combustible ones and noncombustible ones, which include ammonia; the solids are carbon and the other shaly constituents. The richest beds are those made up largely of organic matter, with a minor amount of clay and fine sand.

The Green River formation is 2,000 to 3,000 feet thick and contains three oil-shale zones. An upper one, 500 feet thick, yields only 15 gallons of oil per ton of shale, but a 70-foot section near the bottom yields 30 gallons per ton. This 70-foot section contains some 1-foot beds yielding up to 80 gallons, and some 7 to 9 feet in thickness that yield 55 to 60 gallons per ton of shale. The oil content ranges from

100 to 150 million barrels per square mile and extends over many thousand square miles, Colorado alone having 2,592 square miles. The potential oil reserves are in the hundreds of billions of barrels.

Oil shales are now utilized in places for the extraction of petroleum. They are quite widely distributed and show a great range in petroleum yield. One bed in Estonia, the richest known, yields 100 gallons of oil per ton of shale. Oil shales have been worked in Estonia, Scotland, Germany, France, Spain, Australia, and other places.

THE CYCLE OF CHEMICAL EVAPORITES [1]

Evaporation is a great sedimentary mineral-forming process, familiar in its operation, which supplies homely materials used by the householder, the farmer, the chemist, the builder, the engineer, the manufacturer, and even the birds, the beasts, and the plants. Underground waters have been drawn up to arid surfaces, there to be evaporated, leaving behind valuable minerals once in solution. Lakes have disappeared under the relentless desert sun to leave crusts of salt or borax or soda. Great sections of the ocean have been cut off during slow land oscillations and slowly evaporated to the point that extensive beds of gypsum or common salt have been deposited. Still greater evaporation has yielded rich and valuable potash deposits that supply the chemical industry and agriculture. Evaporation proceeds, of course, most rapidly in warm, arid climates.

When bodies of salty water are evaporated, the soluble salts gradually become concentrated until a point of supersaturation for a given salt is reached, when that salt becomes precipitated. The salts least soluble in the water are precipitated first, and the most soluble salts last. The exact point of precipitation is affected by temperature, time, and the presence of other salts in solution. The effect of temperature and salinity is so great that different salts may be precipitated between summer and winter, and even between day and night. Some of the salts undergo changes after precipitation, as, for example, the change from gypsum to anhydrite. With sea water, which contains about $3\frac{1}{2}$ percent by weight of salts, no deposition will take place until the volume of water is reduced by evaporation to nearly one-half its original bulk; calcium sulfate (gypsum or anhydrite) is deposited when the volume reaches about one-fifth, and sodium chloride (common salt) when it reaches one-tenth. The accompanying table

[1] Sometimes called "evaporates."

gives the composition of sea water, the order of precipitation of its salts, and the reduction in volume at which precipitation starts.

	Percent by Weight	Percent of Total Solids as Salts	Volume	Salt Precipitated
Water	96.2345	0	1.000	
NaCl	2.9424	77.758	0.533	Hematite
$MgCl_2$	0.3219	10.878		Calcite
$MgSO_4$	0.2477	4.737	0.190	Gypsum
$CaSO_4$	0.1357	3.600	0.095	Halite
NaBr	0.0556			Magnesium sulfate
KCl	0.0505			Magnesium chloride
K_2SO_4		2.465	0.039	Sodium bromide
$CaCO_3$*	0.0114	0.345	0.0162	Bittern salts
Fe_2O_3	0.0003			
$MgBr_2$		0.217		

* Includes traces of all other salts.

The sequence of salts given in the table is rarely obtained because evaporation seldom goes far enough to cause precipitation of the last or most soluble salts. The same sequence would not be followed in saline waters whose composition is different from that of sea waters because composition and temperature determine the depositional sequence and the composition of the precipitated salts. The sequences and salts are different, for example, in interior drainage lakes of arid regions, where, instead of sodium chloride waters, there may be sodium carbonate (soda), sulfate, or borate waters. Likewise, different salts are deposited from ground waters and hot springs. Therefore, each is considered separately.

DEPOSITION FROM OCEANIC WATERS

The salts of the oceans come mainly from the land through weathering of the rocks; small amounts are contributed by volcanism, and some are dissolved from the ocean basins. The oceans form a great mixing pot for the diverse contributions of the rivers that annually deliver huge tonnages of salts. The rivers of northern North America contribute excess carbonates, those of the south, excess sulfates.

Water evaporated from the ocean and dropped over the land as rain garners a new load of soluble materials and again reaches the ocean; the cycle repeats itself. Thus salts have continued to accumulate and concentrate in the ocean ever since streams first coursed over the lands. The rivers of the world are estimated to contribute about $2\frac{1}{2}$ billion tons of salts to the ocean each year. The total amount of salts now in the ocean is estimated to be 21.8 million cubic kilometers, or enough to make a layer 60 meters thick over the oceanic basins. The contributions from the land bring in carbonates in excess of chlorides, but in the ocean chlorides exceed carbonates and, as we have already seen under "The Carbonate Cycle," carbonates are continuously being abstracted from the ocean.

The mean depth of the oceans is about 12,000 feet, and the surface would have to be lowered nearly 11,000 feet before salts could be deposited. Consequently the first requirement for deposition of salts is that, to obtain the necessary concentration to induce precipitation, bodies of sea water must become isolated from the ocean in places where evaporation exceeds inflow. Such isolation may be caused by low offshore barriers, over which from time to time new supplies of sea water may be added. Natural cutoffs are not uncommon along coastal areas. Another type of evaporating basin is exemplified by the Gulf of Karabugaz on the eastern side of the Caspian Sea. Here, amid the surrounding deserts, evaporation is so rapid that a current flows continuously from the Caspian Sea through a shallow narrow inlet into the Gulf, augmenting the Gulf waters with some 350,000 tons of salts daily. Gypsum is being deposited on the shallow shores of the Gulf, and sodium sulfate near its center. If the inlet should become entirely closed, sodium chloride and other salts would become deposited eventually.

In such cutoff bodies of salt water the salts tend to become concentrated in the deeper portions of the basins. The salts deposited on the uncovered shores of the receding water body are mostly washed in again by infrequent rains. Thus the salt of a large original volume of water becomes concentrated centrally in a small volume. If the original cutoff body had a volume of 100 cubic kilometers (Karabugaz has 183) and this became concentrated to about 50 cubic kilometers, the iron oxide and calcium carbonate would be deposited. If evaporated to 20 cubic kilometers, the water would then contain about 3,500 million tons of salt, of which about 2,700 million tons would be common salt. At this point gypsum would be deposited, and common salt when the volume reached 10 cubic kilometers. Subsequent evaporation would yield magnesium and potash salts.

CALCIUM SULFATE DEPOSITION. The deposition of calcium sulfate, which occurs in two forms, gypsum (+ water) and anhydrite (− water), depends upon water temperature and salinity. Gypsum will be deposited from saturated solutions below 42° C, and anhydrite above that temperature, which is high for surface waters. Posnjak showed that, when sea water evaporates at 30° C, gypsum will be deposited when the salinity reaches 3.35 times the normal concentration of sea water and will continue to be deposited until the salinity reaches 4.8 times normal, beyond which anhydrite is deposited. Further, about one-half of all the calcium sulfate present will be deposited as gypsum before anhydrite is deposited. Lower temperatures would hardly modify the above ratios. Applying the information to sedimentary beds, it means that, if the lower beds of a calcium sulfate deposit are not gypsum but anhydrite, either the temperature of deposition was above 42° C or the gypsum was converted to anhydrite. Beds of gypsum and anhydrite (and salt) are common the world over, but the rareness of later potash salts is puzzling and must be due to some general cause as yet unknown.

Thick beds of gypsum are a puzzle. A thickness of 250 feet is known in Nova Scotia, 300 feet in Germany, and 1,325 feet in New Mexico. Let us see just why they are a puzzle. The evaporation of a depth of 1,000 feet of sea water yields only 0.7 feet of gypsum; therefore to obtain 300 feet of gypsum would require the evaporation of 425,000 feet of sea water, or a depth of 80 miles. As this is ridiculous, it follows that new supplies of sea water must have been added to the basin during evaporation to supply the volume of water necessary to deposit that quantity of gypsum, and the residual liquors must have become concentrated in subbasins. Another point is that the depth of water in most closed basins should, upon evaporation, also yield a layer of common salt above the gypsum, but in many places this is absent. Further complications are added by the occurrence of salt without underlying gypsum, because gypsum deposition should always precede salt.

How then can one explain the thick beds of gypsum, gypsum beds and no overlying salt, and salt beds with no underlying gypsum? Many theories have been proposed, few have survived, and none are entirely satisfactory. To account for the enormous volumes of water necessary, most students of the subject consider the "bar theory" or some modification of it as the most satisfactory. This postulates partial evaporation of an arm of the sea isolated by a bar over which or through which new supplies of water are added to yield more gypsum deposition. But this theory does not account for the lack of overlying salt in many places. A modification of this theory postulates a series of semi-isolated subbasins separated by low barriers, so that new supplies of water first enter the outer basin and the partially concentrated brines overflow into inner basins, carrying there the salt brines. This gypsum might be deposited without salt. Salt without underlying gypsum might be accounted for by brines from which gypsum has already been deposited being decanted into another basin where salt could be deposited. This could be brought about by coastal tilting. It is obvious that deposition of thick salt beds by evaporation is not a simple problem.

The products that result from calcium sulfate deposition are: (1) beds of relatively pure gypsum or anhydrite from a few feet to several hundred feet in thickness; gypsum beds constitute one of the most valuable of nonmetallic resources for use in plasterboards, plaster, cements, and fertilizer, but anhydrite finds little use because in plaster it absorbs water and swells; (2) gypsum beds with anhydrite impurities; (3) alabaster, a softer and lighter variety of gypsum.

COMMON SALT (HALITE) DEPOSITION. Sedimentary salt beds account for about three-fourths of all salt used. Each person consumes about 12 pounds of salt a year, and the annual world production, amounts to about 40 million tons.

Although the deposition of salt follows that of gypsum, one commonly finds alternations of salt and gypsum beds, indicating that after salt deposition started more sea water was added to supply gypsum. Beds of salt 150 feet thick occur in Ohio, 325 feet in Michigan, and 630 feet in New Mexico. Thicknesses of several thousand feet have been intersected in salt domes of the Gulf Coast and Germany.

These salt domes are unusual features, particularly those of the United States, since prolific oil pools are associated with them. Most of them have been located by geophysical means and explored by drilling. They are dome-shaped, pipe-shaped, or mushroom-shaped (Fig. 10–7). The plug consists of salt and anhydrite, and anhydrite or gypsum forms a caprock 500 to 600 feet thick. The top of the plug may be a few feet to a few thousand feet beneath the surface. No drill has yet penetrated to the bottom of the Gulf Coast salt plugs. They give the impression of being gigantic plugs that have been driven forcibly upward into the overlying strata, rending the strata apart and turning them upward to make oil traps. That is just what is considered to have happened. The salt originally existed in deeply buried sedimentary beds many thousands of feet beneath the surface. Under the pressure of the weight of the overlying rocks the yielding plastic salt was forced upward along some line of weakness, such as a fault. The great thickness of the salt has thus been caused subsequent to its deposition. Barton estimates that ten of the famous Gulf Coast salt domes contain enough minable salt to supply the world's needs for 4,000 years and that all known salt domes would supply the world demand for 300,000 years.

POTASH DEPOSITION. Following common salt deposition, chlorides and sulfates of magnesium and potassium are the chief salts deposited. The potash minerals result from evaporation carried almost to completion but only rarely does that occur. Consequently potash deposits are not common and are known only in the United States, Germany, Alsace, Poland, Russia, and Spain. Most of the world supply of potash is won from sedimentary beds and a minor quantity from brines of saline lakes.

The famed potash deposits of Stassfurt, Germany, illustrate potash deposition and represent the only complete sequence of deposition of oceanic salts. Some 30 minerals are known, of which 5 are important commercial salts, sylvite being the preferred one because of its high potash content and its easy usability. The potash beds contain magnesium and calcium chlorides, and magnesium sulfate. Common salt is present in all the beds, and gypsum and anhydrite beds are present. Deposition started with gypsum and anhydrite, followed by common salt and by potash minerals. A further influx of sea water started the sequence of deposition over again. The total thickness of all the beds is about 1,000 feet. These potash beds are very delicate geologic thermometers, since different minerals are formed with small changes in temperature, and many of the minerals are secondary replacements of primary ones.

Such large potash deposits mean not only almost complete evaporation but also a stupendous concentration of residual bitterns. This must mean a gradual drainage of bitterns, after removal of gypsum and much common salt, into the lowest portions of large evaporation basins. In this way a small area may come to contain the residual constituents formerly dispersed in a great volume of water spread over a large basin. Probably there was some tilting and decanting also.

During the deposition of potash salts common salt and magnesium salts are also deposited, with the result that beds of pure potash salts are rare and some beds are richer than others. Repetitions of beds are not uncommon, indicating that the cycle of evaporation and deposition was interrupted and renewed many times.

Potash deposition took place in the Permian basin of Texas and New Mexico, where a prospective basin of some 40,000 square miles is indicated, and 33 square miles near Carlsbad, New Mexico, is proved to contain commercial deposits of sylvite. The sequence is generally similar to that at Stassfurt, but some 40 potash zones have been recognized, of which the lower sylvite zone, at a depth of 1,000 feet, is 5 to 12 feet thick and consists chiefly of halite and sylvite, averaging over 25 percent potash, the richest in the world. All the beds in this Permian basin have a thickness of 4,000 to 11,000 feet, and the salt series ranges from 1,000 to 2,500 feet in thickness. Repeated sequences of clay, gypsum, common salt, and potash salts indicate a long-continued desiccation with many fluctuations, interruptions, and nonuniform deposition. The richer portions occur in marginal subbasins.

BORATE AND BROMINE DEPOSITION. Some boron minerals are deposited from marine residual liquors, along with potash salts. Bromine is also deposited at the same time. Bromine is remarkably concentrated in the Dead Sea, to the extent of 0.4 percent, compared with 0.0064 percent in ocean water.

DEPOSITION FROM LAKES

In warm semi-arid and arid regions where interior drainage exists, the soluble products of rock weathering are carried by the infrequent rains to depressions without outlet. Here, rapid evaporation brings high concentration of the included salts, and the salinity may be many times greater than that of the ocean. For example, Great Salt Lake of Utah in 1904 had a salinity seven times that of the ocean with salts of oceanic type.

In the ocean, as we have seen, all the different kinds of salts contributed from the diverse drainage areas of the rivers form a common melting pot, uniformly mixed, of one composition. In interior lakes, however, this is not so. Each lake is a unit by itself and collects into itself the soluble salts of its own drainage area, and not of others. Therefore, each lake is different in composition, which is determined by the character of the rocks of its particular drainage basin and by the presence or absence of hot-spring discharge. Thus there are salt lakes of oceanic type, alkali or soda lakes, sodium sulfate lakes, and borate, nitrate, magnesium, and potash lakes. All contain common salt and most of them contain sulfates. The calcium carbonate present in the early stages of the formation of such lakes is soon precipitated in the form of tufa, leaving chiefly sulfates and chlorides. In general

Deposition from Lakes

salt lakes arise from the weathering of sedimentary rocks, and alkali lakes from volcanic rocks. Upon evaporation various kinds of salts are deposited, some of which are important chemical, industrial, and medicinal salts.

SALT LAKES. Salt lakes receive from their drainage areas salts generally similar to those of the ocean. Their salinity varies depending upon season or rainfall, but most of them are much more saline than the ocean. For example, Great Salt Lake, which is a small shrunken remnant of a former great Lake Bonneville, has ranged from 13.79 percent salinity in 1877 to 27.72 in 1904, and is now around 23 percent. It has lost most of its calcium carbonate and now contains some 400 million tons of common salt and 30 million tons of sulfate. The complete evaporation of salt lakes gives rise to deposits somewhat similar to oceanic salts, and these form playas or salt pans, both recent and ancient. The deposits, however, exhibit the fluctuation effects of repeated desiccation and resolution and of recurrent influx of sediments during periods of rainfall and their admixture with the salts.

The Dead Sea is an example of a saline lake that has been subjected to such extreme evaporation that its bittern water resembles brines left over after the deposition of common salt from ocean water. Calcium carbonate is absent, and sulfate is low. Magnesium chloride is the chief constituent, and it is high in potassium chloride and bromine.

ALKALI AND BITTER LAKES. Alkali or soda lakes and bitter lakes are common to regions of sparse rainfall and interior drainage where volcanic rocks, or sedimetary rocks high in sulfate, are undergoing weathering. Volcanic rocks yield much sodium and some potassium. Consequently the lakes obtain sodium carbonate, sodium sulfate, common salt, some potassium carbonate, some magnesium compounds, and smaller amounts of other salts. Some sodium carbonate also results from slow and complex chemical reactions with other sodium and calcium salts; some also is due to the action of algae on sodium sulfate. The potassium carbonate is considered to be the indirect product of the work of organisms.

Alkali or soda lakes contain predominant sodium carbonate, but potassium carbonate may be abundant and common salt is always present. Some examples are Owens and Mono Lakes in California, the Soda Lakes of Nevada, Lake Goodenough in Saskatchewan, and the Natron Lakes of Egypt. The Natron Lakes are alternately wet and dry, and evaporation leaves a layer of natron and salt, bordered by sodium carbonate. The high salinity of Owens Lake varies with seasonal and climatic cycles. It is estimated to contain 8 million tons of sodium carbonate. The Soda Lakes contain one-fifth each of sodium carbonate and sodium sulfate and three-fifths of common salt, but the deposits from it consist of nearly one-half sodium carbonate, one-third sodium bicarbonate, and only a little sodium sulfide and salt. The chief salts deposited from the alkali lakes are the sodium bicarbonate or baking soda, sodium carbonate or soda ash, natron, and trona. Sodium carbonate is commonly deposited around soda lakes as an efflorescence known as "black alkali" because of its effect on vegetation. The efflorescences generally contain some borax and other substances. According to Chatard, fractional crystallization attendant upon evaporation yields (1) trona with some soda, (2) sodium sulfate, (3) common salt, and (4) normal sodium carbonate.

In bitter lakes sodium sulfate predominates, but carbonate and chloride are present also. The sulfate comes from volcanic rocks or buried sulfate beds. Sulfur is contributed from the oxidation of pyrite, by volcanic sources, from hot springs, and, locally, from gypsum beds. If solutions of gypsum and sodium carbonate are mixed, evaporation will give calcium carbonate and sodium sulfate. Bitter lakes are common in arid interior drainage regions of America and Asia. Examples are Searles and Soda Lakes in California, Verde Valley Lake in Arizona, the Downey Lakes in Wyoming, Lake Altai and the Gulf of Karabugaz in Russia. The Gulf of Karabugaz each winter gives deposits of sodium sulfate, which are harvested. Such marginal efflorescences, called "white alkali," are harvested around many lakes. These and the deposits of dried-up lakes consist mainly of the hydrous sulfate, Glauber's salt, and the anhydrous sulfate, thenardite, along with some common salt, natron, and magnesium and calcium sulfates. The deposition of sodium sulfate generally precedes the deposition of common salt and may be controlled by the season of the year. Since the sulfate is more soluble in warm than in cold water, whereas common salt is not, it may be deposited as Glauber's salt in winter and washed upon lake shores. From warm waters thenardite is deposited. Bottom layers may contain beds of epsom salts and soda.

Searles Lake is a semi-solid mass of spongy salt which is sometimes dry. The total salts amount to 35 percent of the very alkaline brine (pH 9.48) and consist of common salt, sodium sulfate (salt cake), sodium carbonate, potassium chloride, bicarbonate of soda, and borate of soda, all of which are recovered. In addition, Searles Lake is now the United States' chief producer of lithium. Beneath the upper crust to depths of 75 to 200 feet is a relatively solid mass of crystals. Hard bottom layers consist of almost pure bicarbonate of soda and some borax.

BORAX LAKES. Borate lakes are not common but are known in western United States, South America, Turkey, Iran, and Tibet. The brines of Searles, Owens, and Borax Lakes yield borax, and borax is deposited in marginal muds. The famed dazzling white surface crusts of borax in Death Valley and the "Twenty-Mule Borax Team" are now of the past.

Boron is a constituent of some igneous minerals and in the form of boric acid and calcium and magnesium borates is well known in volcanic exhalations. Thus in volcanic regions it may become concentrated in lakes, as in Borax Lake, California, where it constitutes 12 percent of the evaporated salts. The boron evaporites are chiefly calcium and sodium borates from which borax is produced. The borate industry has been featured by the successive discarding of one mineral for another more usable one. First borax from efflorescences and lake muds was used, then borax and ulexite were obtained from playas (dried-up lake basins). These impure substances were replaced by bedded deposits of purer colemanite and ulexite from Death Valley; then the most desired mineral, kernite, was found along with borax in older buried lake-bed deposits in California. Subsequent to deposition there have been transformations of some of the boron minerals. Thus kernite is thought to have been a transformation from borax. The solubility of borax and kernite make them preferred. The noteworthy feature of kernite is that it dissolves readily in water and upon evaporation yields normal borax.

About 93 percent of the world borax production is from the United States. It is a well-known household commodity but is of much greater importance in industry. As stated by Schaller, "the cleanser, the pharmacist, the paper and textile maker, the metallurgist, the brazier, and the jeweller all use borax. Hardly any other substance enters into so many diversified lines of manufacture."

OTHER LAKE EVAPORITES. Some lakes contain commercially extractable *potash*, chiefly as chloride, and yield minor production. The Searles Lake deposit contains some 15 potash minerals unknown in the Stassfurt deposits.

A few lakes in western United States, South America, and South Africa contain *nitrates* from which soda niter is won. The nitrogen is apparently derived from volcanic ammonium salts and from the decomposition of organic matter aided by the action of "nitrifying" bacteria, which live in the roots of legumes. Some may originate through electrical discharges in the atmosphere. The soda niter is deposited through evaporation along with other evaporites.

DEPOSITION FROM GROUND WATER

Evaporation from ground water is universal, but in humid regions the evaporites are redissolved and removed. In arid regions the evaporites may accumulate as long as the climate remains arid.

The ground water contains salts in solution similar to those of the rivers and ocean but mostly in very low concentration; in a few places special conditions have provided unusual concentrations of some substances from which deposits originate by evaporation. Those of economic importance in the sedimentary cycle are nitrate salts with iodine, some boron, calcium carbonate, and sodium salts. Deposition follows when evaporation takes place at or near the surface. If the site of evaporation is fed by fresh supplies of ground water, extensive deposits may eventually result. Evaporation will proceed most rapidly where the ground water is supplied relatively close to the surface in arid regions, such as valley bottoms, slopes where hills and valleys merge, and long hill slopes interrupted by gentler or reverse grades.

NITRATE DEPOSITS. The well-known Chile nitrate deposits are the only large natural deposits of commercial nitrates and illustrate the process of formation. The deposits, called caliche, lie in a long narrow belt in northern Chile parallel to the Pacific Ocean and some 13 to 100 miles inland, in one of the most rainless deserts of the world on the east side of the Coast Range. They consist of gravel cemented by sodium nitrate and associated salts at depths of a few inches to a few feet. About one-fourth is sodium nitrate, along with common salt, sulfates, borates, bromides, iodides, phosphates, and lithium and strontium compounds. They yield most of the iodine of the world.

The cycle of nitrogen is only partially known, and the origin of the deposits is highly controversial. The consensus of opinion is that nitrogen compounds were dissolved and transported by ground water and deposited by evaporation toward the base of the Andes. The most probable source seems to be volcanic rocks that mantle the slopes, for there is a close association between their distribution and the nitrate deposits. Moreover, the constant association of nitrates and borates supports this origin. The sources of origin have also been attributed to bird guano, electrical fixation from Andean thunderstorm discharges, and bacterial fixation of former nitrogenous vegetable matter. In arid regions of high evaporation ground water is drawn up toward the surface, and continued deposition of the contained salts yields deposits just beneath the surface.

Other minor deposits resulting from evaporation of ground water are calcium carbonate in the form of so-called caliche or lime, Epsom salts, Glauber's salt, soda, and borax.

HOT-SPRING DEPOSITION

Waters from hot springs upon evaporation build up local deposits, a few of which are of commercial importance; others are scenically beautiful. The deposition is aided by the action of microorganisms. The chief depositions are (1) calcium carbonate in beds of tufa, calcareous sinter, or travertine, (2) siliceous sinter or geyserite, (3) iron oxide in the form of ocher, and (4) manganese wad. The famed White Terrace of New Zealand and the Mammoth Hot Springs of Yellowstone Park are composed of almost pure calcium carbonate. Travertine beds yield a stone much prized in interior decoration because of its pleasing texture and vuggy nature. The type locality, at Travertine, Italy, has been quarried since the time of the Romans.

SELECTED REFERENCES ON SEDIMENTATION

1. Sediments

Principles of Sedimentation. W. T. Twenhofel. McGraw-Hill, New York, 1940. Chaps. 9, 10, 11, and 13. Comprehensive discussion of all phases of sedimentation.

"Solution, Transportation, and Deposition of Iron and Silica." E. S. Moore and J. E. Maynard. *Econ. Geol.* 24:272–303, 365–402, 1929. Theoretical treatment of natural solution and deposition of iron that forms ore beds.

"Wabana Iron Ore Deposits of Newfoundland." A. O. Hayes. *Econ. Geol.* 26:44–64, 1931. Occurrence, mineralogy, and origin.

"Experiments in the Deposition of Iron with Special Reference to the Clinton Iron Ore Deposits." J. R. Castaño and R. M. Garrels. *Econ. Geol.* 45:755–770, 1950. Iron dissolved in stream water is deposited as hematite in sea water in equilibrium with solid calcium carbonate by straight deposition and by replacement.

"Deposition of Manganese." C. Zapffe. *Econ. Geol.* 26:799–832, 1931. Catalytic deposition of manganese from a water supply.

"Mineral Development in Soviet Russia." C. S. Fox. *Min. Geol. and Met. Inst. India Trans.* 34:98–210, 1938. Manganese, gypsum, and iron ores.

Cements, Limes, and Plasters. E. C. Eckel. John Wiley & Sons, New York, 1928. Parts II–VII. Good treatment of gypsum and limestones.

Ore Deposits of the Western United States—Lindgren Volume. Am. Inst. Min. Met. Eng., New York, 1933. Chap. 10. Deposits of uranium, vanadium, radium, copper, manganese, phosphate, and potash.

"The Geologic Role of Phosphorus." Eliot Blackwelder. *Am. Jour. Sci.* 42:285–298, 1916. Details of the phosphate cycle.

"Phosphate Deposits of the United States." G. R. Mansfield. *Econ. Geol.* 35: 405–429, 1940. Occurrence and origin of sedimentary phosphates.

Potash Reserves of the United States. S. H. Dolbear. Am. Potash Inst., Washington, D. C., 1946.

"The Role of Fluorine in Phosphate Deposition." G. R. Mansfield. *Am. Jour. Sci.* 238:863–879, 1940. Methods of transport and deposition.

"Genesis of the Sulphur Deposits of the U.S.S.R." P. M. Murzaiev. *Econ. Geol.* 32:69–103, 1937. Descriptions and origin of sedimentary sulfur.

"Birmingham District, Alabama." E. F. Burchard. *16th Int. Geol. Cong. Guidebook* 2, Washington, D. C., 1933. A concise summary of sedimentary Clinton hematite.

"Vanadium Deposits near Placerville, Colo." R. P. Fischer, J. C. Haff, and J. F. Rominger. *Colo. Sci. Soc. Proc.* 15:115–134, 1947. Flat lenses of uncertain origin in Entrada sandstone.

"Uranium-Bearing Sandstone Deposits of the Colorado Plateau." R. P. Fischer. "Characteristics of Marine Uranium-Bearing Sedimentary Rocks." V. E. McKelvey and J. M. Nelson. *Econ. Geol.* 45:1–11, 35–53, 1950. These papers give carnotite occurrences and occurrences in aluminous and carbonaceous sediments.

Geochemistry. Kalervo Rankama and Thure Georg Sahama. Pp. 900. Univ. Chicago Press, 1950.

Economic Mineral Deposits. 2nd Edit. Alan M. Bateman. John Wiley & Sons, New York, 1950. Sedimentation, 163–182; iron ore, 564–572; manganese, 578–585; vanadium, 603–605; uranium, 619–621; coal, 634–651; petroleum, 652–695; clay, 696–706; structural materials, 711–732; industrial materials, 763–772.

2. EVAPORITES

Geology of Nonmetallic Mineral Deposits. A. W. Grabau. McGraw-Hill, New York, 1920. Vol. I, Principles of Salt Production. Oceanic salts and their precipitation.

"Geology and Origin of Silurian Salt in New York State." H. L. Alling. *N. Y. State Mus. Bull.* 275, 1928. Salt and gypsum deposits and review of theories of origin.

"Gypsum Deposits of the United States." R. W. Stone. *U. S. Geol. Surv. Bull.* 697, 1920. Descriptions of gypsum occurrences.

"Deposition of Calcium Sulphate from Sea Water." E. Posnjak. *Am. Jour. Sci.* 238:559–568, 1940. Chemical controls of deposition of gypsum and anhydrite.

"Geology of Texas." G. R. Mansfield and W. B. Lang. *Univ. of Texas Bull.* 3401:641–868, 1934. Best description of U. S. Permian basin and potash deposits.

"Potash in North America." J. W. Turrentine. *Am. Chem. Soc. Mon. Ser.* 91, 1943. General survey.

"Owens Lake: Source of Sodium Minerals." G. D. Dub. *A.I.M.E. Tech. Pub.* 2235, 1947. Saline minerals and occurrence.

Economic Mineral Deposits. 2nd Edit. Alan M. Bateman. John Wiley & Sons, New York, 1950. Evaporation, 187–201; chemical minerals, 776–799; fertilizer minerals, 800–820.

CHAPTER ELEVEN

Weathering Processes

PRINCIPLES

Weathering is a process that anyone may observe anywhere on the earth's surface. It is familiar but complex. It gives rise to soils, to the constituents of sediments and sedimentary rocks, to the substances dissolved in surface and underground waters; as a result of it valuable new mineral deposits are created, pre-existing substances are concentrated into workable deposits, and existing metallic mineral deposits undergo profound changes.

Under the slow relentless attack of weathering, rocks and enclosed mineral deposits succumb to chemical decomposition and mechanical disintegration. Weathering is a complex operation consisting of distinct processes that may operate singly or jointly. The minerals that are unstable under weathering conditions suffer chemical decay. The soluble parts may be removed, and the insoluble parts accumulate and some of them may form residual mineral deposits. Stable minerals, such as gold or quartz, undergo little or no chemical change but may be freed from their enclosing matrix and accumulate on the surface as residual deposits or be mechanically concentrated into placer deposits by moving water or air.

We generally think of weathering as a double process, one mechanical and one chemical, but both usually operate together. Mechanical disintegration is a physical breaking up of rocks by (1) frost action, (2) expansion and contraction crackling under changes of temperature, (3) abrasion by moving water, (4) gravity, (5) growth of vegetation, or (6) other means. It does not create new minerals but merely frees already formed minerals from their matrix. However, mechanical disintegration facilitates chemical attack and decomposition by reducing the particle size and creating a greater amount of surface for chemical attack. Chemical weathering, however, actually creates new minerals of which some remain stable under surface conditions.

WEATHERING PROCESSES. Mechanical weathering by itself is confined largely to warm, arid regions where rain is sparse and where pronounced annual and daily temperature changes occur, or to very cold regions where surface chemical changes proceed slowly and frost action is vigorous much of the time. Chemical weathering thrives in warm, humid regions where rainfall supplies moisture and nourishes plant life, which in turn supplies humic and organic acids to assist chemical activity. Because deep chemical decay needs deep and long-continued weathering, its products are generally absent from the glaciated regions, as the time since glacial scouring has been too brief to permit much weathering in those regions of cool climate.

The agents of surface chemical decomposition are water, oxygen, carbon dioxide, heat, acids, alkalies, vegetable and animal life, and some of the soluble products of the decomposition of the rocks themselves. Without water there is little decomposition. The absorption of water is general and is important in both decomposition and disintegration. Oxygen permits oxidation, which is almost universal in weathering, and carbon dioxide dissolved in water is a powerful solvent. Acids, such as sulfuric acid, and some sulfates, generated by the oxidation of sulfides, are active agents of decomposition. Vegetation supplies carbon dioxide and organic compounds that decompose rocks and dissolve substances. Bacteria are active in biochemical processes of decomposition and precipitation.

The chemically decomposed rock, ore, and gangue minerals lose many of their constituents in solution or mechanically, with attendant decrease in volume. Carbonates are readily dissolved in carbonated waters and are removed in solution. Vast quantities of limestone thus disappear. Silicate rocks are readily decomposed. Some of the constituents are carried away in solution and form source materials for sediments and evaporites. Others, such as the oxides of iron and aluminum, form insoluble residues that tend to remain behind. During weathering colloidal compounds are formed, some of which go away in solution whereas others are precipitated essentially in place as colloids.

Weathering ordinarily does not extend to depths greater than a few feet to a few scores of feet, although depths of 100 to 200 feet are not uncommon. Sulfides are known to oxidize to depths of more than 3,000 feet.

TEMPERATE VERSUS TROPICAL WEATHERING. There is a pronounced difference, not fully understood, in the nature and results of weathering in temperate climates as compared to tropical and subtropical climates. Essentially, the difference is the effect on silica. In temperate climates most of the silica remains behind; in tropical climates most of it is removed. This difference is of great economic importance—it may mean a common clay soil on the one hand or a valuable bauxite, the ore of aluminum, on the other. In temperate climates most of the silica freed during weathering joins with alumina and water to form clay, along with admixed iron oxides and perhaps grains of quartz. Thus clayey soils are the weathering products of aluminous rocks in all temperate regions.

Tropical-type climates are characterized by alternate wet and dry seasons, by hot weather and warm surface waters throughout the year and, generally, by luxuriant vegetation and an abundant supply of bacterial life and organic compounds. Under these conditions rock decay is carried further, leaching is more complete, the silicates are thoroughly decomposed, but, particularly, silica may be extensively removed in solution. The result may be a *laterite* soil, which is a mixture of hydrous oxides of aluminum and iron with some silica and other impurities. Instead of always being hydrous aluminum silicate (clay) of temperate regions, it may, but not in all cases, be hydrous aluminum oxide (bauxite). Laterites may be so high in iron that they constitute a lateritic iron ore or, conversely, so high in alumina and low in iron oxide and silica that they constitute an ore of aluminum; most laterites are neither. Other physical and chemical conditions also necessary to yield bauxite will be discussed in the section on "Residual Bauxite Formation."

PRODUCTS OF WEATHERING. Weathering results, broadly speaking, in two classes of products, (1) those removed in solution, and (2) those that remain behind. The first class has already been considered in Chapter 10, and the following sections concentrate upon the second class.

Of the products that remain behind, (1) some stay in place to form ordinary soils without mineral value; (2) some consist of residues that constitute workable mineral deposits; (3) some consist of residues of desired heavy and undesired light insoluble minerals from which the light may be separated from the heavy by moving water or air; (4) some of the dissolved products may be reprecipitated at slight depth to form economic mineral deposits. As to which of the four groups results depends upon the nature of the original rock or mineral deposit, the topography, and whether it is temperate or tropical weathering. The second group form one line of descent by purification processes to *residual concentration deposits;* the third group form another line of descent to *mechanical concentrations* or *placer deposits;* the fourth group give rise to *oxidized ore deposits* or *secondary enrichment deposits.*

RESIDUAL CONCENTRATION

Residual concentration is the accumulation of valuable residues after undesired constituents of rocks and mineral deposits are removed through weathering. It is largely a gradual decrease in volume effected by chemical weathering until the residues have accumulated in sufficient volume and purity to make them of commercial importance.

The first requirement for residual concentration of economic mineral deposits is the presence of rocks or lodes containing valuable minerals insoluble under surface conditions, and undesired soluble substances. Second, the climatic conditions must favor chemical decay. Third, the topographic relief must not be too great, or the valuable substances will be washed away as rapidly as formed. Fourth, long-continued crustal stability is essential in order that residues may accumulate in quantity and that the deposits may not be destroyed by erosion. If the conditions are such that the four prerequisites mentioned above coincide, then, for example, a limestone formation free from other impurities except a little iron oxide will slowly be dissolved by surface waters, leaving the iron oxide as a residue. As bed after bed of limestone disappears, the little iron oxide of each bed persists and accumulates until eventually there may be formed a surface mantle of iron ore, which if of sufficient purity and volume would constitute a workable deposit.

The example of iron ore typifies one method of residual concentration wherein the residue is simply the gradual accumulation of a preexisting mineral that has not changed during weathering. In a different mode of formation, however, the valuable mineral is created as a *result* of weathering and then persists and accumulates while the soluble parts are removed. This applies to the accumulations of residual clay or bauxite that are formed from aluminous rocks.

Both temperate and tropical weathering yield residual concentrations, but the tropical and subtropical conditions favor the formation of more kinds of deposits chiefly because of silica removal. Thus the weathering of a limestone with impurities of iron oxide and silica under temperate climate would yield a residue of iron oxide high in silica, which would not be ore, but under tropical conditions the removal of silica would leave a residue of iron ore; a siliceous ironstone would react similarly. Consequently the geographic and climatic location of weathering is important from the standpoint of the formation of residual mineral deposits. Valuable deposits of iron ore, manganese, bauxite, tin, nickel, clays, phosphate, kyanite, barite, ochers, gold, diamonds, and other substances accumulate as economic residual concentrations. The basic principles of weathering and concentration apply to each, but since the processes differ in detail for such widely distinct commodities, it is simpler to consider the cycles of each separately.

RESIDUAL IRON CONCENTRATION

Most rocks contain iron, and the familiar red and yellow soils of chemically weathered areas attest to its presence. Given unusually favorable conditions the iron content of an unweathered rock may accumulate as a residual iron ore. The source materials are (1) lode deposits of iron carbonate or iron sulfide, (2) limestones that have been partly replaced by iron minerals, (3) disseminated iron minerals in limestone free from alumina and silica, (4) ferruginous siliceous sediments, and (5) certain basic igneous rocks. Of these, numbers 1, 3, 4, and 5 are the chief sources of residual iron deposits. In temperate regions iron carbonate deposits and limestones are the only source materials of residual iron ores, since the others leave a residue too high in silica or alumina, or both, to constitute ore. Bodies of iron carbonate weather readily to yield high-grade oxides of iron, as in the valuable residual deposits of Bilbao, Spain. Bodies of iron sulfide only rarely yield residual iron oxides of sufficient freedom from sulfur and other impurities to make good iron ores.

Limestone weathering, as indicated previously, will yield iron ores only from essentially pure limestones containing iron carbonate. Most limestone underlying known residual iron-ore deposits in temperate climates contain only a small amount of disseminated iron and much larger amounts of silica and alumina, so that their weathering would yield, not an ore of iron, but instead a clay soil with some iron oxide. Consequently, residual iron deposits resting on such limestones—and there are several in the southern United States—must have been derived from overlying beds in which the iron was partially concentrated before or perhaps during weathering.

Basic igneous rocks cannot yield residual iron ores during temperate-climate weathering because clay will accumulate in excess of iron oxide. Under tropical weathering, however, nearly all constituents except iron and alumina may be removed in solution, leaving a residue of iron oxide and alumina. If the basic rock originally was much higher in iron than in alumina, an iron-ore deposit may result. The weathering of serpentine has yielded extensive residual iron-ore deposits in Cuba. Iron ores derived from basic rocks will of course be high in alumina but low in silica, phosphorus, and sulfur. However, they may contain as impurities other metallic compounds originally present in the basic igneous rock, such as nickel, cobalt, and chromium.

Ferruginous cherts have been the source of residual iron-ore deposits in various parts of the world. They are originally sedimentary beds, called banded ironstones or cherty ironstones, siliceous ironstones, ferruginous slate, taconite, or other names, and they generally consist of thin alternating layers of iron compounds and chert. Most of them happen to be of Precambrian age (+ 500 million years). The iron is commonly in the form of ferrous silicate or greenalite. This is so in the Lake Superior region, where the greenalite in ferruginous sediments has altered to taconite or ferruginous chert, which in turn is generally considered to have been weathered, with removal of silica leaving rich iron ore. Gruner, however, has advocated that the leaching of silica has taken place by hydrothermal solutions rather than by weathering.

There are many puzzling features about the formation of residual iron ores. In the Appalachian region carbonates were removed and silica and iron remained; in Arkansas silica, iron, and other substances were removed and iron remained; and in the Lake Superior region silica and carbonates were removed and iron oxide and a little clay remained.

In the southern United States examples of residual iron deposits are to be found containing 35 to 55 percent iron but high in alumina, silica, phosphorus, and manganese. Similar ores occur in northeast Texas in the form of layers, nodules, concretions, and lenses derived from greensand. One area contains 150 to 200 million tons of ore that contain up to 48 percent iron and up to 12 percent silica and alumina, illustrating enrichment of iron and accumulation of silica and alumina in temperate weathering. The eastern Cuban ores, on the other hand, represent iron laterite from subtropical weathering. The ores are iron oxides mantling serpentine, from which they were derived. Leith and Mead showed that:

100 pounds serpentine with 1.5 pounds alumina and 10 pounds iron become 17.5 pounds serpentine with 3.8 pounds alumina and 11.7 pounds iron + 2 pounds others

The residue carries 44 percent iron plus some chromium and nickel. These deposits illustrate the amount of shrinkage concurrent with enrichment of iron, nickel, and chrome by removal of silica. Other examples are to be found in Venezuela, Russia, Greece, India, New Zealand, and the East Indies.

RESIDUAL MANGANESE CONCENTRATION

Manganese minerals during weathering in general behave similarly to iron minerals. Manganese and iron are closely related chemically. Manganese accompanies iron but in smaller amounts. Its oxides likewise resist solution and accumulate from the weathering of manganese-bearing rocks to form extensive residual deposits—the most important type next to sedimentary deposits. In fact, outside of Russia, all the chief manganese deposits in the world are residual. Somewhat special conditions are called for to yield large residual manganese deposits.

The source materials are generally similar to those of iron, namely, altered igneous rocks (schists), pegmatite dikes, lode deposits, limestones low in alumina containing manganese carbonate, and limestones containing disseminated introduced manganese minerals. The source manganese compounds are the carbonate, silicate, oxides, and manganese garnet. As do many iron deposits, residual manganese deposits result from silicate rocks only under tropical and subtropical weathering, and few large deposits have resulted from the weathering of limestone because its original low content necessitates so much concentration with such a large reduction in volume of limestone that impurities are apt to accumulate faster than the manganese. The chief residual deposits have been derived from manganese-bearing schists with removal of silica and other compounds and accumulation of manganese and alumina. Lode deposits of manganese carbonate (rhodochrosite) and limestones containing manganese also yield residual deposits of mixed manganese oxides by weathering.

The formation of residual manganese ores apparently takes place with much solution and redeposition, in which colloidal mixtures are formed, which upon crystallization and unmixing yield radiating and colloform structures of intimate mixtures of manganese minerals in bewildering complexity.

Residual deposits of manganese ores are found throughout the unglaciated regions of the world, notably in tropical and subtropical

Residual Bauxite Formation

areas. The large deposits of central India, the Gold Coast, Brazil, Egypt, and Morocco are of residual origin. Other important residual deposits occur in Malaya, the Philippines, Belgian Congo, Angola, and Rumania, with minor ores in United States, east central Europe, the East Indies, Chile, and Puerto Rico.

RESIDUAL BAUXITE FORMATION

Although aluminum is the most abundant metal and the third most abundant element in the earth's crust, it occurs mainly in combinations that are not as yet adapted to commercial extraction of the metal. It is one of the main constituents of clays and soils and of the silicates of common rocks, but so far bauxite, composed of hydrated aluminum oxides, is the only commercial ore of aluminum. Much research and experimentation have been carried on to try to utilize high-aluminous clays, certain igneous rocks, alunite, and andalusite, for the extraction of aluminum, but the high cost of production will defer the use of these materials until bauxite ores are exhausted or inaccessible. All bauxite is formed in residual deposits at or near the surface under special and peculiar conditions of weathering.

Good bauxite ore contains approximately 50 percent alumina (Al_2O_3) and less than 6 percent silica (SiO_2), 10 percent iron oxide (Fe_2O_3), and 4 percent titanium oxide (TiO_2). These substances are accumulated products of peculiar weathering of aluminous silicate rocks essentially lacking in quartz. The silicates are broken down, silica is largely removed, water is added, and the other substances become concentrated in the residuum, which, in syenite rock, occupies about 40 percent of the original rock volume (Fig. 11-1). Bauxite does not exist as a primary mineral substance; it is created for the first time during chemical weathering. Bauxite is not a product of normal weathering in temperate regions; there it is almost lacking from soils because the alumina has joined with silica to form clay. It is restricted to regions of tropical to subtropical weathering under conditions more unusual than those for the formation of residual iron and manganese deposits.

The source rocks of bauxite are mainly nepheline syenite (an igneous rock, mostly of feldspar with no quartz), clay, clayey limestone, clay shales, crystalline gneisses, basalt, and alluvium. The wide variety of rocks is rather striking.

The conditions essential for the formation of bauxite are: (1) a tropical to subtropical climate with seasonal rainfall; (2) suitable source rocks; (3) old erosion surfaces prolonged in their development and marked by gently undulating topography that permits the retention siduum and provides continual and slow downward seep of rain water and a seasonal high-water table; (4) subsurface drainage that permits withdrawal of the rain waters charged with dissolved silica and other substances so that stagnation and lack of removal of waste products do not result; (5) a prolonged period of weathering; and (6) preserval of the bauxite from the ravages of erosion or destruction by encroaching marine invasions. The last feature requires fortuitous

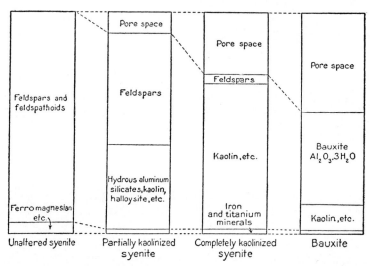

FIGURE 11-1. Diagram showing changes of volume and of minerals in the weathering of syenite to form bauxite. (W. J. Mead, *Econ. Geol.*)

circumstances, such as burial during quiet sedimentation, followed by freedom from dissection, or preservation on undissected remnants of old erosion surfaces.

The chemical processes of bauxite formation are imperfectly understood. It would appear that the ordinary reagents available in tropical weathering have been the effective agents of alteration. These would include tepid rain water, carbon dioxide, and humic acids; carbonic and organic acids break up silicates, and the resulting alkali carbonates are competent solvents for silica. Bacteria also aid solution and redeposition of alumina. A temperature of 20° C favors chemical processes by which silica goes into solution and iron oxide and alumina remain behind. Accumulation of free carbon dioxide on the surface is hindered during the wet season, which is the one of formation of the oxides of iron and aluminum, and the dry season is the one of leaching away of silica from these oxides. The bauxite apparently undergoes some solution and reprecipitation during the weathering process. This is affected by the hydrogen-ion concentration (pH) of the solutions. Aluminum remains dissolved in acid solution with pH less than 4 and also in alkaline solution with pH greater than 9, and bauxite is precipitated around the neutral point. If carbon dioxide, which keeps the solutions acid, escapes, the solution turns neutral and bauxite is precipitated, but silica still stays in solution. Thus purification of the bauxite and removal of silica take place.

Bauxite and clay are generally associated. The deposits of Arkansas mostly grade through clay into underlying syenite (Fig. 11-2);

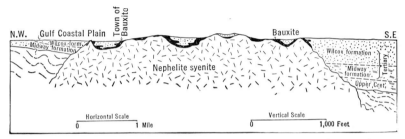

FIGURE 11-2. Cross section of bauxite deposits in Arkansas, showing bauxite (black) lying on weathered surface of syenite rock. (After G. C. Branner, *16th Int. Geol. Cong.*)

the deposits of southern Europe derived from limestone all rest on a clay base; those of the Guianas and Gold Coast also rest on clay. Herein is the basis of some controversy—whether clay is formed first and then under later tropical or subtropical conditions the clay is converted to bauxite, or whether the clay is only an intermediate step in the formation of bauxite and all under tropical conditions, or whether later clay was formed from bauxite by resilication. There are advocates for all three ideas.

To suggest that the clay was formed under temperate weathering and at a later period changed to bauxite would require rather improbable fortuitous climatic changes at just the right periods. It seems most probable that the clay is a transitory intermediate step in the continuous alteration of original rock to bauxite; i.e., underlying conditions favor clay formation, and overlying conditions favor the changeover of clay to bauxite. This idea is strengthened when one considers that the bauxite belt of southern Europe has been formed not from feldspathic rocks but from limestones. The solution of impure limestone left behind a residuum of clay upon which bauxite rests. Obviously, bauxite cannot form from limestone; therefore it must have formed from the underclay. Moreover, I have noted in British Guiana that "icicles" of pure bauxite penetrate downward into the underclay and the boundary between the two is highly irregular, indicating encroachment of bauxite into clay. In addition, deep residual clay without bauxite occurs near the bauxite deposits, and in fact I saw it in many parts of British Guiana. Therefore residual clay formation is characteristic of this tropical region, and bauxite formation is the exception. This would indicate that bauxite forms from the clay. This idea does not preclude the possibility that some later clay may form from bauxite by resilication of bauxite.

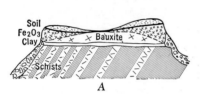 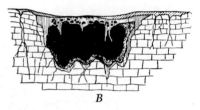

FIGURE 11–3. A, section of Gold Coast bauxite deposits (after W. G. G. Cooper, *Gold Coast Geol. Survey*); B, section of pocket bauxite, Istria. (A. Desio, *Geologia.*)

All the commercial bauxite deposits have been formed in past geologic ages (mostly Mesozoic and Tertiary) and in part under climatic conditions different from those existing today in the Arkansas–Alabama, southern Europe, and other deposits that occur in what are now temperate climates. These geologic periods apparently supplied optimum conditions for bauxite formation.

Bauxite deposits always lie on ancient erosion surfaces and are generally a few tens of feet thick (Fig. 11–2). They occur as (1) almost horizontal blankets, (2) pocket deposits or irregular masses sheathed in clay, and (3) as interstratified lenses and beds within sedimentary formations. The last type represents in part bauxite that has been formed in place and then covered by later sediments or, in some cases, bauxite that has been washed from its site of accumulation into adjacent areas of accumulation of sediments, and thus is sedimentary. Most of the deposits are covered by soil or later thin formations of sedimentary beds.

The large bauxite deposits of the world are found in Arkansas and Alabama–Georgia (low grade), Gold Coast (Fig. 11–3), Surinam, and British Guiana as blankets and interstratified deposits beneath

Residual Clay Formation

partly eroded thin covers of sediments. The greatest bauxite belt of the world lies in southern Europe, paralleling the Mediterranean Sea, in France, northern Italy, Dalmatia, Yugoslavia, Hungary, and Greece, where pocket deposits are numerous. The deposits of France and Dalmatia lie on erosion surfaces and are overlain by folded limestones. Fairly extensive deposits occur in Russia, the Netherlands Indies, Celebes, Rumania, Jamaica, Haiti, and Dominican Republic. Smaller deposits occur in Germany, Spain, Ireland, Poland, Gulf of Guinea, Australia, and Nyasaland.

RESIDUAL CLAY FORMATION

As we have seen under "Residual Bauxite Formation," clay results wherever aluminous rocks undergo chemical weathering. Most of it is carried away by rain wash to form the sedimentary deposits described in Chapter 10; part remains as soil. Under suitable conditions of weathering, topography, and source rocks, high-grade clays may persist as residual deposits. Most of the high-grade clays, such as kaolin or china clay, which are used for the famous porcelains and chinaware or for making fine paper or refractory materials, are residual clays.

The high-grade residual clays occur chiefly as blanket deposits, veins, and bedded deposits. The chief source rocks are silicic granular igneous rocks rich in feldspars and low in iron, such as syenite, granite, gneiss, and some pegmatite dikes. Clayey limestones, shale, or dark igneous rocks yield mainly low-grade clays.

Residual clays result from normal weathering in which water, oxygen, carbon dioxide, and organic acids take part. It is noteworthy that good white clays commonly underlie swamps or former swamps. The organic compounds of swamps serve to remove coloring matter, leaving white clays. They change iron from the insoluble ferric oxide to soluble ferrous oxide, permitting its removal and thereby bleaching the clay. High-quality fire clays commonly underlie coal beds. The formation of clay from feldspar is essentially a breakdown of the feldspar to form hydrous aluminum silicates, such as kaolinite, and soluble silica and alkalies which are removed in solution. Grains of quartz remain behind in the clay and must be washed out to obtain pure clay. Kaolin clay is also produced by the action of sulfates and sulfuric acid. There are a number of the "clay minerals," and different ones impart different properties to the clay. A thorough removal of alkalies gives rise to kaolinite, and a less complete removal favors the formation of another clay mineral (montmorillonite). Kaolinite is the chief mineral of the fine kaolins.

It is known that several of the clay minerals are formed by hydrothermal solutions, and because of this many investigators have claimed that kaolin deposits have resulted from hydrothermal action. This has been held particularly for the famed Cornish china clays of England. Most kaolin deposits, however, can be traced downward through partially kaolinized into fresh unweathered rocks, and there is little doubt that most residual clays result from weathering.

Residual clay deposits assume roughly the form of the source rock. Dikelike deposits are derived from pegmatite dikes and are up to 300 feet wide and as deep as 120 feet. Those derived from feldspar rocks occur as mantles that may be many acres in extent and several tens of feet in depth; the Cornish kaolins have been worked to a depth of 300 feet. Some residual clays are covered by later formations.

The color and purity of clays vary widely. Most of the kaolins derived from feldspar rocks consist of 10 to 50 percent of white clay minerals, the remainder being grains of quartz, fragments of rock, and other decomposition products that must be removed by washing.

Geographically, residual clays are much more restricted than sedimentary clays. The best kaolins come from temperate humid regions and are lacking in arid regions. Where they occur, ceramic centers of world renown have sprung up, and fine chinaware and porcelains made from them find international markets. Fine chinaware has been made in Kiangsi Province, China, since A.D. 220, and a million people have been employed in the industry. China clays from Cornwall and Devon have world markets for chinaware and high-gloss papers. Kaolins of Allier, Drôme, and Brittany have made France famous for its fine Limoges and Sèvres porcelains. Large deposits at Dresden, Halle, and Kemmlitz in Germany supply the famed Dresden ware and other high-quality wares, and paper clays. The clays of Pilsen and Zettlitz give rise to famed chinaware and supply fine paper clay. North America has few high-grade kaolins. Fine paper clays come from Georgia and the Carolinas, with smaller deposits in some of the eastern states. Nonresidual clays, of course, are widely distributed.

RESIDUAL NICKEL CONCENTRATION

Certain basic igneous rocks contain small quantities of nickel in some unknown form, but presumably held in crystal structures. Under tropical and subtropical weathering such rocks lose silica and yield compounds of nickel and magnesium. In several places the

Other Residual Concentrations

nickel, in the form of the mineral garnierite, has undergone sufficient residual concentration on the surface to yield workable deposits of nickel ore, as in New Caledonia, Cuba, Celebes, Venezuela, and Brazil.

The deposits of New Caledonia, formerly the world's chief source of nickel, illustrate the mode of formation of residual nickel deposits. Part of the island is underlain by deeply weathered serpentine, and nickel occurs throughout the weathered mantle. Here and there residual concentration of nickel has taken place to form many small deposits from 25 to 35 feet thick, localized in six centers. The largest deposit contains 800,000 tons. The deposits lie on slopes rather than on ridges, indicating that seepage waters have been important in the formation of the ore. Migration of nickel during concentration is indicated, since nickel ore penetrates cracks in the underlying serpentine. The ore carries 1.5 to 3.5 percent nickel and a little cobalt and chromium and is high in silica, magnesia, and iron. There has been a large loss of alumina and calcium, reduction in silica, and a very high concentration in nickel, cobalt, and chromium.

At Nicaro, Cuba, traces of nickel in serpentine have been concentrated to 1.4 percent in the surface mantle 20 to 50 feet thick. A huge plant was built to treat this ore from 1943 to 1946, and was reopened in 1951.

OTHER RESIDUAL CONCENTRATION PRODUCTS

Massive residual boulders of *kyanite* are won from a surface mantle overlying kyanite-bearing rocks in India. *Barite* occurs in residual lumps and masses embedded in residual clay in Missouri, the source being small hydrothermal deposits in the underlying rocks. *Phosphate* pebbles in a matrix of sand and clay have been weathered out of the underlying formation at Bone Valley, Florida, and worked over by an advancing sea. The deposits support a large industry. *Tripoli,* an earthy abrasive of nearly pure silica, is a residual product of the weathering of chert, cherty limestone, and siliceous limestone, from which the calcium carbonate has been removed. Several such deposits are worked in the southern states. *Mineral paints* or *ochers* are varied colored materials consisting of iron oxides mixed with clay and in places with manganese, the colors depending upon the proportions of each. Such natural paints have been used since early man was first attracted by their colors and spreading ability. They are residual products of weathering of some rocks. There are ochers, umbers, and siennas. Indian red (barn paint), vermillion red, Persian red, Spanish red, and Venetian red are composed mostly of residual hematite. Ochers contain limonite in addition, and umbers have a high content of manganese, which gives it the brown color. Various greens and the whites are weathering residues of different kinds of rocks.

SELECTED REFERENCES ON RESIDUAL CONCENTRATION

Mineral Deposits. W. Lindgren. 4th Edit. McGraw-Hill Book Co., New York, 1933. Chap. 21:344–378. Residual deposition in general.
Iron Ores, Their Occurrence, Valuation and Control. E. C. Eckel. McGraw-Hill Book Co., New York, 1941. Various residual deposits.
"Lake Superior Iron Region." W. O. Hotchkiss and others. *16th Int. Geol. Cong. Guidebook* 27, 1932. Résumé of geology, origin, and references.
"Brown Iron Ores, Tennessee." E. F. Burchard. *Tenn. Geol. Surv. Bull.* 39, 1932.
Aluminum Industry. J. D. Edwards, F. C. Frary, and Z. Jeffries. McGraw-Hill Book Co., New York, 1930. Vol. I, Chap. 4, by E. C. Harder. Résumé of bauxite deposits.
"Bauxite Deposits of Arkansas." W. J. Mead. *Econ. Geol.* 10:28–54, 1915. Occurrence and origin by removal of silica.
"Relations of Bauxite and Kaolin in Arkansas Bauxite Deposits." M. Goldman and J. I. Tracey, Jr. *Econ. Geol.* 41:567–575, 1946. Claim that bauxite is derived from syenite and not from kaolin.
"Formation of Bauxite from Basaltic Rocks of Oregon." V. T. Allen. *Econ. Geol.* 43:619–626, 1948. Alteration first to clay minerals, then to bauxite.
Clays and Other Ceramic Minerals. C. W. Parmelee. Edwards Bros., Ann Arbor, 1937.
"Geology of Some Kaolins of Western Europe." E. R. Lilley. *A.I.M.E. Tech. Pub.* 475, 1932. Good description and discussion of origin of residual clays.
"Residual Kaolin Deposits of Spruce Pine District, North Carolina." J. M. Parker. *N. Car. Dept. Cons. Dev. Bull.* 48, 1946. Occurrence and origin.
"Tripoli Deposits of the Western Tennessee Valley." E. L. Spain. *A.I.M.E. Tech. Pub.* 700, 1936.
"Nickel-Silicate and Associated Nickel-Cobalt-Manganese Deposits, Goiáz, Brazil." W. T. Pecora. *U. S. Geol. Surv. Bull.* 935–E, 1944. Residual nickel-cobalt deposits.
Economic Mineral Deposits. 2nd Edit. Alan M. Bateman. John Wiley & Sons, New York, 1950. Residual concentration, 201–226; aluminum, 553–560; manganese, 578–586; clay, 696–708; pigments, 726–728.

MECHANICAL CONCENTRATION—PLACERS

The weathering of rocks and included mineral deposits leaves in the residual mantle those substances that resist solution. Some of these are light, some are heavy; some are brittle, some are durable.

These substances may undergo mechanical concentration, which is a natural gravity separation of heavy, durable minerals from light or friable minerals by means of moving water or air, and their concentration into deposits called placer deposits.

Concentration can occur only if the valuable minerals possess three properties: heavy weight, chemical resistance to weathering, and durability (malleability, toughness, or hardness). Placer minerals that have these properties are gold, platinum, tinstone, magnetite, chromite, ilmenite, rutile, native copper, gemstones, zircon, monazite, phosphate, tantalite, columbite, and, rarely, quicksilver. Those of greatest economic importance are gold, platinum, tinstone, ilmenite (titanium ore), diamonds, and rubies.

The placer minerals are derived from commercial lode deposits, such as gold veins; noncommercial lodes, such as small gold-quartz stringers or veinlets of tinstone; sparsely disseminated ore minerals in rocks, such as grains of platinum or diamonds; rock-forming minerals, such as magnetite, ilmenite, or zircon; and from former placer deposits, either surface ones or ancient buried ones.

PRINCIPLES INVOLVED

In the winning of ore minerals, man mines, crushes, and concentrates the desired ore minerals. Nature has done the same on a grand scale in releasing placer minerals from their matrix through weathering. The comminuted materials are washed slowly downslope by rain wash to the nearest streams or seashore. Moving water then sweeps away the lighter matrix, and the heavier placer minerals sink to the bottom or are moved downstream relatively short distances. Waves and shore currents also separate heavy minerals from light ones and coarse grains from fine ones. From thousands of tons of debris the few heavy minerals in each ton are gradually concentrated in the stream or on the beach until they accumulate in sufficient abundance to make workable deposits. Eventually the gold of countless thousands of tons of rock is concentrated into relatively small volume. These may form rich placer gold deposits such as gave rise to the journey of the Argonauts to seek the Golden Fleece, to the great California gold rush of 1849, or to the Klondike stampede to Yukon and Alaska. Billions of dollars in gold have been won from such deposits, and gold-seeking pioneers have opened up vast new lands.

There are a few basic principles underlying mechanical concentration that rest chiefly on the difference in weight, size, and shape of particles as affected by the velocity of a moving fluid. First, in a body of water a heavier mineral sinks more rapidly than a lighter one of the same size, and water accentuates the difference in weight. For example, the specific gravity (weight compared to water) of gold is 19, quartz is 2.6, water is 1, and the ratio of these in air as compared to water is:

$$\frac{\text{Gold in air, } 19}{\text{Quartz in air, } 2.6} = \frac{7.3}{1} \text{ whereas } \frac{\text{Gold in water, } 19-1}{\text{Quartz in water, } 2.6-1} = \frac{11.2}{1}$$

Thus water accentuates the difference in specific gravity. Secondly, the settling rate in water is affected by the amount of surface of particles. Of two spheres of the same weight but different size, the smaller sinks faster because of its lesser surface and friction. Thirdly, the shape of a particle affects the rate of settling. A spherical pellet has less surface than a thin platy disc of the same weight, and therefore will sink more rapidly.

Now, add to these principles the effect of moving water. The ability of moving water to transport a solid depends upon the velocity and increases at a much faster rate than the velocity. Everyone is aware that a racing floodwater can carry substances that a quiet stream cannot. When the velocity is doubled the transporting power is increased about four times. Conversely, if the velocity is halved much of the transported load is dropped. Hence placer minerals being transported will be dropped where the current slackens. Faster-moving water increases the settling differences between minerals, and if particles of gold and quartz are dropped into moving water the gold might drop directly to the bottom and the quartz be swept downstream. Also, those with larger surface are more rapidly swept away.

Again, add another factor, that a particle in suspension is more rapidly swept away than one at rest. As one stirs up sugar from the bottom of a teacup, so stream eddies lift light minerals from the bottom and enable currents to swish them away. Moreover, the swirls and eddies of streams create a jigging action by which gold particles scattered through gravels become concentrated on the bottom, even though the gravels are thick. This is a familiar step in artificial ore concentration.

These various factors operate together to separate light and fine minerals from heavy and coarse ones, and with long-continued action placer minerals present eventually may become concentrated into workable deposits. It follows that the water velocity must be favorable for placer action; if too low the lighter minerals will not be removed from the heavier; if too great, they will all be swept away and perhaps dissipated. A slackening of velocity causes deposition and accumulation at the place of slackening. Thus the changing of a stream gradient, meandering, spreading, or obstacles all produce slackening and permit heavy minerals to drop and accumulate.

The scraping and pounding of placer minerals during transportation and jigging eventually crushes brittle minerals to a powder; it rounds off sharp corners of durable ones, and compacts and flattens the malleable ones. The extent of these changes is an indication of the distance of travel. Sharp, angular, placer gold, for example, is not far from its source, and the wily prospector uses this principle in his search for the "mother lode."

Another essential for concentration is a continued supply of placer minerals. This means that the most favorable regions are those of deep weathering and topographic relief, the weathering to create supplies and the topographic relief to permit debris to move streamward or beachward. A flat plateau could not supply much debris or velocity to move it. The most favorable areas are where uplift of the land causes new valleys to be cut into old ones, bringing about rewashing and reconcentration of earlier gravels. The more of this, the better is the degree of concentration.

When weathering takes place on a hill slope, the heavier particles move downslope more slowly than the lighter ones, giving a rough concentration into *eluvial* placers. Those concentrated in streams give *stream* (*alluvial*) placers; those on beaches give *beach* placers; those by wind give *eolian* placers. The basic principles enumerated above apply to each of the different kinds of placers, but special features contribute to individual groups, so that we will consider each separately.

ELUVIAL PLACER FORMATION

Eluvial placers may be considered an intermediate or embryonic stage in the formation of stream or beach placers. They are formed without stream action from materials released from weathered lodes that crop out above them (Fig. 11–4). The heavier, resistant minerals accumulate just below the outcrop, whereas the lighter products of decay are washed downhill or blown away by the wind. This brings about a partial concentration by reduction in volume, which continues with the downslope creep. Generally, fairly rich lodes are necessary to yield workable deposits by this incomplete concentration. Gold and tinstone are the chief eluvial deposits, but there are minor ones of tungsten, kyanite, barite, and gemstones.

Many eluvial gold deposits have been mined in Australia, New Zealand, western United States, Mexico, Kenya, Sierra Leone, and British Guiana. Some of the large gold nuggets have come from such deposits.

Tinstone is obtained in abundance from eluvial deposits in the Netherlands Indies, Malaya, Siam, Belgian Congo, and other places

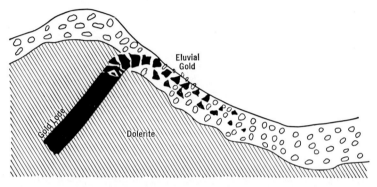

FIGURE 11-4. Eluvial placer gold (black) formed immediately below the outcrop of gold lode. San Antonio vein, Chontales district, Nicaragua. (After N. Carter, *A.I.M.E.*)

where weathering of low granite domes seamed with veinlets of tin ore releases particles of tinstone that hang back on the slopes. Eventually the tinstone reaches streams, where greater concentration occurs.

STREAM PLACER FORMATION

Stream placers are the chief type of placer deposits and have yielded the greatest quantity of gold, tinstone, platinum, and diamonds. It was on such deposits that primitive mining undoubtedly started because the ease of extraction and the richness of some deposits made them as eagerly sought in early times as in recent times. They have caused some of the great gold and diamond "rushes" of the world and led to the opening up of former wilderness in California, Alaska, Australia, and Africa.

Gold placers are a "poor man's" type of deposit. The hardy miner, single-handed, needs only a gold pan and a shovel to obtain the gold, and in a rich deposit a fortnight may bring fabulous riches. The California gold rush of 1849 and the perilous stampede to the Klondike and Alaska in 1897 initiated ephemeral placer mining that gave way to more stable lode mining. Behind them in most places came settlement, agriculture, industries, and the development of great new countries.

The California gold rush of 1849 amidst the hardships of overland travel and the danger of savage Indians, or around Cape Horn in inadequate ships, was, according to Rickard, only the modern counterpart of an ancient gold rush made classic by the Greek legend of the Golden Fleece.

Today the production of placer gold is small compared with its romantic past. The easily discoverable placers have been found, and fewer strikes or rushes can occur in the future. Outside of Russia, only the slightly explored new lands hold hope of yielding new placer discoveries of consequence. Placer diamond mining, however, is at its height and today yields about 85 percent of the world supply of diamonds. Over 75 percent of world tin production comes from placer deposits.

REQUISITES FOR STREAM PLACERS. Given the source materials mentioned in a preceding paragraph, the concentration of placer minerals is effected by stream water, which, although ever flowing downstream, does so with irregularity. It rushes through canyons, sweeping everything before it; it slackens in wide places; it swirls around the outside of bends, creating back eddies on the inside; it laps up over bottom projections, forming quiet eddies on the lee side. In the slack waters the heavy substances, both pebbles and placer minerals, drop to the bottom. The jigging action jumps the gravels up and down, permitting the placer minerals to seek the bottom part of the gravels. Further concentration takes place by the rubbing and abrading of the gravels, the resulting powder being swept away and the gravels being reduced in thickness. During dry seasons of low stream velocity the placer minerals remain at rest, but in flood time they and the enclosing gravels may all be swept further downstream and reconcentrated on bars, stream margins of flood plains, or other favorable places. Minute particles of gold (colors), however, may be carried far downstream.

The most favorable sites for placer deposition are the middle and lower-upper reaches of streams where the water is neither sluggish nor too rapid to sweep everything away. There, deposition takes place where slack water is to be found. This occurs at river bars, at the confluence of streams, where streams widen, where obstacles pond the water. Ribs on the bedrock form natural riffles that trap gold and the other placer minerals.

The accumulation of placer minerals in stream channels requires a nice, long-continued adjustment between stream velocity and gravel accumulation. The gravels cannot be too thick and must slowly move downstream and must be water-soaked to permit jigging action. The stream gradients of placer streams range from about 30 feet up to 150 feet per mile. The presence or absence of these various conditions means rich "pay streaks" in some streams and sparsely disseminated minerals in others.

The optimum conditions for high concentration of placer minerals are those where a first concentration has occurred in stream chan-

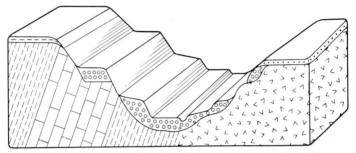

FIGURE 11-5. Section across valley, showing location of bottom and terrace gold placer deposits.

nels and in benches along the riverside, and where the region then becomes uplifted, with resultant incision of the river in its bottom, digging a new and steeper-sloped valley beneath the older, broader one. This causes all the side or bench gravels to be rewashed into the lowered stream bottom (Fig. 11-5), and reconcentration takes place. The highest-grade gravels result from more than one such reconcentration.

CHARACTER OF DEPOSITS. Gold and other placer minerals tend to occur in concentrated pay streaks that are likely to be narrow and relatively rich. The coarser gold is deposited in the upstream reaches of a "run" and the finer in the lower reaches. The pay streaks are irregular in outline; they split and branch and are absent in places. They may not always lie in the center of the stream but at one side, and, if the stream shifts its course through meandering, the pay streak may lie buried away from the stream. Some gold is also generally scattered throughout the body of the gravels. In many places the richer pay streaks have been mined by hand, and the lower-grade gold and tinstone gravels are mined by means of huge power dredges.

Stream Placer Formation

Most placer gold is in the form of fine specks called dust, but to the joy of the miner some larger lumps called nuggets are found. They mostly range from the size of small shot to a half inch across, but rare ones have been found up to 2,280 ounces in weight. Most gold particles have become flattened into disc shapes, but tinstone becomes rounded and the diamond, because it is the hardest substance known, may show little or no rounding. An astonishing feature of placer gold is its divisibility into minute scales, called flour gold. Hite estimated that the flour gold of the Snake River averages about 5,000 particles (colors) to equal 1 cent in value. The smaller colors require 7 to 8 million of them to equal 1 cent. Most placer gold is about 800 fine (1,000 is pure gold); some is 950 fine, which is higher than the gold in nearby veins, indicating purification during placer formation.

Most placer gold rests within a foot or so of bedrock, but rarely it may extend 15 to 20 feet above. One deposit in the Klondike averaged $4.13 per cubic yard in the lower 6 feet of gravel, and the next 6 feet averaged only $0.18 per cubic yard. Sporadic small pockets may yield several thousand dollars. The richer gravels of the Klondike ranged from $9.00 to $50.00 per cubic yard, but the dredging gravels of Alaska yield only $0.17 per cubic yard, and those of California from $0.25 to $0.34.

Tinstone and diamonds, being lighter than gold, are not so closely concentrated and may be more widely distributed in the gravels. Tin gravels may be dredged with only a half pound of tinstone per cubic yard. Diamond gravels in the Congo average 0.6 carat per cubic yard, of which only 5 percent are gemstones.

In addition to bottom gravels, placer gravels may also occur as bench gravels perched on valley sides. Also, they may become submerged by the deposition of barren gravels on them. Some gold gravels near Fairbanks, Alaska, lie at depths of 300 feet, and rich channels have been found at the same depth at Ballarat, Australia. Ancient hardened gold placer gravels nearly ½ billion years old have been uncovered near the Homestake mine in South Dakota.

The "high-level gravels" of California are particularly unusual and interesting, and have been made classic by Lindgren. In the Sierra Nevada, gold gravels accumulated in Cenozoic time (early Tertiary) in stream channels that traversed a gently sloping country. These were buried by lean gravels and by volcanic tuffs and breccias to a depth of 1,500 feet. The range was elevated and new streams eroded canyons to depths of 2,000 to 3,000 feet, or an average of about 2,600 feet below the early Tertiary stream bottoms, and in a direction of flow at an angle to the courses of the early streams. The present valley walls disclose the elevated cross sections of the earlier stream valleys with gold gravels in their bottoms. The eroded parts have supplied much of the present-day stream gold.

Perhaps the most remarkable placer diamond deposits are those of Lichtenburg, South Africa, which witnessed a great stampede of 25,000 "runners" in 1926. Here, on a flat erosion plain carved across dolomitic limestone, stream gravels occur as a sinuous embankment resting above the plain, marking the meandering course of a former stream that once had been entrenched in the dolomite. During erosion the dolomite surface was lowered by solution, and the stream disappeared into underground caverns. Diamond-bearing gravels left behind protected the underlying dolomite from solution while the adjacent region was dissolved scores of feet deep. This left the gravels projecting up as a ridge on the lowered surface; some gravels slumped into sinkholes, and a few of these pockets proved fabulously rich. In 1927 these gravels yielded over 2,100,000 carats of diamonds.

DISTRIBUTION OF PLACERS. Gold placers have been mined in most parts of the world. The most notable ones, however, are those of California and other western states, Yukon-Alaska, Australia, New Zealand, Siberia, British Columbia, Mexico, and the Andean region of South America. Large-scale mining is presently going on in California, Alaska, Siberia, Colombia, Central Africa, and New Guinea.

Platinum placers occur in the Urals, Colombia, Australia, and South Africa, with minor deposits in Alaska, Ethiopia, Japan, Panama, Papua, and Sierra Leone.

Diamonds in stream placer deposits are found extensively in Belgian Congo-Angola, Sierra Leone, and other west coast and central African areas, South Africa, and Brazil. Most rubies, sapphires, chrysoberyls, aquamarines, zircons, and spinels are washed from stream gravels. Their home was in pegmatite dikes or metamorphosed rocks, and they are found in India, Ceylon, Burma, North Carolina, Georgia, and other places.

BEACH PLACER FORMATION

In the formation of beach placers, the waves and shore currents separate the lighter from the heavier minerals. Pounding waves throw up the materials on the beaches, and the undertow carries out the lighter and finer material, which is moved alongshore by currents. The placer minerals are derived by weathering and stream delivery and by wave action on rocky shores or elevated beaches.

Beach placers yield gold, diamonds, ilmenite, rutile, monazite, zircon, magnetite, chromite, garnet, and quartz. Beach *gold* is mined at Nome, Alaska, where there are six beaches, two submerged and four elevated, that have yielded over 75 million dollars in fine gold. *Ilmenite, monazite,* and *rutile* have undergone unusual beach concentration at Travancore and Quilon, India, where commercial sands long supplied most of the world's titanium. These sands contain 50 to 70 percent ilmenite—an amazing concentration—with 2 to 5 percent monazite, desired for radioactive thorium. Similar, but less valuable, beach sands occur in Brazil, Florida, North Carolina, Ceylon, Senegal, Argentina, Australia, and New Zealand. *Magnetite* or black sands are found on the coasts of Oregon, California, Brazil, Japan, and New Zealand. Chrome sands are mined in Japan.

One of the most remarkable beach concentrations is that of the diamond-bearing gravels of Namaqualand, South Africa, which in 1928 yielded 50 million dollars. The diamonds, of unusually fine quality, occur in marine gravels in wave-cut terraces 20 to 210 feet above sea level, extending 3 miles inland and 200 miles along the desert coast south of the Orange River. By means of diagnostic oyster shells the old beach lines could be traced along the coast. The diamonds are presumed to have come from the weathering of diamond pipes in South Africa and to have been delivered by the Orange River to the coast, where they were sorted and distributed by waves and the prevailing southerly shore currents. They are found irregularly distributed in coarse beach shingle from which the sand has been blown away.

EOLIAN PLACER DEPOSITS

Wind instead of water has been the agent of concentration of placer minerals in some rainless areas. In Australia, desert weathering has released gold from its matrix in small gold-quartz veins. The lighter material has been blown away, but the heavier gold particles remained behind and thus became concentrated into workable deposits. Similar concentration of gold has taken place in Lower California, Mexico, where gold is recovered by "dry washing," utilizing wind instead of water.

SELECTED REFERENCES ON PLACER DEPOSITS

"Tertiary Gravels of the Sierra Nevada." W. Lindgren. *U. S. Geol. Surv. Prof. Paper* 73, 1911. A fascinating correlation of Tertiary gravel formation and physiographic development of the region, resulting in remnants of elevated buried gravels.

"New Technique Applicable to the Study of Placers." O. P. Jenkins. *Calif. Dept. Nat. Res.*, 31, 1935.

"Klondike Gold Fields." R. G. McConnell. *Can. Geol. Surv. Ann. Rept.* 143, 1905; and "The Klondike High-Level Gravels." *Can. Geol. Surv.*, 1907. Occurrence and origin of rich Klondike placers.

"Gold." J. M. Maclaren. *The Mining Journal*, London, 1908. Descriptions of important placer deposits of the world.

"Law of the Pay Streak in Placer Deposits." J. B. Tyrrell. *Inst. Min. and Met. Trans.*, London, May 16, 1912. Development of pay streaks in stream gravels.

Diamond-Bearing Gravels of South Africa. A. F. Williams. *3rd Empire Min. and Met. Cong.*, S. Africa, 1930. Descriptions of various types of placer gravels.

Ore Deposits of the Western United States—Lindgren Volume. Am. Inst. Min. Met. Eng., New York, 1933. Chap. 10. Thumbnail sketches of many western gold districts.

Gemstones. Imp. Inst., London, 1933. A brief geographical survey of gemstones in all countries; good bibliography.

Industrial Minerals and Rocks. 2nd Edit. Edited by S. H. Dolbear and Oliver Bowles. Am. Inst. Min. Met. Eng., New York, 1949. Chap. 35, by S. H. Ball. Concise summary and interesting history; good bibliography.

"The Diamond Industry." S. H. Ball. Annual Jewelers' Circular. A concise up-to-date summary.

"Geologic and Geographic Occurrence of Precious Stones." S. H. Ball. *Econ. Geol.* 17:575–601, 1922; also 30:630–642, 1935.

"A Visit to the Gem Districts of Burma and Ceylon." F. D. Adams. *Can. Inst. Min. Met. Bull.*, Feb., 1926, pp. 213–246. A popular description of interesting gem localities.

Economic Mineral Deposits. 2nd Edit. Alan M. Bateman. John Wiley & Sons, New York, 1950. Mechanical concentration, 227–245; gold, 428–435; tin, 546–552; gemstones, 834–854.

OXIDATION AND SECONDARY ENRICHMENT

When ore deposits undergo weathering a natural chemical laboratory is started. Oxygen and water are powerful chemical agents when combined with the chemical constituents of the ore deposits, and the results they produce are profound and drastic.

The first effect of weathering is the joining of the oxygen from the atmosphere or from surfaced waters with elements of the ores. This is called oxidation, and it yields new minerals and solvents that dis-

Oxidation and Secondary Enrichment 227

solve other minerals. An ore deposit thus becomes oxidized and generally leached of many of its valuable minerals to as great a depth as oxidation can proceed, and this level is generally the water table. The oxidized part is called the *zone of oxidation*. The effects of oxidation, however, may extend far below the zone of oxidation. As the cold, dilute leaching solutions trickle downward, some or all of their metallic content may be precipitated within the zone of oxidation and form oxidized ore deposits—a familiar group of deposits readily accessible to mining because of their shallow depth, and one made colorful in the glamorous beginnings of many mining districts. But when the down-trickling solutions penetrate beneath the water table they enter a zone where oxygen is normally absent, and their metallic content may be precipitated in the form of metallic sulfides or related compounds to form a zone of *secondary (supergene) sulfide enrichment*. This zone gradually merges downward into the *primary zone*. This zonal arrangement, shown in Figure 11-6, is characteristic of many mineral deposits that have undergone long-continued weathering.

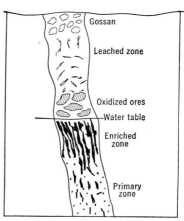

FIGURE 11-6. Diagram of zones of a weathered copper vein, showing the relation of oxidation and enrichment to the water table.

Of course the arrangement does not occur in all ore deposits because some do not contain minerals that undergo oxidation or that yield the necessary solvents, or the climatic conditions do not permit oxidation. Also, in places the secondary enrichment zone may be absent, and usually in glaciated countries there is no oxidized zone. Without the formation of an oxidized zone there can be no secondary enrichment zone. Special conditions of time, climate, topography, and amenable ores are necessary to yield the results mentioned above, but they are sufficiently common for oxidized and enriched supergene ores to occur in most nonglaciated lands of the world. Many of the great mining districts would not be in existence had oxidation and enrichment not taken place.

The effect of oxidation on ore deposits has been profound. It has either rendered barren the upper parts of many ore deposits or has changed the primary ore minerals into more usable or less usable forms or has even made rich bonanzas. The effect of secondary sulfide enrichment has been even more far-reaching, for it has added much where previously there was little; leaner parts of veins have been made richer; and unworkable primary material (called *protore*) has been enriched to workable ore.

Oxidation and supergene enrichment go hand in hand. Without oxidation to supply solvents by which metals may be dissolved and carried downward, there can be no precipitation in either the oxidized zone or the secondary enrichment zone. The process of oxidation and enrichment resolves itself, therefore, into three distinct phases: (1) oxidation and solution in the zone of oxidation; (2) deposition in the zone of oxidation; and (3) deposition of sulfides in the secondary enrichment zone. Each will be considered separately.

OXIDATION AND SOLUTION IN THE ZONE OF OXIDATION

Through oxidation the minerals are mostly altered, and the structure is largely obliterated. The metallic substances for the most part are leached or are altered to new compounds that require quite different metallurgical treatment for the extraction of their metals from that employed for the unoxidized ores. Compact ores are made cavernous. Yellow or red limonite obscures everything and imparts to the oxidized zone (gossan) that familiar rusty color, which from earliest times has attracted the curiosity of the miner. One can only infer what may lie beneath this gossan (p. 231), but in this inference one can be guided by the character of the oxidation products themselves.

CHEMICAL CHANGES INVOLVED. To understand the changes that take place it is necessary to delve a little into the chemical reactions that occur.

Everyone is familiar with oxidation, even though he may not call it by that name. A piece of iron pipe left out in the open turns rusty. That is oxidation. The iron joins with oxygen and water to make hydrous iron oxide, or the mineral limonite, which is the rust. In the oxidation of ore deposits the same thing takes place. Water and oxygen are powerful oxidizing agents, and carbon dioxide also plays a part. Locally, chlorides, iodides, and bromides enter in. Another strong oxidizing agent is iron chloride, which results when sulfuric acid reacts with sodium chloride (common salt) to yield hydrochloric acid, and the hydrochloric acid with iron gives iron chloride. The sulfuric acid, as will be shown below, is one of the

Oxidation and Solution

early solvents formed in the ore deposit. Two chief chemical changes occur in the zone of oxidation: (1) the oxidation, solution, and removal of certain minerals; and (2) the transformation in place of some of the minerals into oxidized compounds.

The common mineral pyrite (FeS_2) is the spark plug of most of the chemical reactions, and it is present in nearly all ore deposits. It readily gives up part of its sulfur to form iron sulfate and sulfuric acid as is generally indicated by reaction [1]:

[1] $$FeS_2 + 7O + H_2O = FeSO_4 + H_2SO_4$$
Pyrite + oxygen + water = ferrous sulfate + sulfuric acid

The ferrous sulfate by further reaction yields ferric sulfate, thus:

[2] $$2FeSO_4 + H_2SO_4 + O = Fe_2(SO_4)_3 + H_2O$$
Ferrous sulfate + sulfuric acid + oxygen = ferric sulfate + water

or [2a]

$$6FeSO_4 + 3O + 3H_2O = 2Fe_2(SO_4)_3 + 2Fe(OH)_3$$
Ferrous sulfate + oxygen + water = ferric sulfate + ferric hydroxide

The ferric sulfate so formed plays several roles: first, by reacting with sulfide minerals in the ore deposit it regenerates itself continuously by forming new ferrous sulfate, which in the presence of oxygen immediately changes over to ferric sulfate; second, with water it yields ferric hydroxide, which changes over to limonite and hematite, the iron oxides, which remain behind as such; third, it is the chief solvent for most of the metallic minerals of the ore deposit. For those who desire it, the following three reactions illustrate the end results of these three roles:

[3] $$Fe_2(SO_4)_3 + FeS_2 = 3FeSO_4 + 2S$$
Ferric sulfate + pyrite = ferrous sulfate + sulfur

[4] $$Fe_2(SO_4)_3 + 6H_2O = 2Fe(OH)_3 + 3H_2SO_4$$
Ferric sulfate + water = ferric hydroxide + sulfuric acid

[5] $$CuS + Fe_2(SO_4)_3 = CuSO_4 + 2FeSO_4 + S$$
Copper sulfide + ferric sulfate = copper sulfate + ferrous sulfate + sulfur

Thus it can be seen that, if pyrite (or an equivalent sulfide) is present, the powerful solvent ferric sulfate is generated, and once formed continues to regenerate itself and manufacture the rusty-looking iron oxides. If pyrite is absent, no solvent for the various metals is generated and no secondary sulfide enrichment can take place.

Now let us see how solution of metals takes place. We may illustrate it by choosing four common minerals that occur separately or together in many ore deposits, namely, chalcopyrite [6] the copper iron sulfide, sphalerite [7] the zinc sulfide, galena [8] the lead sulfide, and native silver [9]. The end products, without intermediate steps, are indicated in reactions [6] to [9].

[6] $2Fe_2(SO_4)_3 + CuFeS_2 = CuSO_4 + 5FeSO_4 + 2S$
Ferric + chalcopyrite = copper + ferrous + sulfur
sulfate   sulfate   sulfate

[7] $2Fe_2(SO_4)_3 + ZnS + 4H_2O = ZnSO_4 + 8FeSO_4 + 4H_2SO_4$
Ferric + sphalerite + water = zinc + ferrous + sulfuric
sulfate   sulfate sulfate acid

[8] $Fe_2(SO_4)_3 + PbS + H_2O + 3O = PbSO_4 + 2FeSO_4 + H_2SO_4$
Ferric + galena + water + oxy- = lead + ferrous + sul-
sulfate   gen   sulfate sulfate furic
                                                 acid

[9] $Fe_2(SO_4)_3 + 2Ag = Ag_2SO_4 + 2FeSO_4$
Ferric sulfate + silver = silver sulfate + ferrous sulfate

Note in these reactions how more ferrous sulfate is generated and that each of the four metals goes into solution as a sulfate. Generally similar reactions apply to many more metals and minerals. In the case of silver, chlorides, bromides, and iodides of silver are commonly formed in arid and semi-arid regions where these substances occur in the soils and surface waters.

The sulfates of the metals, formed as indicated above, are, except for lead, all readily soluble. They trickle slowly downward through the ore deposit and, as we shall see further on, later give up their metals.

The general tendency of the chemical changes in the zone of oxidation is to break down complex minerals and form simple ones. In general, the metallic minerals that lack oxygen (e.g., sulfides) are the ones most susceptible to oxidation, and those that already contain oxygen are little affected. Among the gangue minerals, carbonates are readily decomposed, most of the silicates are broken down into a few stable forms, quartz is unaffected, but silica set free by decom-

Oxidation and Solution 231

position of silicates is taken into solution and may join with some of the metals to form silicates of the metals, e.g., copper silicate.

Another chemical effect during oxidation is the separation of mixed metals in an ore deposit. The same four minerals used above to illustrate solution, along with pyrite, commonly occur together as "mixed ores" in deposits in limestone. As shown above, the pyrite is dissolved, and the copper, zinc, and silver sulfates are taken into solution, which migrates downward, but the lead sulfate, being insoluble, remains behind as the oxidized compound. Thus, the oxidized part

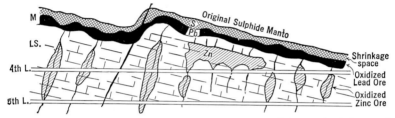

FIGURE 11–7. Sketch of longitudinal section of an ore bed (manto) of Encantada mine, Sierra Mojada, Mexico, illustrating the separation of metals during oxidation. Original ore body (M) was composed of sulfides of iron, zinc, and lead. Oxidation altered the iron sulfide to iron oxide and removed the zinc, which was transported and deposited below, and left the lead as lead carbonate (Pb) in the original ore body, the shrinkage being represented by S.

of the deposit will be enriched in lead and will be freed of the other metals, and complex metallurgical treatment is simplified by nature, rendering the deposit of greater value than the original. If the downward-migrating zinc sulfate comes in contact with limestone farther down, it reacts to form zinc carbonate. There are many places in the United States and Mexico where zinc and lead have been separated in this manner, as is shown by Figure 11–7. Similarly, gold is separated from sulfide matrix and remains behind as "free milling" gold.

Before following the trickling sulfates of the metals in their downward journey, let us stop to consider the rusty gossan.

GOSSANS AND THEIR INDICATIONS. The rusty gossans that mark the outcrops of oxidized zones are signboards that point to what lies beneath, provided it can be interpreted. They arrest attention and incite interest as to what they may mask. Most ore deposits, save in glaciated regions, are capped by gossans, generally barren of valuable metals; hence the finding of one may herald the discovery of hidden wealth. But, alas, noncommercial mineral bodies also yield gossans, and unfortunately there are many more of them. To distinguish between them, therefore, may lead either to discovery of new resources or to the saving of needless expenditure of money. This involves experience, knowledge, and careful observation. The distinctions involve delicate differences of color and form, and oftentimes the only metal compound one can see in the gossan is limonite.

Limonite is universally present in gossans and occurs in a variety of positions, forms (called limonite boxworks), and colors, each characteristic being significant. Limonite is not a mineral but a mixture

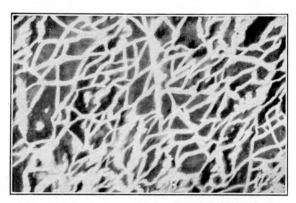

FIGURE 11-8. Triangular boxwork of limonite derived from bornite, Ruby mine, California. Photo magnified 5 times. (R. Blanchard and P. F. Boswell, *Econ. Geol.*)

of oxides of iron with generally some sulfates. There are two kinds, called *indigenous* and *transported*. Indigenous limonite is fixed at the site of the parent sulfide, and transported limonite has been carried in solution and deposited away from the parent sulfide—a vital difference. The indigenous limonite is thrown down in the insoluble ferric state and indicates the former presence of copper. It remains in the void vacated by the parent copper sulfide, and the color and shape of its boxwork (see Fig. 11-8) can be interpreted to indicate what the predecessor sulfide was. Thus, it is a favorable indication that ore and copper ore minerals may be expected beneath a gossan that displays indigenous limonite. Transported limonite on the other hand indicates lack of copper or a high ratio of iron to copper. Generally it means only pyrite, which is not an ore mineral. The transported limonite leaves the voids quite empty and permeates the surrounding minerals, commonly forming halos of rusty rock around the voids. Such indications are unfavorable for the expectation of copper ore beneath the gossan.

Transported limonite may give rise to *false gossans*. Since it is precipitated from solution by rocks that react with it, the transported limonite may give rise to an iron-stained area in formerly barren rock that resembles a true gossan in many respects. The false gossan is distinguished from the true by the lack of indigenous limonite and sulfide voids and by the character of the transported limonite. Such false gossans do not overlie ore deposits, but their presence may indicate former sulfides not far distant.

There are still other features of gossans that permit inferences as to what may lie hidden beneath them. Their shape and size approximate that of the underlying mineral deposit, except that the limonite may mushroom out around the original ore body. Therefore only indigenous limonite can be used to indicate the shape and size of the original ore body. Voids vacated by sulfides show by their abundance the amount of sulfides formerly present, and thereby the amount that may be expected beneath. Also, their shapes may be molds of known sulfides. Lack of voids shows, of course, that no predecessor ore minerals existed. Relict sulfides that have survived because they are tightly enclosed in pieces of impervious quartz give direct clues to expectable underlying minerals. The character and intensity of the hydrothermal rock alteration surrounding the gossan gives an indication of the character and intensity of the primary metallization.

Stains of other oxidation minerals may be present along with limonite and so tell what metals may have been originally present, for example, the black stain of manganese, the blues and greens of copper, the brilliant pink of cobalt, the apple green of nickel, the yellow of bismuth and cadmium (in zinc), and the bright yellow of molybdenum.

FACTORS CONTROLLING AND LIMITING OXIDATION. Oxidation is a surficial phenomenon and ceases in depth. Certain factors, therefore, control its localization, depth, and cessation. Of these, climate, especially temperature and rainfall, is one of the chief factors. Oxidation is obviously inhibited in regions of perennial ground frost and is not appreciable in cold arctic climates. High temperatures certainly accelerate it, and low temperatures retard it. It seems to proceed fastest in warm humid climates with evenly distributed rainfall.

The water table is perhaps the next most important factor affecting oxidation. The rock pores above the water table are alternately occupied by air and rainwater, and this condition is the home of oxidation.

Oxidation may extend right down to the water table, but only under rare circumstances does it extend beneath, for here free oxygen is not available. The persistence of ferrous salts in the ground water indicates that oxidation is not taking place. The bottom of the zone of oxidation is, therefore, generally a gently undulating base with an abrupt passage into unoxidized ore beneath. In the rare exceptions pendants of oxidation project down into unoxidized ore along large faults where circulating water carries down entangled oxygen, as at Cananea, Mexico.

The water table normally moves slowly downward, except under low-lying flat lands, in company with the downward progress of erosion

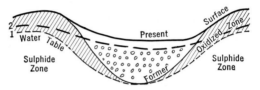

FIGURE 11–9. Sketch illustrating rising water table and a drowned oxidized zone (shaded) at Miami-Globe, Arizona; 1 is former water table corresponding to former land surface (light line). The valley became filled by 2,000 feet of gravel, giving a new surface and raising the water table to 2, drowning part of the oxidized zone.

of the surface. Thus, as the water table is depressed, the oxidized zone encroaches upon the sulfide zone and likewise progresses downward. When the water table comes to a standstill, so does the zone of oxidation; if it remains stationary, oxidation eventually ceases. A change of climate from humid to arid also causes a lowering of the water table and of the zone of oxidation. If the water table should be depressed very rapidly, as where steep canyons are being excavated, oxidation may not be able to keep up with it, and the bottom of the zone of oxidation becomes stranded well above the water table. This is just what has happened in the mining district of Bingham Canyon, Utah.

The water table may also rise in response to opposite processes, i.e., a change of climate from arid to humid, or the filling up of valleys by deposition of sediments. For example, in the Globe-Miami mining district of Arizona the valley has been filled by 2,000 feet of gravels. This raised the water table by a like amount and submerged the oxidized zones of the adjacent ore deposits, as shown by Figure 11–9. Also in the Rhodesia copper belt a change from arid to humid climate raised the water table some 2,000 feet or more, and oxidized copper ores lie deep beneath the high-water table. Such drowned oxidized zones are common in the western United States. Faulting may also lower an oxidized zone beneath the water table.

Time and rate of erosion also affect oxidation. It is a slow process and needs lots of time. The interval since the glacial period is too short to have permitted much oxidation. The rate of erosion and the ensuing lowering of the water table cannot be too rapid, else oxidation cannot keep pace with it. A very slow rate of erosion would cause an almost stagnant position of the water table and little oxidation.

A very important factor affecting oxidation is the nature of the enclosing rocks—their structure and physical and chemical properties. Oxidation proceeds fastest in permeable or crackled rocks and slowest in dense, unfractured, impervious rocks. Fault zones, shear zones, and inclined bedding planes may permit deep penetration of oxidation. Faults that have thick impervious clay gouge on their walls may act as dams and prevent down-seeping oxidizing waters from reaching unoxidized ores beneath them.

CESSATION AND DEPTH OF OXIDATION. Since oxidation requires oxygen, it ceases when the supply of oxygen ceases. Cessation is brought about by (1) the water table; (2) a rise of the water table; (3) refrigeration of the ground in Arctic regions, as in the Kennecott mines of Alaska, where the ground is frozen to a depth of 700 feet and partial oxidation extends to a depth of more than 2,000 feet; (4) depletion of available oxygen above the water table by its being consumed by abundant sulfides; (5) burial by sediments or lavas, as at United Verde Extension mine, Arizona.

The depth of oxidation may range from a few feet, which is uncommon, up to 3,000 feet, as at the Lonely mine, Rhodesia. At Kennecott, Alaska, and at Tintic, Utah, it is over 2,000 feet. Its general range is from a few tens to a few hundreds of feet. It is generally shallow in humid regions and deep in arid regions, and very shallow in humid flat regions. It is moderate to deep in mountain regions and is generally lacking or negligible in glaciated regions.

ORE DEPOSITION IN THE ZONE OF OXIDATION

We return now to the down-trickling sulfate solutions of the metals formed in the zone of oxidation. We have abundant evidence that such solutions are in reality present within the zone of oxidation. In copper mines, trickles and pools of green copper sulfate eat into the steel rails and the nails of miners' boots. Efflorescences of sulfates of copper, iron, zinc, manganese, magnesium, cobalt, and other metals are found on the roofs and walls of underground workings, indicating down-seeping sulfate solutions that dried out. A story is recorded of the early days in Butte, Montana, that a "tin" can fell into discharged mine water draining through a miner's yard and became converted to copper, whereupon the miner unobtrusively obtained a lease upon the water, ran it over scrap iron, and became wealthy producing "cement" copper. The copper sulfate water plus iron equals iron sulfate plus copper. In many copper mines, much copper is recovered from waste dumps by running water over them, which by reaction forms copper sulfate from which copper is precipitated by scrap iron.

The metal-bearing solutions may lose their metallic content within the zone of oxidation by redeposition chiefly as native metals, oxides, carbonates, and silicates. The chief metals so deposited are copper, lead, zinc, silver, vanadium, and uranium, of which there are many large and valuable deposits. Of lesser importance are oxidized deposits of manganese, iron, cobalt, and a few other substances. For the most part the ores are deposited in the lower part of the zone of oxidation.

CAUSES OF DEPOSITION. Several different chemical factors bring about deposition of oxidized ores in the zone of oxidation. Most important is the reaction between sulfate solutions and wall rocks or gangue minerals. Solutions of copper, zinc, or iron sulfate react with carbonate rock to form malachite and azurite, the carbonates of copper, smithsonite, the carbonate of zinc, and siderite or ankerite, the carbonates of iron. These solutions react with silica released in the weathering of silicates, to form the silicates of copper and zinc, namely chrysocolla and hemimorphite. Lead sulfate, being insoluble, does not migrate, but carbonated waters that come in contact with it may change the sulfate (anglesite), to the lead carbonate (cerussite). Silver sulfate does not form carbonates or silicates.

The sulfate metal solutions may intermingle with other solutions and bring about deposition. Carbonated waters react with sulfate solutions and cause deposition of the metal carbonates mentioned above, and also oxides of copper and zinc, and native copper. Native copper and native silver are also formed by reaction between the sulfates of these metals and iron sulfate. Common salt is a widespread constituent of the soil and ground water of arid regions. Salt solutions react with metal sulfates and other compounds to form chlorides

Oxide Ore Deposition

of copper, silver, lead, and other metals. Silver chloride or cerargyrite, a very important ore of silver, is formed as follows:

$$Ag_2SO_4 + 2NaCl = 2AgCl + Na_2SO_4$$
Silver sulfate + salt = silver chloride + sodium sulfate

Bromides and iodides of silver are similarly formed but in much less abundance. Most of the early-mined silver in the western United States was won from such ores.

Evaporation also causes deposition of metal compounds from sulfate solutions. In the greatest single copper deposit of the world, at Chuquicamata, Chile, tens of thousands of tons of ore per day have been mined for over 35 years from oxidized ores consisting of various sulfates of copper deposited by evaporation. Oxidation and hydration also yield oxides and hydrous oxides of several of the metals.

CHARACTER OF DEPOSITS. Many of the important mineral deposits of the world consist of oxidized ores formed in the manner described above. Copper carbonates and silicates have given rise to large and rich ore bodies at Bisbee, Globe, and Morenci in Arizona, and in Northern Rhodesia and the Belgian Congo. Oxidized ores of lead and zinc have been formed at Tintic, Utah, Leadville, Colorado, and other western districts and have contributed largely to the high Mexican production of lead and zinc. Other examples of oxidized ores are world wide.

Some of the ore minerals formed in the zone of oxidation are also formed by hydrothermal processes, but many are formed only through oxidation and are, therefore, diagnostic of it. The carbonates, silicates, and sulfates of copper and zinc originate only in the zone of oxidation. This is also true of the oxides of copper, cobalt, antimony, molybdenum, bismuth, and of the chlorides, iodides, and bromides of silver, and of many other minerals.

When it is established that ores are oxidized, certain broad conclusions may be drawn regarding them, such as: (1) that the ores are bound to change in character with depth; (2) that there is likely to be a pronounced change in tenor or grade in depth; (3) that in most cases only shallow depth may be expected; (4) that different metallurgical treatment will be required for the underlying ores; (5) that more adequate transportation is necessary for the exploitation of oxidized ores than for unoxidized ores, because most of them are shipped directly to smelters without concentration.

Oxidized ores, like gossans, may be indicators of sulfide ores that lie beneath them. Therefore it is desirable to distinguish indigenous from transported ores. If they are indigenous, unoxidized ores may be expected beneath them. If they are transported, they may be distant from their source and so may not have sulfide roots; their source is more likely to be upward and lateral, as has been found in a number of copper and zinc deposits (see Fig. 11-7). If transported, the size of the ore body may have no relation to the size of the source body.

SECONDARY SULFIDE ENRICHMENT

The metals in solution that escape capture in the zone of oxidation trickle down to where there is no available oxygen, generally to the water table, and are there deposited as secondary (supergene) sulfides. The metals removed from above are thereby added to those already existing below, thus bringing about an enrichment of the upper part of the sulfide zone. This forms the *zone of secondary enrichment,* or, as it is more generally referred to, the *supergene sulfide zone.* It in turn is underlain by the *primary* or *hypogene zone.* As progressive erosion brings about deeper oxidation, after a time the supergene sulfides themselves become oxidized and their metal content is again transferred to the downward-progressing enrichment zone. This may be repeated again and again until the primary ore may be enriched to as much as ten times its original metal content. Rich ores are made richer, lean ores are made more valuable, and noncommercial primary material or *protore* is built up to commercial grade. Many great copper camps such as those of the "porphyry coppers" owe their start to supergene enrichment of what was then valueless protore. The process is, therefore, not only of great scientific interest but of far-reaching economic importance to the mineral industry.

Favorable conditions must exist for supergene sulfide enrichment to take place, but they are sufficiently common so that enrichment deposits are widely scattered over the nonglaciated areas of the earth. The process is of greatest importance to copper and silver deposits, and the following remarks deal largely with them.

The following are prerequisites for supergene sulfide enrichment: (1) foremost is oxidation to yield the metals in solution—oxidation may occur without sulfide enrichment, but enrichment cannot take place without oxidation; (2) suitable primary minerals are necessary to yield the solvents, and iron sulfides are essential; (3) the deposit beneath the oxidized zone must be permeable to permit penetration of the trickling sulfate solutions—impervious deposit and pervious wall

Secondary Sulfide Enrichment

rocks cause dissipation or loss of the solutions; (4) there must be freedom of precipitants in the oxidized zone, else the metals will be captured there and will not be available for secondary sulfides; (5) there must be a zone of no available oxygen; (6) there must be available precipitants in the unoxidized zone, and this means sulfides, since the means of deposition is entirely by replacement of pre-existing sulfides; if there are no sulfides in the oxygen-free zone, then no supergene sulfides are formed.

MODE OF FORMATION. If the metals are taken into solution in the oxidized zone, why don't they continue to stay in solution? The answer to this rests on a simple chemical law, namely that, if a metal in solution meets a sulfide more soluble than itself, the more soluble one is dissolved and the less soluble one is precipitated at its expense. The order of solubility of the common metal sulfides is: mercury, silver, copper, bismuth, lead, zinc, nickel, cobalt, iron, and manganese, the last being the most soluble. Therefore, if copper in solution meets any of these sulfides of lower solubility, copper sulfide will be precipitated at the expense of the more soluble sulfide. Mercury would be deposited by each one below it; manganese, being the most soluble of this group, would not be precipitated by any of them. The deposition in nature takes place only by replacement of the other sulfides, and it is volume-for-volume replacement. Hence balanced chemical equations depicting such changes do not correctly represent what actually takes place. They do, however, indicate the trend of the exchange. Using copper sulfate as an illustration, let us glance at a few such equations to see how copper sulfides are formed at the expense of other sulfides.

[1] Galena to covellite:

$$PbS + CuSO_4 = CuS + PbSO_4$$
Lead sulfide + copper sulfate = covellite + lead sulfate

[2] Sphalerite to covellite:

$$ZnS + CuSO_4 = CuS + ZnSO_4$$
Zinc sulfide + Cu sulfate = covellite + zinc sulfate

[3] Pyrite to chalcocite:

$$5FeS_2 + 14CuSO_4 + 12H_2O = 7Cu_2S + 5FeSO_4 + 12H_2SO_4$$
Iron sulfide + Cu sulfate + water = chalcocite + ferrous sulfate + sulfuric acid

[4] Pyrite to covellite:

$$4FeS_2 + 7CuSO_4 + 4H_2O = 7CuS + 4FeSO_4 + 4H_2SO_4$$

Iron sulfide + Cu sulfate + water = covellite + ferrous sulfate + sulfuric acid

[5] Chalcopyrite to covellite:

$$CuFeS_2 + CuSO_4 = 2CuS + FeSO_4$$

Copper-iron sulfide + Cu sulfate = covellite + ferrous sulfate

[6] Bornite to chalcocite:

$$5Cu_5FeS_4 + 11CuSO_4 + 8H_2O = 18Cu_2S + 5FeSO_4 + 8H_2SO_4$$

Copper-iron sulfide + Cu sulfate + water = chalcocite + ferrous sulfate + sulfuric acid

[7] Covellite to chalcocite:

$$5CuS + 3CuSO_4 + 4H_2O = 4Cu_2S + 4H_2SO_4$$

Copper sulfide + Cu sulfate + water = chalcocite + sulfuric acid

Secondary copper sulfides similarly replace many other minerals. The end product is always chalcocite or covellite, although chalcopyrite and bornite may form as temporary transition products. Certain reactions also yield native copper, and this mineral is common in the chalcocite ores of Chino, New Mexico.

Silver sulfate reacts similarly to copper, forming silver sulfide or native silver at the expense of some other sulfides. Less is known about other supergene sulfides. Lead does not form a mobile sulfate; therefore supergene lead sulfide is not known. Zinc should be expected to form supergene sulfides, as it is higher in the solubility scale than a few other sulfides, but no unchallenged supergene zinc sulfide has been found. Both nickel and cobalt form soluble sulfates, and one might expect supergene sulfides of these metals. Supergene sulfides of nickel have been recorded, but none of cobalt. Gold may be dissolved by ferric sulfate in the joint presence of sodium chloride and manganese dioxide and precipitated if the ferric is reduced to ferrous sulfate. However, few occurrences of supergene gold have been recorded; it tends to stay behind in the oxidized zone.

Supergene sulfides exhibit all degrees of enrichment, from incipient replacement, where other sulfides are thinly tarnished or lightly veined with microscopic cracks of supergene sulfides, to a more advanced stage, where the primary sulfides are left as fragments in a

Secondary Sulfide Enrichment

sea of supergene sulfides (Fig. 11-10), to almost complete enrichment, with only a few microscopic residuals of the original sulfides. The incipient phase marks either the bottom of the zone of enrichment or weak enrichment. Supergene enrichment may also be selective in that only certain minerals may be changed over to supergene sulfides and adjacent minerals may remain untouched.

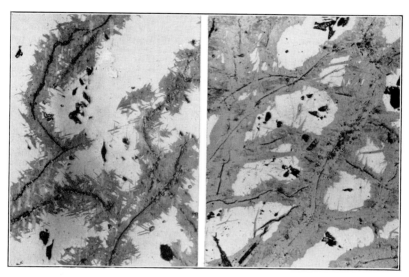

FIGURE 11-10. Photographs through a microscope of polished surfaces of enriched ore, magnified 180 times, showing two stages of sulfide enrichment. Left, initial stages of enrichment of chalcopyrite (white) by feathery covellite (gray) growing outward from cracks (black lines). Right, more advanced enrichment of same, in which the covellite has widened and coalesced, leaving residuals of unreplaced chalcopyrite; cracks still visible. Chalcopyrite contains 34% copper, and covellite 66%. (Photos by author.)

Sulfide enrichment starts at the water table and extends downward far below it, generally diminishing downward. Its upper surface, being controlled by the water table, is generally smooth and undulating and roughly conforms to the topography that existed at the time enrichment took place. Its bottom is highly irregular and is a gradual transition to primary ore. Only under rare circumstances does supergene sulfide enrichment take place above the water table, as at Bingham, Utah, where the oxidized zone was stranded above a rapidly lowered water table. Many enrichment zones are now out of adjustment with the present topography and water table. They are either depressed or elevated, and their tops are related to former water tables that often can be correlated with older erosion surfaces. Hence a study of sulfide enrichment must go hand in hand with a study of topographic development of the region. For example, sulfide enrichment zones rarely occur beneath steep canyons where canyon cutting went on faster than oxidation and sulfide enrichment could keep pace. In such development, enrichment zones may occur beneath upland remnants of mature topography lying between steeply incised valleys, being related to the development of the older topography. Many enrichment zones now lie deeply buried beneath later rock formations.

FACTORS INFLUENCING SULFIDE ENRICHMENT. A high stationary water table, such as occurs in flattish, humid regions, means a thin zone of oxidation, little metal dissolved, and a thin but well-enriched sulfide zone, as at Ducktown, Tennessee. A deep water table, especially one being slowly depressed, favors a thick zone of oxidation, release of much metal, and a thick and well-enriched supergene sulfide zone. The ideal condition for optimum enrichment is active erosion, with progressive depression of the water table at a rate such that oxidation and enrichment keep pace with it.

Supergene sulfide enrichment is dependent upon the enriching solutions coming in contact with suitable precipitants in the primary zone. If they do not, no secondary sulfides are precipitated, and the metal removed in the overlying oxidized zone is lost. This rarely occurs in widespread, permeable, disseminated copper deposits, but, in inclined impervious veins enclosed in much shattered wall rocks, the enriching solutions may trickle vertically downward and pass off into barren wall rocks. The commercial width of the enriched zone is commonly much greater than the underlying primary zone. This is because the primary metallization is commonly accompanied by wall rock alteration with attendant formation of pyrite, and within the enriched zone the pyrite of the wall rocks may be converted to chalcocite, thus giving a wide ore zone.

The wall rocks control the character of enrichment, particularly in disseminated deposits. If they are brittle and crackled they allow easy ingress of the enriching solutions, but if they are dense and pliable their imperviousness keeps the enriching solutions away from the primary sulfides, and enrichment may be weak or lacking. Intense kaolinization of wall rocks showing evidence of primary wall rock alteration, accompanies enrichment. Carbonate rocks prevent enrichment because they precipitate the metals in the oxidized zone. Faults and shear zones in the rocks may permit deep prongs of enrichment, or, if they contain impervious gouge, inclined zones and faults may shield the underlying primary sulfides from contact with down-seeping enriching solutions.

The relation of sulfide enrichment to topography, climate, erosion, and length of time is the same as that given for oxidation. Even more than with oxidization, a long period of time is necessary to yield appreciable enrichment. It is not a question of time alone, but rather the length of time that there has been exposure to oxidation, without covering by later sediments or lava flows. A "porphyry copper" deposit might have been exposed to erosion in early Miocene time, which was long enough ago to permit adequate enrichment, but, if it was buried by middle Miocene lava flows and only recently exhumed, there would have been insufficient time to yield much enrichment. It is quite essential that there be sufficient erosion and oxidation to release enough metal to make appreciable addition to the primary ore.

As was mentioned earlier, the most complete enrichment is obtained by the continuing abstraction of metals, during a lowering water table, from the newly exposed top of the enrichment zone, and the addition of these metals below. The very completeness of enrichment, however, may bring about its own demise. If the enrichment has gone on to the stage where, for instance, all the primary sulfides, pyrite included, are replaced by chalcocite, then the zone of enrichment, lacking in pyrite, when exposed to oxidation will generate no solvents for copper. The chalcocite will then be converted to copper carbonates *in situ*, and no copper will be released to be carried below. Enrichment ceases, and a zone of disseminated oxidized ores results. This, I think, is what has happened in parts of Northern Rhodesia, Miami, Arizona, Bagdad, Arizona, and other places. Cessation of enrichment is also brought about by other means, such as burial, submergence, base leveling of a region causing stationary water table, and bottoming of the primary ore. The enriched zones may reach thicknesses up to 1,500 feet, but a few hundreds of feet is more common.

ENRICHMENT DEPOSITS. Supergene sulfide deposits are characterized by the three zones. It may happen that the primary zone is lacking, if enrichment extends to its bottom, or, in a few cases, the oxidized zone may be largely removed by erosion. However, they generally have a gossan, and enriched ores leave evidence of themselves in the gossan for interpretation by those who can make it. Such evidence consists of removal of metals from the gossan, voids, indigenous limonite boxworks, and kaolinization of the wall rocks. Evidence from gossans opposed to sulfide enrichment would be rich oxidized ores in the oxidized zone, transported rather than indigenous limonite, limestone country rock, and lack of kaolinization of minerals susceptible of being kaolinized.

The minerals of supergene sulfide deposits are not always diagnostic of enrichment, since, except for sooty chalcocite, most of them are also formed by hydrothermal processes. Association of minerals is more diagnostic than are single minerals. Microscopic evidence is generally necessary to make sure that a given deposit has been secondarily enriched. It is characteristic of enriched copper deposits that the copper content is negligible in the oxidized zone, jumps suddenly at the top of the enriched zone, and slowly tapers off toward its bottom. The iron content shows a reverse curve. It is fairly high in the oxidized zone, drops off sharply at the top of the enriched zone (because copper has replaced iron), stays less than copper (in disseminated deposits) to the bottom of the enriched zone, and then rises above copper in the primary zone (Fig. 11–11).

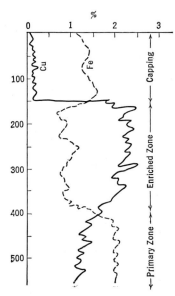

FIGURE 11–11. Assays of copper and iron content of ore from boreholes. In enriched ore, the iron content in the oxidized zone is about constant (indigenous iron), but its curve crosses that of copper at the top of the enriched zone and recrosses at the bottom because copper replaces the iron. The copper content rises in the enriched zone to about 2½ times that of the primary ore.

There are many examples of supergene sulfide-enriched deposits in various parts of the world, except in glaciated regions. All the great "porphyry copper" mines were initiated on sulfide-enriched ores, although today their copper content is contained largely in primary minerals. Few copper deposits of western United States and Mexico lack enriched sulfide ores. The great copper deposits of Chile likewise underwent a high degree of supergene sulfide enrichment. Those of Rhodesia exhibit little supergene sulfide enrichment. Enriched silver deposits are likewise widespread.

SELECTED REFERENCES ON OXIDATION AND ENRICHMENT

"The Enrichment of Ore Deposits." W. H. Emmons. *U. S. Geol. Surv. Bull.* 625, 1917. Older, comprehensive treatment of supergene sulfide enrichment.

Copper Resources of the World. Vols. 1 and 2. 16th Int. Geol. Cong., Washington, D. C., 1935. Brief details of individual deposits.

Ore Deposits of the Western United States—Lindgren Volume. Am. Inst. Min. Met. Eng., New York, 1933. Chap. 9, "Supergene Enrichment," by W. H. Emmons. Later developments.

Leached Outcrops as Guides to Copper Ore. A. Locke. Williams and Wilkins, Baltimore, 1926. Outcrop characteristics of oxidized and enriched deposits.

"The Hydrated Ferric Oxides." E. Posnjak and H. E. Merwin. *Am. Jour. Sci.*, Vol. 47, 1919, and *Am. Chem. Soc. Jour.*, Vol. 44, 1922. Iron oxides, their composition and formation.

"Oxidation Products from Sulphides." Group of papers by R. Blanchard and P. F. Boswell. *Econ. Geol.*, Vols. 22, 25, 29, 30, 1927–1935. Various types of limonite boxworks.

"Oxidation at Chuquicamata, Chile." O. W. Jarrell. *Econ. Geol.* 39:251–286, 1944. Excellent discussion of processes of oxidation of large copper deposits.

"Oxidation and Enrichment in the San Manuel Copper Deposit." G. M. Schwartz. *Econ. Geol.* 44:253–277, 1949. Good example.

Economic Mineral Deposits. 2nd Edit. Alan M. Bateman. John Wiley & Sons, New York, 1950. Chaps. 5–8, oxidation and supergene enrichment, 245–289; silver, 459–462, 464–473; copper, 486–495, 498–503, 507–515.

CHAPTER TWELVE

Metamorphic Processes

Metamorphism is a process by which rocks and minerals respond to a change in their environment by taking on new forms that are stable under the changed conditions. The chief agencies involved are temperature, pressure, and sometimes a little water or carbon dioxide. The source materials are rocks. The metamorphic agencies bring about recrystallization or recombination of the existing rock constituents, as was explained under contact metamorphism in Chapter 8, but no new mineral constituents are added as in contact metasomatism. The changes produce new nonmetallic minerals, which in some cases constitute valuable mineral deposits. The chief deposits formed by metamorphism are graphite, asbestos, talc, soapstone, sillimanite-kyanite-andalusite, garnet, and abrasives.

Metamorphism operates deep within the earth where overlying rock mass supplies pressure, to which may be added additional pressure resulting from movements within the earth. The heat is that existing within the earth, and we know that earth temperatures are at least sufficiently high to form magmas. Additional heat may be added from igneous intrusions or from movements within the earth, such as folds and faults. The moisture is that existing in the rocks or bound up in rock minerals; in metamorphosed igneous bodies some remnants of magmatic juices were probably present. In general, the agencies and forces are the same as those that produce common metamorphic rocks.

In view of the diversity of the resulting mineral deposits and the somewhat different conditions that were involved in the formation of each, the origin of each group will be considered separately.

GRAPHITE FORMATION

Graphite or "black lead" is one of the many forms of carbon that occur as *crystalline* graphite, consisting of thin, nearly pure black flakes, and so-called *amorphous* graphite, a minutely noncrystalline impure variety. True graphite marks paper and is referred to as the "lead" of lead pencils. Graphite is formed in several ways, of which metamorphism is only one. It is formed by hydrothermal solutions as vein graphite, by contact metasomatism, and as original crystallization in igneous rocks; that produced by metamorphism is found in marble, gneiss, schist, quartzite, and altered coal beds. Most of the crystalline variety of metamorphic origin occurs in minute flakes disseminated through metamorphic rocks, and it may have an organic or inorganic origin.

Coal beds have been intruded by igneous rocks at Sonora, Mexico, Raton, New Mexico, and other places, and the heat of the intrusion has driven off the volatile constituents of the coal and converted the residual carbon into crystalline graphite, resulting in commercial deposits. In Rhode Island some anthracite coal has been metamorphosed almost to graphite. These occurrences are clearly of organic sources.

Concerning the disseminated crystalline graphite in metamorphic rocks, two views of origin are held: one, that the graphite is altered organic matter formerly present in the sediments; and the other, that it results from the breakdown of calcium carbonate. Black carbonaceous limestones, when metamorphosed, yield white marbles with disseminated graphite. Either the original hydrocarbons have been broken up, giving rise to precipitation of the carbon, or they have been converted into carbon monoxide and carbon dioxide, which in turn were deoxidized and the carbon precipitated. In either case the formerly distributed carbon has moved into concentration centers. There are other examples of organic origin.

The inorganic concept is that carbonates are broken down, yielding their calcium, magnesium, or iron to join with silica and form silicates, at the same time releasing carbon monoxide and carbon dioxide, which in turn become deoxidized to form graphite. A. N. Winchell suggests that two reversible chemical reactions may take place under geologic conditions:

$$C + H_2O \leftrightarrows CO_2 + 2H_2$$

$$C + CO_2 \rightleftarrows 2CO$$

If so, either could account for free carbon.

The fact that graphite occurs in Precambrian rocks before much life is supposed to have existed suggests an inorganic rather than an organic origin for the carbon. It is also possible that the graphite of igneous rocks, pegmatite dikes, and veins may have been picked up from underlying carbonate rocks.

The chief metamorphic graphite deposits have been formed in Bavaria, Austria, Czechoslovakia, Korea, Madagascar, Russia, and the United States. Those of central Europe, long the most productive in the world, occur as large lenses in a gneiss that contains layers of schist and limestone, all surrounded by granite. The highly productive deposits of Korea also occur as layers and lenses in schist not far from granite contacts. The large deposits of Russia, which for long were the chief supply of pencil graphite, occur in schist and syenite. The Madagascar graphite lies in pockets, layers, and masses in schists and gneisses and constitutes a belt 400 miles long—an almost inexhaustible supply. The chief United States occurrences are near Ticonderoga, New York, in quartz-schist (5–7 percent graphite) associated with garnet gneiss and limestone, and in Alabama in mica schists with included pegmatites, carrying 2.5 to 3.5 percent flake graphite.

ASBESTOS FORMATION

Asbestos is unique among materials of the mineral kingdom in that it consists of delicate, flexible fibers, so soft and silky that they can be spun readily into threads and woven into cloth. Because of its resistance to fire, acids, and electricity, it has been much sought ever since the time of the Romans, who used it for lampwicks that would not burn out and for cremation cloths that could be cleansed by fire. Its uses are manifold.

Asbestos is a commercial name applied to a group of fibrous minerals of which the chief ones entering industry are the serpentine asbestos, chrysotile (the silky variety), crocidolite or "blue asbestos," and the long harsh variety, amosite. The serpentine asbestos, which is the most valuable variety, consists of hydrous magnesium silicate of the identical composition of pure serpentine rock, generally called serpentinite.

To understand the controversial origin of serpentine asbestos, let us first glance briefly at its mode of occurrence. It always occurs in serpentine, which in turn is chiefly an alteration product of ultrabasic igneous rocks; a minor amount is an alteration product of magnesian limestone or dolomite. The asbestos fiber is chiefly in veinlets with cross fibers oriented at right angles to the walls of the veinlets, so that the length of the fiber is the width of the veinlet. Poorer-quality asbestos occurs as slip fiber and mass fiber with random orientation.

Asbestos Formation

The cross fibers range up to 4 or 5 inches in length, rarely 8 inches, but mostly less than 1 inch. The fiber may make up 2 to 20 percent of the rock. The veinlets are commonly short, discontinuous, and criss-cross in networks. Less commonly they occur in parallel veinlets, forming the "ribbon rock" (Fig. 12-1) of the Transvaal. The deposits make up only a small part of the igneous rock masses, and those in Quebec are as large as 800 by 200 feet in dimension.

FIGURE 12-1. Bands of asbestos in serpentine forming ribbon-rock structures. (M. J. Messel, *A.I.M.E.*)

The limestone serpentinite is in layers parallel to the limestone bedding, with alternating layers of serpentine and unserpentinized rock. The fiber layers occur as elongated lenses within the layers of serpentine and with the fibers oriented at right angles to the bedding. The limestones lie next to sills of basic igneous rock.

In both types of occurrence the fibers occur only in the serpentinite, but serpentinite may occur without asbestos. In some places the entire igneous mass may be serpentinized. The serpentinite always occurs along fractures, indicating it is later in origin than the igneous rock. Serpentinite is generally considered to have been formed by the hot magmatic juices left over upon final consolidation of the igneous rock and to be a late magmatic effect. The origin of the serpentine in magnesian limestones is obviously later than the sedimentary rock. Unquestionably, magmatic emanations given off from the cooling igneous sills introduced silica, which combined with the magnesia of the limestone to form the hydrous magnesium silicate, or serpentine.

The origin of the asbestos is puzzling. With the same composition as serpentine, how did the fibers form and how did they become emplaced? Some investigators think that the fiber commenced to crystallize from the serpentine along tight cracks and that the growing fibers pushed the walls apart, aided by tension, to make the veinlets. However, the network type of fiber veinlets and those in horizontal limestones with undisturbed bedding and with much overlying weight do not support this view. Other investigators think they are hydrothermal fissure fillings, and still others that they represent recrystallization of serpentine, a variety of replacement, outward from tight cracks. Fissure filling does not seem tenable, because how could the cracks remain open, particularly the horizontal ones such as those of the Transvaal and those in limestones, under such heavy loads of overlying rocks? The most probable explanation is that the serpentine along cracks recrystallized into a fibrous form, probably aided by circulating solutions that had undergone some slight change, such as a change in the hydrogen-ion content.

The crocidolite and amosite varieties, called amphibole asbestos, occur in slates, schists, and banded ironstones over an extensive area in South Africa. They also occur as cross fibers, but in lengths up to 12 inches. Peacock thinks the crocidolite originated by molecular reorganization, without essential transfer of materials or constituents of the enclosing banded ironstones. Deep burial is thought to have supplied the heat and pressure that resulted in the metamorphism of the rock constituents into this peculiar blue fiber. This is strongly suggested by its wide distribution in similar rocks generally unassociated with igneous intrusives.

There is still much to be unraveled about the origin of asbestos deposits.

Large asbestos deposits have been formed in Quebec, Russia, Rhodesia, South Africa, and Swaziland, with minor deposits in Cyprus, the United States, Korea, Finland, Czechoslovakia, and Australia. The amphibole varieties come chiefly from South Africa, and the high-quality material from Quebec, Rhodesia, and Russia.

TALC AND SOAPSTONE FORMATIONS

Talc, the softest of all minerals, is known to everyone because of talcum powder. The pure, soft mineral in trade is called talc, but this term also includes block talc or steatite, a massive variety, and soapstone, which includes other minerals than talc. Pyrophyllite is often included with talc because it is used for the same purposes.

Talc, a hydrous magnesium silicate, and soapstone are restricted to metamorphic areas and occur in metamorphosed ultrabasic igneous intrusives or in dolomitic limestones. The very best quality is formed in metamorphosed dolomite limestone as beds and lenses that attain widths of 125 feet. The deposits in igneous rocks are more numerous but smaller than those in altered limestones, and are associated with serpentine.

Talc is a deep-seated alteration of the magnesian minerals of the two rocks by carbon dioxide and water under metamorphic conditions of elevated temperature and pressure. The magnesia is largely, if not entirely, derived from the rocks in which the talc occurs. The formation of talc and soapstone is thus analagous to the formation of serpentine. Talc is always late in the mineral sequence of alteration products.

The commercial talc and soapstone deposits are all confined to regions of metamorphism. In the United States, the country of largest production, it is found in six states. The New York variety goes to paint, ceramic, and paper industries, North Carolina talc to the rubber industry, and Vermont talc to the roofing industry. Soapstone comes from Virginia and Georgia. The finest grades of toilet talc come from the Pyrenees and Alps, and steatite talc from India and Sardinia.

SILLIMANITE GROUP MINERALS

Three interesting minerals, sillimanite, kyanite, and andalusite, being born under high-temperature conditions, are able to withstand high temperatures and change over to mullite, and are eagerly sought for high-grade refractories (e.g., spark plugs) and ceramic uses. The three have identical composition, being composed of aluminum oxide and silica. Sillimanite is formed, as a result of high-temperature metamorphism, as slender prisms in aluminous crystalline schists. It comes chiefly from India where it is associated with corundum in sillimanite schists.

Kyanite is a common mineral of metamorphic rocks, although commercial deposits are few. These consist of disseminated crystals or small masses in gneisses and schists. Kyanite is considered to have been formed from mica schists or other aluminous silicate rocks by dynamothermal metamorphism, perhaps accompanied by magmatic emanations. The only commercial occurrences are in India, Kenya, and the United States. In each locality it occurs in kyanite schists or quartz-kyanite schists in disseminated crystals or masses. The Appalachian occurrences and those of California have been formed from aluminous schists.

Andalusite has its home in metamorphosed mud rocks, where it occurs along with other metamorphic minerals such as garnet, tourmaline, and corundum. The largest deposit known is in White Mountain, California, where it occurs enclosed in schist as a result of metasomatic processes.

Another generally similar mineral but one containing boron, dumortierite, is commonly included among the sillimanite minerals, as it is used for similar purposes. However, it occurs in pegmatite dikes or quartz veins that cut across aluminous schists, and its origin is probably due to magmatic emanations. During a 10-year period the Champion Spark Plug Company made 350 million spark plugs from andalusite and dumortierite.

OTHER METAMORPHIC PRODUCTS

Garnet of abrasive quality (almandite and rhodolite) is won from commercial deposits in gneisses and schists, where it has been formed as a result of metamorphism. There are also five other varieties of garnet that originate in pegmatite dikes, in deep-seated, ultrabasic, igneous rocks, and in contact-metasomatic zones. The largest commercial deposits, in upper New York State, occur in a garnet gneiss in grains and crystals mostly from $\frac{1}{4}$ to 1 inch in diameter. Crystals up to 1 foot in diameter are quite common, and a few even attain 3 feet in diameter.

Emery, which is a mixture of magnetite and corundum with hematite or spinel, is formed by metamorphic processes, mostly contact metamorphism. The chief deposits are in Turkey, Greece, and the United States.

Abrasive stones, consisting of millstones, grinding pebbles, and sharpening stones, although made in part from sandstone, consist mostly of metamorphosed sandstone and siliceous mudstones forming quartzite, which possesses great hardness. These are merely metamorphic rocks.

Roofing slates are metamorphosed mud rocks in which thin cleavage has been developed.

Anthracite coal is metamorphosed bituminous coal.

SELECTED REFERENCES ON METAMORPHISM

"Asbestos." Oliver Bowles. *U. S. Bur. Mines Bull.* 403, 1937. Summation of geology, occurrences, and industry.

"Asbestos Deposits of Thetford Area, Quebec." H. C. Cooke. *Can. Geol. Surv. Sum. Rept.* 1931:1–24; 1932:41–59; *Roy. Soc. Can. Trans.* 29:7–19, 1935. Geology, origin. Also in *Econ. Geol.* 31:355–376, 1936.

"Geology of the Shabani Mineral Belt, Southern Rhodesia." F. E. Keep. *Southern Rhodesia Geol. Surv. Bull.* 12, 1929. Description of Rhodesian occurrences.

"Geology of Quebec," Vol. 2. J. A. Dresser and T. C. Denis. *Quebec Dept. Mines Geol. Rept.* 20:413–442, Quebec, 1944. Occurrence and origin.

"Asbestos in the Barberton District, South Africa." A. L. Hall. *South Africa Geol. Soc. Trans.* 1921:168–181; 1923:31–49. Unusual chrysotile occurrence.

"An Arizona Asbestos Deposit." Alan M. Bateman. *Econ. Geol.* 18:663–683, 1923. A limestone type formed by replacement.

"Alabama Flake Graphite in World War II." H. D. Pallister and R. W. Smith. *A.I.M.E. Tech. Pub.* 1909, 1945. Occurrence and developments.

"Graphite for Manufacture of Crucibles." G. R. Gwinn. *Econ. Geol.* 40:86, 1945.

"Origin of Talc and Soapstone." H. H. Hess. *Econ. Geol.* 28:634–657, 1933.

"Origin of Talc and Soapstone Deposits of Virginia." J. D. Burfoot. *Econ. Geol.* 25:805–826, 1930. Problems of origin.

"Andalusite and Related Minerals, White Mountain, California." P. F. Kerr. *Econ. Geol.* 27:614–643, 1932. Occurrence and origin.

"Kyanite Deposits of North Carolina." J. L. Stuckey. *Econ. Geol.* 27:661–674, 1932. Geological features.

"Occurrences of Sillimanite in North Carolina." C. E. Hunter and W. E. White. *N. Car. Dept. Cons. Dev. Inf. Circ.* 13, 1946. General occurrence.

"Mining and Treatment of the Sillimanite Group of Minerals and Their Use in Ceramic Products." F. H. Riddle. *A.I.M.E. Tech. Pub.* 460, 1932. Technology and uses.

Industrial Minerals and Rocks. 2nd Edit. Edited by S. H. Dolbear and Oliver Bowles. Am. Inst. Min. Met. Eng., New York, 1949. Chap. 2, "Asbestos," by G. F. Jenkins. Concise geology and technology; good bibliography. Chap. 19, "Graphite," by G. R. Gwinn. Good survey and bibliography. Chap. 42, "Sillimanite Group," by F. H. Riddle and W. R. Foster. Brief summary of mineralogy and geology, distribution, and uses. Complete bibliography. Chap. 48, "Talc and Ground Soapstone," by A. E. J. Engel. Comprehensive; four pages of bibliography.

Economic Mineral Deposits. 2nd Edit. Alan M. Bateman. John Wiley & Sons, New York, 1950. Metamorphism, 289–300; graphite, 738–742; asbestos, 749–756; talc, 758–761; abrasives, 823–833.

CHAPTER THIRTEEN

Ground-Water Processes

    Ground water or subsurface water ranks as a mineral resource of highest importance. From earliest time in many parts of the world the development of civilization has been dependent upon it. Oscar Meinzer remarked that one of the truly great achievements of civilized man has been his ability to provide adequate and reliable supplies of good water on those parts of the earth where surface waters are lacking or unusable. Despite our familiarity with water, the principles of occurrence and laws of ground water are inadequately known to many, and with few other subjects have so much superstition and misinformation been handed down through the centuries. Those who have been fortunate enough to live in regions of abundant supplies of surface water little realize how extensive are the areas entirely dependent for life and agriculture upon underground supplies of this essential substance. Too often the supplies are meager—a meagerness due as much to lack of knowledge of ground water as to a lack of supply. It has been estimated by Cox that 20 million acres in India alone are irrigated by subsurface water, an average comparable to the total irrigated (both by subsurface and surface water) area of the United States. In arid regions the world over pumping subsurface water is one of the major industries. Witness also the withering of once-flourishing agriculture and civilization with the changing of a humid to an arid climate in the Valleys of the Nile, the Euphrates, and the Tigris.

    The demand for water supplies is mounting steadily with advancing civilization, and the average per capita consumption of water in the cities and towns of the United States is now about 100 gallons a day. Growing metropolitan areas are reaching farther and farther back into rural areas for available surface waters, with overlapping from adjacent cities. The rural areas are now facing lessened supplies for their own growth. High and growing demands are made upon water supplies by large industrial plants, and in several coastal areas the limit of withdrawal of surface water has already been reached or exceeded. It is necessary in such places to force recharge of discarded water back into the ground to build up the supply of ground water.

The ocean is the great residual reservoir of water, and most of this water has probably come from within the earth throughout geologic time. The land-surface water in lakes and ponds, in streams, and in soils is ever returning to the ocean. Water is evaporated from the ocean into the atmosphere, from whence it falls upon the lands as rain and snow. Much of this runs over the surface and returns to the ocean again; part of it sinks into the ground and forms a temporary reservoir of ground water, which in turn eventually seeks an ocean outlet; a part is returned to the atmosphere by vegetation. Thus the ever-recurring hydrologic cycle continues, and under normal conditions ground water should be present.

In this chapter surface water is not considered, and only the bare hydrologic features of ground water will be dealt with. It is hoped that these may help to dispel the great amount of ignorance regarding this everyday substance and clarify many erroneous conceptions of underground water.

HOW GROUND WATER OCCURS

As everyone realizes, a part of the rainfall and snowfall sinks into the ground, and the flatter and more heavily vegetated the region, the more sinks in. This goes to supply the ground water or subsurface water that fills all pores and openings in the soil and rocks within the zone of water saturation. Ground water is not, as is often popularly supposed, an unlimited underground stream or pool, and can be no greater than the amount of pore space in the rocks and soils. The top of the zone of saturation is called the *water table*, which generally lies a short distance below the surface. In humid regions it is shallow, but in arid regions it may be very deep. Between the water table and the surface is a zone that is alternately dry and wetted by descending rain water. In dry seasons the water table sinks a little and in wet seasons it rises again, forming a zone of fluctuation between the two. If the water table rises to the surface, a swamp is formed. If a depression is made beneath the water table, the ground water occupies and stands in the depression. If the depression is a natural closed one, a pond or lake results. If the depression is an open-ended valley, the ground water seeps into its side and feeds a stream. If, in a dry season the water table sinks below the bottom of a closed depression, the pool or pond or lake goes dry; if it sinks below the bottom of a well, the well goes dry—but not if the well is below the zone of fluctuation. If water is withdrawn from a well, the water table is temporarily lowered at that point until it is re-established by the ground water slowly seeking its own level. Ground water, like surface water, merely follows the little-appreciated elementary principle that the laws of gravity determine its movements. However, it moves more slowly than surface water because of the friction exerted by the small rock pores through which it flows.

In a flat area the water table parallels the surface, but in a rolling humid region the water table only roughly parallels the hills and

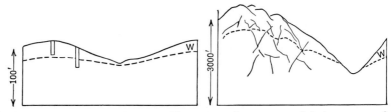

FIGURE 13–1. Relation of water table W to surface in an undulating topography (left) and in mountainous topography (right) where large fissures carry water to discharge in lower regions.

vales, being deeper under the hills and shallow under the valleys (Fig. 13–1). The motion of the ground water from hill toward valley is continuous and slow and would reach a level were it not that surface recharge takes place. In a humid mountainous region the valley slopes may intersect the water table, giving valley-side seeps and springs. Wherever the ground surface intersects a sloping water table, a spring must result.

The movement of ground water in a given area is from places of higher elevation to lower places, so that it continually moves downslope from intake to point of discharge, whether that point be a natural seep or spring, or plant evaporation, or a well or sea level. The rate of movement is determined by the difference in height between the highest point and the lowest, and by the permeability of the rock or soil. In a mountainous region the movement may be relatively rapid; in a flattish region with low head it is slow. Its path may be short or scores of miles in length, and its time of travel may be short or many centuries. Field tests have shown the rate to be as high as 420 feet in a day, which is very rapid, and the lowest test was 1 foot in 10 years. However, in well-defined water-bearing strata,

How Ground Water Occurs

which are called aquifers, the natural rate is generally not greater than 5 feet a day nor less than 5 feet a year.

Artesian is one of the most misused terms relating to water. Any common spring is often miscalled artesian. Most wells drilled into the water table are mistakenly called artesian wells, and water obtained from such wells is erroneously referred to as artesian water. Generally it is not; it is simply common ground water. Real artesian conditions exist where water is confined under a "head," and tends to rise by pressure to the approximate height of its intake. Thus a city

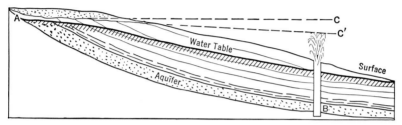

FIGURE 13-2. Artesian water conditions. Water is carried under pressure in permeable aquifer overlain by impervious strata. AC is level of intake; AC' is level to which water will rise because of friction and leakage; and BC' is the head under which water would be forced from the aquifer into a well at B. This would be a free-flowing well, but if the surface were at C the well would not be a flowing one.

water supply from a high-standing reservoir confined within water mains is an artificial artesian supply, and the higher the head, the greater the pressure of flow. In nature an artesian circulation is set up when ground water is confined within an inclined permeable stratum that is overlain and underlain by impervious strata, and when the ground-water intake is higher than a possible outlet, as may be seen from Figure 13-2. If such an inclined aquifer is punched by a cased drill hole, the water will rise under pressure to a height that is a little below the height of the intake, the difference being due to friction and some leakage. If the top of the well is substantially below the height of the intake, there will be a flowing artesian well. If the top of the well is at or above the height of the intake the well will not be a flowing one, but the water will rise up in the well to almost the height of intake, and it will still be an artesian well. Contour maps are made on such artesian aquifers, depicting the elevation of the aquifer and at the same time showing the topographic contours. Thus, if the map at a given point shows a surface elevation of 2,000 feet and an aquifer elevation of 1,200 feet, it means that a well drilled at that point will cut the aquifer at 800 feet depth. Further, water-pressure contours (piezometric) may also be added to the map, and, if at the same point this is 2,100 feet, it means that water in a cased well will flow out of the well with pressure. Such maps are made by the Water Resources Branch of the United States Geological Survey, and permit a landowner to determine the depth, head, and expected flow of artesian water that might underlie his land. Notable artesian areas lie in the states surrounding the Black Hills and in Florida.

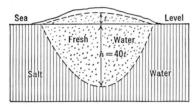

FIGURE 13–3. Fresh water below sea level under a marine island. According to Herzberg formula, fresh water rides on salt water to depth h, which is 40 times the height t of the water table above sea level.

In coastal areas, such as along the low-lying shores of New Jersey, Long Island, and Holland, a knowledge of the occurrence and laws of ground water is particularly pertinent. In such places fresh water rides on top of salt water, and, if withdrawn too rapidly, will cause encroachment of salt water and ruination of ground-water supplies. One ordinarily does not think that fresh water can be obtained from sands below sea level, but it can. Fresh water falling on a sea island floats on top of and depresses underlying salt water by an amount equal to about 40 times the height of the water table above sea level. The significance of this law (Herzberg formula), which rests on the difference in specific gravity between fresh water (1) and salt water (1.025), is too little appreciated. It is shown in Figure 13–3. Translated into everyday language, it means that on Long Island, for example, if the water table lies 20 feet above sea level there is a depth of 800 feet of fresh water below sea level. This is very reassuring. However, it can be disturbing too, because the law works both ways. If by too rapid withdrawal the water table should be lowered 10 feet, it follows that there would be a corresponding rise of 400 feet of salt water—one-half the ground-water supply would be ruined. Hence in such areas great care has to be exercised that the rate of withdrawal does not exceed the rate of intake.

In arid regions slightly different water conditions rule. Here, because of small rainfall intake, the water table is apt to stand very low, except in piedmont valleys. The intake is small in the areas between streams because of the lack of vegetation and because brief arid-region showers do not penetrate rapidly or deeply into the air-filled pores of desert regions. Most of the intake is from throughflowing streams that receive their water in distant mountain areas. In consequence of this the water table is higher under arid-region

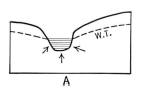 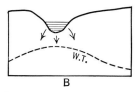

FIGURE 13-4. *A* is a stream (effluent) fed by ground water in a humid region, and *B* is ground water charged by a stream (influent) in an arid region. (After O. E. Meinzer, U. S. Geol. Survey.)

streams and lower away from them, just the reverse of the water table under humid-region stream valleys (Fig. 13-4). The feeding of water into the Tennessee River is shown by Figure 13-5.

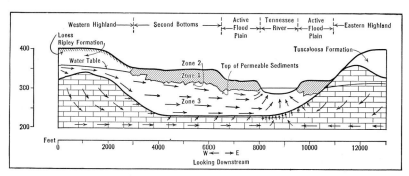

FIGURE 13-5. Geologic section at Kentucky dam site, Tennessee River, showing flow of ground water, confined beneath impervious sediments (Zone 1), into river under slight artesian head. (R. Rhoades, *Econ. Geol.*)

Changes in the water table, either up or down, during geologic time, and their causes, have already been considered in Chapter 11 under "Oxidation and Secondary Enrichment."

WITHDRAWAL AND RECHARGE OF GROUND WATER

The ground water is a great underground water storage that extends downward from the water table. Although theoretically its lower limit is presumed to be the depth where rock openings cannot persist, deep mine workings disclose that there is no visible unconfined water at depths below 2,000 and 3,000 feet.

FIGURE 13-6. Kucher's concept of the origin of springs and rivers. The whirlpools mark openings in the sea bottom into which the water rushes and then rises through subterranean channels into caverns in the mountain tops, where it issues as springs and returns as rivers to the sea. (From Kucher, *Mundus Subterraneus*, 1678.)

A well extending beneath the water table will contain standing water at the level of the water table and will not be affected by drought, provided it extends below the lower limit of the zone of fluctuation. When a well is pumped the water table is lowered around the well in the form of an inverted cone of depression (Fig. 13-7). If pumping is continuous the cone of depression deepens and spreads out laterally, and inflow becomes slower. Continued "drawdown" gradually lowers the water table at a decreasing rate, and nearby shallow wells go dry. If pumping exceeds inflow the water storage is drawn upon until it is exhausted. A *safe yield* is indicated if the recovery by inflow between periods of pumping is complete.

Therefore it is important that the rate of pumping be controlled unless the storage is so great that the supply is almost limitless. Meinzer points out that at a winter garden area in Texas the Carrizo sand transmits 24 million gallons a day, at a rate of 50 feet a year; accordingly, a mile of the outcrop of the sand at this rate contains storage for a century.

Natural recharge goes on continuously in humid regions by intake from rainfall and snow. This is aided, of course, by forest and vegetation, which retards surface runoff and facilitates water seepage into

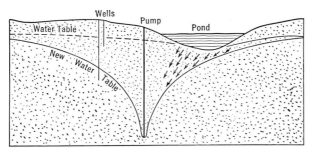

FIGURE 13–7. Cone of depression of water table formed by pumping, causing one well to go dry and lowering water in pond. (F. M. Veatch *et al.*, *U. S. Geol. Survey*.)

the soil. On deforested and plowed land, on the other hand, the proportion of run-off becomes greater and recharge is less. It is often erroneously thought that in humid regions recharge takes place from natural streams and lakes; on the contrary, they are fed by ground water. In semi-arid regions the scarce vegetation causes greater run-off and less run-in, and most recharge is from through-flowing streams and ponding. Artificial recharge is now practiced in many places where withdrawal of ground-water supplies is heavy. Recharge is accomplished by drilling recharge wells or by impounding surface waters over spreading grounds to permit in-seeps. In one impounding basin of 47 acres of the Los Angeles water system about 106 cubic feet of water per second seeped into the ground at a rate of 3 to 10 feet a day and reached the North Hollywood pumping station. Overpumping in Brooklyn, Long Island, brought such serious encroachment of salt water that recharge into wells had to be practiced.

GROUND-WATER SUPPLIES

The United States has some 62,000 large irrigation wells drawing on ground-water supplies, in addition to the countless other large wells for municipal and private industrial purposes. About 6,500 public waterworks also draw supplies from ground water, which likewise furnishes domestic supply for more than half the population of the United States. Ground-water supplies are, therefore, of great economic importance in many parts of the world.

The recovery of ground water is mainly through wells, which may be dug, driven, bored, or drilled. In many large wells, horizontal tunnels or collecting galleries are driven from the bottom. In places like Hawaii tunnels are driven into hillslopes beneath the water table and serve as collecting and discharge galleries. In some arid regions shafts are sunk to bedrock on either side of a dry stream channel and their bottoms connected by a tunnel, which serves to trap all the hidden ground water that flows down the valley on the top of bedrock.

The quality of ground water is of vital concern if it is to be used for domestic purposes or for agriculture. But an abundant supply of impure water is more beneficial than a sparse supply of pure water, because the impurities can often be removed. Impurities consist of pollution and contamination. However, bacterial pollution is generally absent from ground water because its percolation through strata is a most effective means of natural filtering and purification. The contamination is mainly from salt, carbonates, sulfates, iron, manganese, and dissolved gases. Of these, salinity is the most common, particularly in coastal and arid-region ground water. It is generally expressed in parts per million (ppm), sea water being 35,000. Some of the tolerances are: 400, no taste of salt; 1,000–2,500, strong taste but bearable; 3,300, usable domestically; 5,500, unfit for human use; 6,250, horses thrive on it; 7,800, horses can live on it; 9,385, cattle can live on it; 16,000, beyond limit for grasses. The carbonates impart *temporary hardness,* which can be removed by water softening or utilizing lime and soda or exchange silicates. Other contaminants can be removed by base-exchange silicate minerals or by aeration.

One of the most urgent problems in ground-water supplies is the determination of the rate at which water strata will supply water to wells. This involves the determination of additions to supply and discharge, called ground-water inventory. Some water strata are merely reservoirs, and others are water conduits conducting water from other underground reservoirs. It is necessary also to determine the safe yield, or the practicable rate of withdrawal without depletion. The factors affecting increase in ground water are: (1) rainfall penetration to the water table; (2) natural in-seepage from surface waters;

Long Island, New York, Ground-Water Supply   263

and (3) inflow of ground water from outside the area. Some factors affecting decrease are: (1) seepage and flow of free ground water from springs; (2) seepage and spring discharge of artesian water; (3) leakage from aquifers; (4) evaporation, transpiration, and artificial removal; and (5) subsurface discharge. These data are supplied by special measurements.

The character and variability of ground-water supplies may perhaps be understood best by considering briefly three different areas, namely, Long Island, the Dakota sandstone province, and Florida.

LONG ISLAND, NEW YORK, GROUND-WATER SUPPLY

Long Island, 115 miles long and up to 20 miles wide, is inhabited by nearly 5 million people who depend largely on ground water for

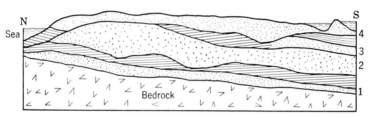

FIGURE 13–8. Generalized section across Long Island, near Jamaica, showing water-bearing sand marked 1 to 4. 1, 2, Cretaceous sands, 4, Pleistocene sands. (Adapted from D. G. Thompson, F. G. Wells, and H. R. Blank, *Econ. Geol.*)

their water supply. In addition there is a heavy industrial demand in the western or Brooklyn end. The island is underlain by Cretaceous sands, clays, and gravels containing three aquifers that incline southward toward the ocean, as shown in Figure 13–8. The maximum elevation is 420 feet. The surface deposits contain a great reservoir of water supplied only by a rainfall of 42 inches a year, of which one-third to one-half reaches the water table.

A thick body of fresh water floats on salt water and sustains pumping of over 200 million gallons a day, with concentrated pumping at the western end. A delicate balance is thus maintained between fresh and salt water, but overpumping brought some contamination. According to Legget, the water table under Brooklyn dropped to an elevation of 40 feet above sea level, and much water there is now unfit for drinking purposes. Natural recharge near Brooklyn is restricted because of the large area of paved streets that prevent in-seep and channel the rainfall into sewer discharge. Now compulsory artificial recharge of industrial water into wells is practiced. This small area is an excellent example of concentrated utilization of ground water floating on salt water, and of recharge.

A similar occurrence of a floating fresh-water supply is in the coral limestones and sands of the Bahama Islands.

FLORIDA ARTESIAN PROVINCE

Florida, a low-lying state, is largely underlain by a vast storage of fresh and nearly fresh water in several aquifers, notably the Ocala

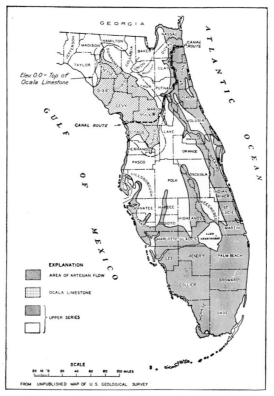

FIGURE 13–9. Area of artesian flow in Florida. (S. Paige and V. T. Stringfield, *Econ. Geol.*)

and the Hawthorne, the Ocala limestone being the important one because of its many large and small caverns. It is warped into a gentle dome with its crest near Ocala, from where it dips seaward to the east, south, and west. This great artesian aquifer extends at depth under much of the Florida coast, and even out beneath the sea.

Dakota Sandstone Artesian Province

Excellent geologic-hydrographic maps have been made that show the area underlain by artesian water (Fig. 13-9), contours on the buried Ocala limestone, and another one showing the head or pressure of the artesian water in all localities. There are three high intakes from which the water flows out radially in all directions, one in central Florida, one to the north, and one near Tampa. This great aquifer discharges into the sea, and a back-up of discharge causes large inland springs. When intersected by wells this aquifer yields copious flows of fresh water, but in the south, where it is about 1,000 feet deep, the water is mineralized and unfit for use. This may be due to infiltration of sea water or to incomplete flushing of original sea water laid down with the marine limestone. Miami, therefore, has to obtain its water supply from wells tapping ground water in surface gravels supplied by rainfall.

This great Florida water supply typifies both artesian water and a supply of fresh water above salt water down to about 2,000 feet beneath sea level.

THE DAKOTA SANDSTONE ARTESIAN PROVINCE

The third example has its intake in the domed-up strata of the Black Hills of South Dakota which supplies one of the greatest aqui-

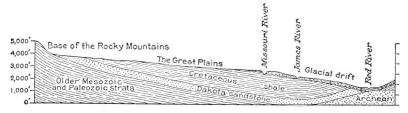

FIGURE 13-10. Section across the Great Plains, showing the Dakota sandstone artesian basin. (H. E. Simpson and W. Upham, *U. S. Geol. Survey.*)

fers, the permeable Dakota sandstone. This sandstone is 100 feet thick and inclines in all directions outward from the Black Hills, beneath the Dakotas, Nebraska, Iowa, and Wyoming (Fig. 13-10). Its impervious shale cover permits intake water from the Black Hills to be transmitted under artesian head for hundreds of miles, and it is punched by thousands of wells to supply artesian water for domestic, rural, and urban use. Owing to the large discharge the head has dropped appreciably in many places, and much water is being drawn from storage. A study of this great aquifer led Meinzer to conclude that compaction of the Dakota sandstone takes place upon withdrawal of water, and much of the early discharge is water squeezed out by such compaction.

> How can we know, we have not seen,
> But we believe the sands will yield
> Their water to the thirsty field,
> And all the desert turn to green.

SELECTED REFERENCES ON GROUND WATER

"Occurrence of Ground Water in the United States, with a Discussion of Principles." O. E. Meinzer. *U. S. Geol. Surv.* W.S.P. 489, 1923. General résumé.

"Outline of Ground-Water Hydrology." O. E. Meinzer. *U. S. Geol. Surv.* W.S.P. 494, 1923. Comprehensive outline and definitions.

"Outline of Methods for Estimating Ground-Water Supplies." O. E. Meinzer. *U. S. Geol. Surv.* W.S.P. 638–C, 1932. Full treatment of methods of estimation.

"Present Status of Our Knowledge Regarding the Hydraulics of Ground Water." O. E. Meinzer and L. K. Wenzel. *Econ. Geol.* 35:915–941, 1940. A theoretical treatment.

Ground Water. C. F. Tolman. McGraw-Hill, New York, 1937. First comprehensive textbook on this subject; principles, occurrence, and western examples.

Practical Handbook of Water Supply. 2nd Edit. Frank Dixey. Murby, London, 1950. Occurrence, principles, recovery, with application entirely to African conditions.

"Artesian Water in the Florida Peninsula." V. T. Stringfield. *U. S. Geol. Surv.* W.S.P. 773–C, 1936. A fine example of an artesian basin.

"Recent Geologic Studies on Long Island with Respect to Ground Water Supplies." D. G. Thompson, F. G. Wells, and H. R. Blank. *Econ. Geol.* 32:451–470, 1937.

"Hydrology in Relation to Economic Geology." O. E. Meinzer. *Econ. Geol.* 41:1–12, 1946. Scope of hydrology and relation to engineering.

"Artificial Recharge of Ground Water on Long Island, N. Y." M. L. Brashears, Jr. *Econ. Geol.* 41:503–516, 1946. Recharge by wells and pits to prevent overdevelopment.

"Problems of the Perennial Yield of Artesian Aquifers." O. E. Meinzer. *Econ. Geol.* 40:159–163, 1945. Procedures of investigation.

"Ground-Water Investigations in the United States." A. N. Sayre. *Econ. Geol.* 43:547–552, 1948. Problems of water supply, control, and conservation.

Recent Groundwater Investigations in the Netherlands. W. F. J. M. Krul and F. A. Liefrink. Elsevier Pub. Co., New York, 1946. Ground-water problems in a country near sea level.

Hydrology. Edited by O. E. Meinzer under auspices of National Research Council. Dover Publications, Inc., New York, 1949. A complete reference on hydrology by twenty-four experts.

Applied Hydrology. Ray K. Linsley, Jr., Max A. Kohler, and Joseph L. H. Paulhus. McGraw-Hill, New York, 1949. A text-reference book for students and engineers on general data, theory, and applications.

CHAPTER FOURTEEN

Controls of Mineral Localization

There is an old Cornish saying about ore: "There it is where it is." In a way it expressed futility, as it implied that the cause of ore's being in one place and not in another was undecipherable. A modern version, changed to "why ore is where it is" has stimulated thought as to the cause of localization of economic mineral deposits and directed observation to those geological features that control the location of mineral bodies. Many of these have been found to be structural controls, and to such an extent have they been used successfully in ore-finding that the tendency has been to assume that structural controls are almost the sole cause of ore localization. As in petroleum geology, structural traps were for long the only kind of control searched for, until the realization that there are also prolific stratigraphic traps. So, in ore-finding, there are several other kinds of ore controls for localization of ore, although structural controls dominate.

Controls of ore deposition obviously are of most importance in the localization of epigenetic deposits, or those formed later than the enclosing rocks, and these constitute the bulk of the ores of the American mountain regions. However, controls also play a part in the localization of certain types of sedimentary and secondary enrichment deposits. A number of ore controls have been discussed or referred to in Chapters 6 to 12.

STRUCTURAL CONTROLS OF MINERAL LOCALIZATION

Structural controls are both regional and detailed. The regional controls determine the broader localization of ore belts or mineral districts within wide areas barren of economic minerals. The relationship is broadly clear but vague in detail. The detailed features determine the immediate localization of ore and are the ones observed in careful studies of the deposits.

Regional controls of a broad type are mountain ranges or the roots of eroded ranges. They are sites of thick sedimentation, crustal movements, dislocations, and igneous intrusions that yield ore-forming fluids. Thus mountains are the home of mineral deposits. In a large way also, belts of igneous intrusives give rise to belts of ore deposits (see Fig. 14–1). Commonly mineral deposits occur in or are clustered about the larger types of intrusives. Few mineral deposits of the North American mineral areas are far removed from intrusives.

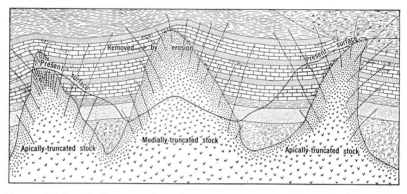

FIGURE 14–1. Ideal section through an igneous intrusive showing places where veins are formed, the parts of veins removed, and the parts left after erosion had advanced to the present surface. The content of eroded veins might form placer deposits. (After B. S. Butler, A.I.M.E.)

Large regional faults may serve as master channelways for uprising mineralizing solutions, from which they may become distributed into subordinate connecting channelways. The great Keweenaw fault in Michigan is believed to have served as a major channelway for the 300-mile length of the Lake Superior copper belt. In a broad way, folding, igneous intrusion, faulting, and metallization are related and occur in definite sequence.

Detailed structural features are the prime cause of the immediate localization of ore in one place rather than in another. They are mostly various types of openings in rocks caused by structural processes, such as fissures, shear zones, fold openings, brecciation, and other types described and illustrated in Chapters 8, 9, and 11.

SEDIMENTARY CONTROLS OF MINERAL LOCALIZATION

Features inherent in sediments are of the greatest importance in the localization of oil, gas, and water, but they enter into the localiza-

Sedimentary Controls 269

tion also of sedimentary ores, as shown in Chapter 10, as well as some other types of deposits. Stratigraphic controls are likewise regional and detailed.

Regional controls may be broad basins of sedimentary deposition in which, along with the sedimentary rocks, there may be sedimentary mineral deposits such as coal, iron ore, manganese, phosphates,

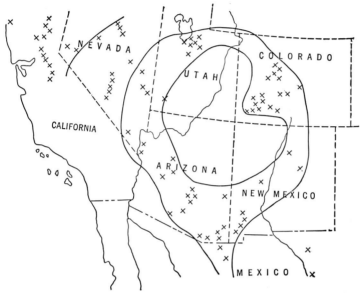

FIGURE 14–2. Barren Colorado Plateau and concentration of mineral deposits and mineral districts (marked by x) surrounding it. Heavy lines denote boundaries between stationary areas and those that have undergone deformation. (After B. S. Butler, A.I.M.E.)

salines, clays, and others discussed in Chapter 10. Plateau margins are another type of broad regional control. Some plateaus remain relatively stationary, and around their margins are areas of extensive sedimentation followed by folding and faulting, intrusion, and mineralization. This has been strikingly shown by B. S. Butler for the Colorado Plateau, with its surrounding halo of innumerable mining districts, as shown by Figure 14–2. Unconformities, or places where later sediments overlie eroded rocks—commonly, tilted ones—are sites of accumulation of residual and placer deposits, as discussed in Chapter 11.

Detailed sedimentary controls are commonly the immediate cause of localization of various types of mineral deposits and mineral fluids.

Bedding planes serve as channelways, and favorable beds localize replacement deposits (Fig. 14–3), contact-metasomatic deposits and gas, oil, and water. Included carbonate lenses may be replaced to form ore lenses, or sandstone lenses may be saturated with oil. Impervious shale beds may serve as covers to ascending mineralizing solutions, giving rise to blanket mineral deposits beneath them, or to oil or gas pools. Similarly, impervious basement rocks may prevent

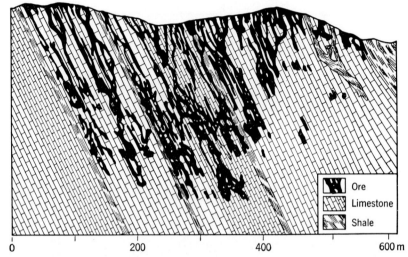

FIGURE 14–3. Bedding planes localizing lead replacement deposits (black) in tilted limestone beds. Monteponi, Sardinia. (G. Zoppi, *Mem. descr. of geol. map Italy.*)

the downward movement of secondary solutions and localize deposition of oxidized ores or supergene sulfides. Downarched impervious beds may also hold water or oil.

PHYSICAL AND CHEMICAL CONTROLS

The physical and chemical properties of host rocks play an important role in helping to localize mineral bodies, but the exact properties that aid or cause mineralization are not always evident.

Among the physical factors, permeability permits the flow of solutions along bedding and is particularly important in such types of deposits as the copper lava beds of Lake Superior or the copper deposits of Rhodesia. Brittleness permits cracking of rocks and aids permeability, as in the "porphyry coppers."

Chemical factors have long been ascribed an important role in the localization of epigenetic ore deposits. Deserving as is this role, it

Igneous Rock Controls

has also been overrated, occult attributes having been assigned to it. Some rocks, particularly the carbonate rocks, are much more congenial host rocks than are others, and often for no observable reasons. In district after district in North America, fissures carry ore in limestone but not in other rocks, a selective replacement due to chemical control. Many other rocks also display relative preferences for ore, as has been noted in the Canadian gold district, at the Homestake

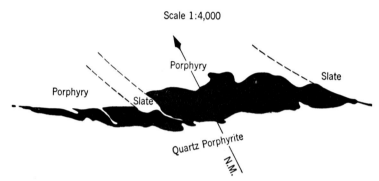

FIGURE 14-4. Massive pyrite-copper replacement deposit at Rio Tinto, Spain, illustrating selective replacement in porphyry as compared to slate. Mina de S. Domingos. (After A. B. Edge, *C. R. 14th Int. Geol. Cong.*)

mine, South Dakota, at Rio Tinto in Spain (Fig. 14-4), and in the gold deposits of Australia. In oxidized and secondary enrichment ores the character of the wall rocks is of vital importance.

IGNEOUS ROCK CONTROLS

There have long been recognized, even before the realization of the magmatic source of mineralization, certain associations between specific igneous rocks and specific kinds of ores. Some examples of these are: (1) diamonds with ultrabasic igneous rock (kimberlite); (2) magnetite, ilmenite, and chromite with basic igneous rocks (gabbro, peridotite); (3) nickel-copper ores with norite; (4) silver-cobalt ores with diabase; (5) copper ores with intermediate igneous rocks (quartz-monzonite, granodiorite); (6) corundum with nepheline syenite; (7) tin, tungsten, and uranium with granite. Most of the examples mentioned are found only with those types of rocks, but copper also occurs with basic igneous rocks in places.

It cannot always be assumed that the ore originated with the igneous rock in which it is now found, as both probably came from the same magmatic reservoir. Too often an igneous associate is assumed for a given deposit, and much unnecessary exploration work has been carried on under such assumptions. A genetic relationship is generally clear in (*a*) magmatic concentrations, and (*b*) contact metasomatism. Hydrothermal emanations offer the greatest difficulty in establishing genetic associations, as the source rock may not even be exposed to view (Fig. 14–5).

The following criteria aid in establishing a genetic association: (1) magmatic concentrations enclosed in a specific intrusive, as chromite

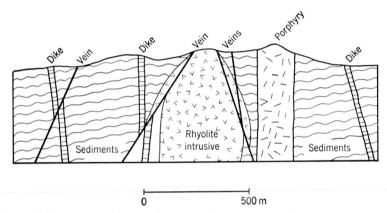

FIGURE 14–5. Structural control of veins by igneous contacts. Section across Tepezalá district, Mexico. (After A. Wandke and T. G. Moore, *Econ. Geol.*)

in peridotite, (2) magmatic injections within or near specific intrusives, such as magnetite deposits in and about anorthosite; (3) contact-metasomatic deposits adjacent only to a specific intrusive; (4) deposits confined to the periphery but absent from the interior of a specific intrusive, e.g., "porphyry coppers"; (3) spatial distribution of deposits with respect to one or more similar intrusives, e.g., tin deposits of Cornwall, England; (6) world-wide associations of specific ores and specific kinds of intrusives, such as those given in the preceding paragraph; and (7) zonal distribution of ores about a single intrusive, such as occurs at Bingham, Utah. (See subsection on "Zonal Distribution of Mineral Deposits" below.)

LOCALIZATION OF ORES IN PARTS OF BATHOLITHS

A batholith is the largest type of igneous intrusive body. Supposedly it has no bottom, and it may be scores of miles wide and scores or hundreds of miles long. The tops are very irregular, projecting upward into the overlying rocks as gentle domes or conical or elongated cupolas. W. H. Emmons and others have pointed out that the magmatic fluids would tend to rise up from within the batholith and become concentrated in these cupolas, from which they would stream upward into the outer frozen part of the cupolas and into the overlying cover rocks, as shown in Figure 14–6. The cupolas would thus be the centers of distribution and, in that way, the localizers of deposits. If erosion has reached down to the roof over the cupolas or into the cupolas themselves, mineral deposits might be

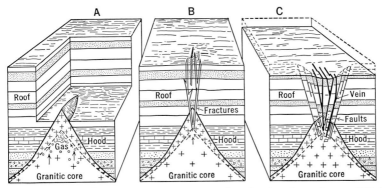

FIGURE 14–6. Cupolas on igneous intrusives and roof veins above them. (W. H. Emmons, *Gold Deposits of the World.*)

expected to be found there. But if erosion has extended deep into the batholith, then mineral deposits would not be expected to show on the surface; any that might have formed would be removed by the erosion of the cover and cupola. In a broad way it is found that mineral deposits rarely occur in the centers of large batholiths but are localized about the peripheries. It should be pointed out, however, that the cupolas would be the first to solidify, and probably before the hydrothermal solutions are formed during the later phase of consolidation of the batholith. In this case cupolas might not be localizers of ores any more than any other part of the upper periphery.

ZONAL DISTRIBUTION OF MINERAL DEPOSITS

It has been noted that in many mineral districts there is a zonal arrangement of ore minerals about and outward from an intrusive or from a hot center of the intrusive. This arrangement is marked by the formation of higher-temperature ore minerals nearest the hot spot and lower-temperature ore minerals outward and upward from it.

Thus, at Butte, Montana, there is a central zone within the Boulder batholith in which higher-temperature copper minerals occur. This is surrounded by an intermediate zone, both outward and upward, of lower-temperature copper minerals with some zinc minerals. This in turn is surrounded by an outer shell of lower-temperature zinc-silver-manganese ores (Fig. 14–7). Again at Bingham, Utah, is a conical-shaped intrusive that contains the great Utah Copper mine—the greatest "porphyry copper" deposit of the United States. Adjacent to the

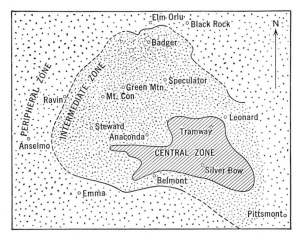

FIGURE 14–7. Zonal arrangement of ore minerals at Butte, Montana, at an elevation of 4,600 feet, showing the three zones; circles denote position of chief mines. (After R. Sales, A.I.M.E.)

porphyry in the invaded rocks are copper-zinc-lead ores; farther out are silver-lead ores, each outward zone being of lower-temperature formation. Again, at Cornwall, England, tin lies below copper ores, and copper below lead-zinc ores—an upward gradation. At Zeehan, Tasmania, tin lies in granite; a zone of iron, tungsten, and copper lies outside of this; lead-zinc-copper is next outside, and farthest out are iron carbonate, lead, zinc, silver, antimony, and manganese, the outermost zone being of the lowest temperature of formation.

This zonal arrangement is considered to be due largely to a temperature drop of the mineralizing solutions as they move outward from the hot intrusive into colder rocks. The neatness of the arrangement has led some investigators to assume that it is of general occurrence with hydrothermal deposits, and that where lower-temperature ores are found higher-temperature ores may be expected beneath; i.e., if lead-zinc-silver ores are present these should be underlain by copper ores, or even by tin-tungsten ores. However, so many discrepancies of arrangement occur that such conclusions would be unsafe and unwise. Also, the cause may not be entirely due to temperature decrease.

METALLOGENETIC EPOCHS AND PROVINCES

In a broad way metallogenetic epochs have served as a localizer of ores in time. Most mineral deposits affiliated with igneous activity represent but one event within a larger period of igneous activity. The periods of igneous activity in turn are generally associated with periods of crustal deformation that give rise to folding, faulting, or mountain building. Therefore such mineral deposits are related in time and place to periods of crustal and igneous activity that have occurred at definite periods in the earth's history. These are termed metallogenetic epochs and are not to be confused with areas within which specific metals have been relatively concentrated during one or more periods. The latter are metallogenetic provinces.

Metallogenetic epochs have been recognized on each continent, and in many cases the same suites of ores are repeated in the same metallogenetic epoch of each continent. For example, the Precambrian the world over was a period of great and varied mineral formation, probably in part because of the great length of time involved—at least three times as great as all succeeding geologic history. Three-quarters of the world production of gold, including that of South Africa and Canada, comes from deposits of this age, as well as great deposits of iron, titanium, chromium, nickel, copper, tungsten, silver, zinc, and some lead. All the Precambrian shields are areas of great mining activity. Other examples are shown on the Geological Time Table in Appendix A. Eight distinct metallogenetic epochs are recognized. Of those later than the Precambrian, two outstanding epochs, early Tertiary and late Tertiary, compass most of the great mineral deposits of the western Americas. These include the great copper deposits of the world and many rich areas of gold, silver, lead, zinc, mercury, and antimony deposits, as well as other metals. The late Tertiary is rather characteristic of a belt surrounding the Pacific Ocean but particularly evident in the western part of the two Americas.

Of the many metallogenetic provinces of the world, some of the most distinctive are the southwestern United States copper province, the Andean copper province and the central African copper province; the Precambrian gold provinces of Canada and Africa; the diamond province of central and South Africa; the Mississippi Valley lead-zinc province; the Lake Superior copper and iron provinces; the rich silver province of the central and southern Andes; the distinctive tin province of the Malayan belt and its southern and northern extensions; the prolific silver-lead and gold provinces of Mexico; the Scandinavian province of iron and nickel; the Ural province of chrome, iron, and platinum; the rich gold province of Australia; the chrome and manganese provinces of central-southern Africa; the lead-zinc-silver area of Broken Hill, Australia; and many other lesser but nevertheless important provinces in other areas. In some of the areas, such as the southwestern copper province of the United States, copper appears in different geologic ages—in the Precambrian, late Mesozoic, early Tertiary, and late Tertiary.

Metallogenetic provinces raise a puzzling problem of origin. Why should the Arizona province be so rich in copper or the Canadian shield in gold? It must mean that the intrusives of the Arizona region were particularly rich in copper, those of Canada and the Sierra Nevada in gold, and those of Idaho in silver and lead. Vagaries of erosion and depth of exposure can be eliminated. One can understand that the constant association of tin with granite or chromium with peridotite is a matter of differentiation whereby tin collects with the silicic residual magma and chromium with the basic portion. But the predominance of such diverse metals as gold, silver, copper, lead, zinc, molybdenum, and others in genetic association with identical host rocks cannot be a matter of simple differentiation associates. The underlying magmas must have been rich in one or other of the metals. Why this should be is an unsolved problem that awaits further research.

SELECTED REFERENCES ON MINERAL CONTROLS

Ore Deposits of the Western United States—Lindgren Volume. Am. Inst. Min. Met. Eng., New York, 1933. Chap. 6, "Structural and Stratigraphic Controls," by B. S. Butler *et al.;* broad features. Chap. 7, "Ores and Batholiths," by W. H. Emmons. Chap. 8, "Rocks and Associated Ores," by A. F. Buddington; relationships. Chap. 12, "Utilization of Geology by Mining Companies"; mine methods, structural controls.

Structural Geology. B. Stočes and C. H. White. Van Nostrand, New York, 1935. An elementary textbook with many mining examples of ores and structures.

"Structural Control of Ore Deposition." C. D. Hulin. *Econ. Geol.* 24:15–49, 1929. Relation of mineralization in time and place to structural deformation.

"Structural Control of Ore Deposition in the Uncompahgre District, Colorado, with Suggestions for Prospecting." W. S. Burbank. *U. S. Geol. Surv. Bull.* 906–E, 1940. An excellent discussion.

Selected References

"Physical Chemistry and Stratigraphic Problems." G. R. Mansfield. *Econ. Geol.* 32:533–549, 1937. Stratigraphic controls of nonmetallic mineral deposits.

"Structural Environment of Bendigo Goldfield." J. B. Stone. *Econ. Geol.* 32:867–895, 1937. Saddle-reef type of structural control of gold ores.

Copper Deposits of the World. 16th Int. Geol. Cong., Washington, D. C., 1936. Many fine examples of structural control.

"Structural Control of Ore Deposition in the Red Mountain, Sneffels, and Telluride Districts of the San Juan Mountains, Colorado." W. S. Burbank. *Colo. Sci. Soc. Proc.* 14, No. 5:141–260, 1941.

"Factors in the Localization of Mineralized Districts." C. D. Hulin. *A.I.M.E. Tech. Pub.* 1762, 1945. Larger structural controls for ore.

Ore Deposits as Related to Structural Features. Edited by W. H. Newhouse. Princeton Univ. Press, Princeton, N. J., 1942. Excellent examples of structural controls.

Structural Geology of Canadian Ore Deposits—A Symposium. Geol. Div. Can. Inst. Min. and Met., Montreal, 1948. Comprehensive treatment of all Canadian mining districts or mines from control standpoint.

Mining Geology. H. E. McKinstry. Prentice-Hall, New York, 1948. Chaps. 11, 12, and 13. Stratigraphic and structural controls as ore guides.

Economic Mineral Deposits. 2nd Edit. Alan M. Bateman. John Wiley & Sons, New York, 1950. Rock openings, 96–163; Chap. 6, "Controls of Mineral Localization," 303–326; Chap. 7, "Folding and Faulting of Mineral Deposits," 327-354.

CHAPTER FIFTEEN

Exploration and Exploitation of Mineral Deposits

Most of the easily discoverable mineral deposits have already been found, and it seems more and more difficult to discover new ones as the unprospected areas are gradually eliminated. Scientific prospecting is now the vogue, and it is necessary to become much more scientific if the mineral resources that we are rapidly depleting are to be replaced. Geophysical prospecting has advanced by leaps and bounds, particularly in relation to oil finding, but it has lagged somewhat with respect to ore finding. In the exploration, development, and mining of known mineral bodies, geology is playing a more and more important role, and in the valuation and appraisal of mineral deposits a knowledge of the origin, occurrence, and behavior of mineral deposits is essential. In these various phases the function of the geologist coordinates with that of the mining engineer and the geophysicist. The subject matter of these fields is extremely broad, and each would need a large book to cover them. Consequently, this chapter is merely a brief survey of these subjects in order to integrate mineral deposits with their finding and extraction.

GEOLOGICAL EXPLORATION

In geological exploration for mineral deposits, aside from geophysical exploration, the knowledge presented in the preceding chapters is applied in a practical manner to ore finding. Although geological and geophysical exploration must be utilized together for proper interpretation of results, either one may be carried on independently of the other. Generally, geological exploration is the first step, and geophysical exploration is a follow-up exploratory check on the geological findings. The geophysical findings then have to be interpreted by geologists in the light of the known geology.

Homage is due the old-time prospector who braved the hardship of wilderness and desert in the search for minerals. His discoveries of the past constitute the lion's share, and many famous mines perpetuate his memory. His industriousness has eliminated areas of obvious discovery, and his own opportunities have thereby been diminished to the point that he is rapidly becoming nonexistent. It is now common practice to supplement him or supplant him with trained geologists or mining engineers for areal prospecting. Formerly, one or two prospectors traveled afoot, with pack animal or canoe, but modern prospecting has become more organized and scientific. A practical prospector and a geologist are often paired together, or groups of paired prospectors operate in different localities under supervision of geologists who check their results at intervals or move them to more promising geological localities. In many places all prospecting is now done by trained geologists who are serviced by trucks, as in open African regions, or by airplane, as in Canada and Siberia.

FOR LODE DEPOSITS. Field exploration is initiated by a study of geological maps and reports, if such exist. Work is directed to regions where the geological conditions are favorable for mineralization. In the investigation for primary lode deposits, intrusive igneous bodies should be sought, as they may be a source of mineralizing solutions. Features that direct control of mineralizing solutions, such as fissures, faults, shear zones, or drag folds, should then be looked for and followed. If limestones are intruded by or lie near intrusives or are cut by fissures, they should receive careful attention. If deposits are known in the area the mineral habit of the region can then be utilized. Particular attention should be paid to rock alteration of the type that accompanies hydrothermal solutions. Observations are made of minerals that have been introduced into the rocks. Oxidized areas are sought to see if there are any croppings or gossans that can tell a story of the character and abundance of pre-existing sulfides.

If no geologic maps exist the formations may be broadly outlined by reconnaissance trips, and reconnaissance geologic maps are made, as was done in the large-scale geologic prospecting programs carried on in Northern Rhodesia with such beneficial results. It is common practice today to have overlapping aerial photographs made upon which to plot the geology and from which to make ground maps.

Much information can be gained by carefully examining "float" and deducing its source or direction of source. Tracing of glacial erratic boulders containing copper-gold ore followed in the direction of the glacial striae led to the discovery of important mineral deposits in Scandinavia. From the earliest time the panning of stream gravels has been utilized to trace minerals upstream to their source. It has led to lode gold deposits in many places and to rich diamond deposits in central and South Africa, and even helped in the finding of the large manganese deposits of the Gold Coast.

Geochemical investigations of soils and the leaves of shrubs have been used to detect the presence of metals in certain areas. The detection of trace metals in streams and ground water may also offer clues as to the whereabouts of possible mineral deposits.

Reinvestigation of older mining districts has proved very productive in mineral exploration. Such districts are known to be centers of metallization as distinct from the great barren areas, and intensive geologic search of known mineral controls has led to the discovery of many new ore bodies and has greatly prolonged the life of many mining camps.

After discovery comes geologic appraisal, as mere discovery does not indicate whether a deposit is of value. That must be determined by careful exploratory work to outline the character, shape, size, grade, and prospective tonnage of a deposit. Few discoveries survive this stage; mostly they are proved to be noncommercial. If they do survive and hold hope of becoming economic deposits, still further exploratory work is called for to delimit the deposit to determine the size of operations and plant capacity they can support. Upon the completion of the exploratory program the property is developed for commercial production, which is then followed by mining operations.

During the exploratory stages of lode deposits careful attention is given to detailed, large-scale geologic mapping of the rocks and structure to arrive at the origin of the deposit, cause of ore localization, shape, behavior, probable extent, and possible sites for the finding of similar ore bodies. For this purpose the mapping and study of folds, faults, intrusives, congenial host rocks, ore outlines, wall rock alteration, localization of ore shoots, particularly ore controls, and other geologic features, provide data for drawing suitable conclusions and inferences regarding ore dislocations and expectable ore.

FOR OIL AND GAS. Geological exploration for oil and gas requires different information and methods, and fundamental geological knowledge is essential. Surface expressions of petroleum such as oil and

Geological Exploration 281

gas seeps, asphalt, and bitumens in the soil are only rarely present, but of course when present are direct evidences of the existence of nearby hydrocarbons. Rather, one has to rely on geologic inferences based upon the knowledge of the origin and occurrence of petroleum. First, impossible areas are eliminated, such as areas of igneous and metamorphic rocks, and basins of continental sedimentation. Attention is concentrated on basins of marine sediments that contain both possible source beds and suitable reservoir beds. The geologic column of an area and the age of the strata must be learned. The various strata must be deciphered and the conditions of sedimentation inferred. Unconformities must be recognized, and the changing conditions of sedimentation with an advancing and retreating shore line. Structural and stratigraphic traps are sought. Such work, however, rarely by itself results in the direct finding of oil. It merely constitutes the reconnaissance to provide opportunity for more detailed investigation.

If a selected area shows merit, then more detailed geological exploration is carried on as a preliminary to geophysical work or drill testing. A geologic map is made, generally utilizing aerial photography, and the geologic formations are plotted. Fossils are collected for age determination, and the structure is carefully depicted to yield evidence of structural traps for oil at depth. If any drill holes exist in the region the cores are examined to determine the character and age of the formations passed through. Information is gleaned of the microfossils, character of sand grains, heavy minerals, porosity of beds, water, and other features. Should the geologic features look favorable for oil finding, the next step is either geophysical testing or a test drill hole (wildcat hole). Since drilling is costly, the geologic possibilities and geophysical indications are arrived at before drilling is started.

After the data from one or more drill holes become available the exploration stage continues, utilizing the subsurface data from the well logs and well cuttings. These data may yield information regarding structures that may not be revealed at the surface, or regarding potential stratigraphic traps, and on the position and character of possible reservoir and source beds. Studies are made of porosity, permeability, pressure, drill-hole temperatures, water salinity, and electrical and radioactive logs. Structure contours are drawn on key strata, and locations for new holes are spotted with respect to subsurface structure and stratigraphy. Commonly one to two years of preliminary geological exploration may precede drilling. The stage is now set for geophysical investigation.

GEOPHYSICAL EXPLORATION

The witch stick or divining rod was the forerunner of modern geophysical instruments. From the time of the ancients the quest for buried treasures has stimulated man's inventiveness to find them, and this has culminated today in the development of scientific instruments and methods whose foundation, unlike that of the divining rod, is physics and mathematics. They are used, and with great success, to locate mineral substances or favorable mineral environments that lie within the earth. Such rapid advances have been made in geophysical prospecting, particularly for oil but also for metal, that it has grown to be a science in itself, with a voluminous literature. The methods are also utilized in engineering and engineering geology.

The chief methods and adoptions used for exploration of oil and metals are outlined here in principle only, and details of instruments, procedure, and interpretation will be found in the extensive literature on this subject. The various methods may be grouped as follows:

1. Magnetic
2. Electrical
3. Electromagnetic
4. Gravimetric
5. Seismic
6. Radiometric

MAGNETIC METHODS

The principle of the well-known horizontal and vertical pull of the earth's magnetic field upon the compass needle is utilized in magnetic types of geophysical prospecting. It was early learned that magnetic bodies in the earth caused local deviation of the compass needle. Conversely, a local deviation of the needle indicates a disturbing magnetic body.

The simplest magnetic survey uses refined horizontal needles and vertical dip needles. Readings are taken at specific intervals along measured parallel lines, and the position of an attracting body is indicated and plotted. This simple but very effective method is used only for substances that themselves exert magnetic attraction or that are enclosed in rocks that, because of their mineral composition, exert a stronger magnetic attraction than surrounding rocks lacking such minerals. Thus bodies containing magnetic minerals such as magnetic iron ore, pyrrhotite, nickel, and cobalt may be detected; also rocks

Magnetic Methods 283

high in iron, such as basalt lava that may contain copper. A refinement of the common magnetic needle is the Hotchkiss Superdip, which is a much more delicate detector of magnetic substances.

MAGNETOMETER. A more widely employed magnetic method utilizes not the absolute magnetic value, as does the magnetic needle, but the variations in the strength of the horizontal and vertical com-

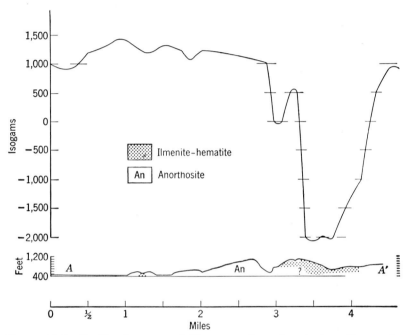

FIGURE 15-1. Profile of magnetic field across a large ilmenite-hematite ore deposit in basic igneous rock at Allard Lake, Quebec, made by "flying magnetometer" at a height of 500 feet. (Private survey, 1948.)

ponents of the earth's magnetic field. Different rocks show different magnetic susceptibilities that can be measured by the delicate instrument called the magnetometer. Magnetic permeable rocks such as basic igneous rocks (paramagnetic substances) apparently increase the earth's magnetic field. Others, such as many sediments (diamagnetic substances), decrease the earth's magnetic field, and it is these variations that are measured. The magnetic susceptibility of basic igneous rocks is 10 to 100 times stronger than that of granite and some sediments, and that of some dolomites is 20 to 50 times less than the magnetic susceptibility of some shales. These relative differences are measured both for the vertical and horizontal components and can be detected deep beneath the surface.

Readings are easily made along parallel lines or on a squared network, but considerable calculations are required. The determined local vector is resolved into horizontal and vertical vectors, and these are plotted on a map or cross section (Fig. 15–1). The features that cause the anomalies lie mostly at depths of 2,000 to 15,000 feet, and the horizontal effect is felt within a distance of about twice the depth. This method is utilized for detecting mineral bodies beneath the surface and for detecting faults, rock contacts, and folded rocks. It is utilized, therefore, both for metal and petroleum exploration. A magnetic anomaly, of course, does not necessarily mean a mineral deposit.

FLYING MAGNETOMETER. A variation of the magnetometer method utilizes a specially designed magnetometer towed behind an airplane, giving thereby a continuous record along prescribed parallel lines of flight at desired elevations. The procedure is rapid, but the calculations required are lengthy. The method eliminates the necessity of cutting out and measuring ground lines. It is extremely effective in disclosing underground mineral bodies, faults, rock contacts, rock differences, and structure and is now widely employed in metalliferous and petroleum exploration.

ELECTRICAL METHODS

Electrical methods are chiefly valuable in exploration for metallic mineral deposits, but some are utilized for petroleum exploration, logging oil wells, deciphering structure, and in engineering for determining the depth of bedrock. Several methods are based upon the conductivity of minerals, rocks, and fluids. In some, natural earth currents are measured, and in others, electric currents are introduced into the ground and measured.

NATURAL CURRENT METHODS. This method, also called self-potential and spontaneous polarization, requires no outside energizing force. Ore bodies in the ground set up small natural currents that can be measured and plotted, and the presence of hidden mineral bodies is thereby detected. The upper oxidized part of an ore body is chemically more active than the lower part, and this upper part sets up electric currents that flow down through the ore body and up through the surrounding rock, spreading outward because of the resistance of the rocks, as in Figure 15–2. The currents can be traced on the surface by points in the ground connected by a sensitive galvanometer. Curves can be drawn connecting points of equal potential, the negative center indicating the position of the ore body. The method is not well suited to arid regions or to regions where the soil is deep. There must be present in the ore body at least 5 percent of conductive metallic sulfides.

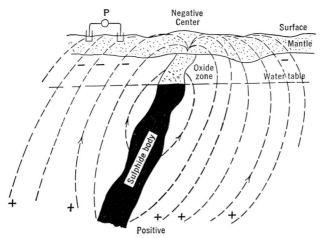

FIGURE 15-2. Self-potential or spontaneous polarization produced by oxidizing sulfide body. Currents are measured and plotted.

EQUIPOTENTIAL METHOD. This simple electrical method has several variations. A common one is to lay out two grounded bare wires, as in Figure 15-3, connected to a generator (or batteries and a buzzer), and to pass current from one bare wire to the other. If the intervening ground is uniformly conducting, the current drop will give the equal potential dotted lines of Figure 15-3. However, if a good conducting ore body is present, the lines of equal potential will be distorted, as shown by the solid black lines. Two operators stick steel rods connected by a short cable into the ground, and earphones record no buzzing when both rods are on points of equal potential. The points are pegged and surveyed. Two grounded points can be used instead of parallel grounded wires. This method has been quite successful in ore exploration, notably in Sweden, Canada, and Newfoundland.

In the "leapfrog" variation of Eve and Keyes the drop in potential between parallel base wires is measured by means of two porous pots inserted in the ground and connected by a 100-foot cable. First, one pot is placed on the bare wire and the other in the ground 100 feet

FIGURE 15–3. Equipotential method, using parallel grounded wires connected to current supply. Dashed lines represent normal lines in absence of a conductor, and solid lines are the measured distortions by an underground conductor.

away, and the drop in potential is noted; then the first pot is swung 100 feet beyond the second one, and the next drop recorded. Repetition of this gives lines of equal potential.

RESISTIVITY. This much-used method is applied to ore finding, logging oil wells, and determining depth to bedrock for foundations, and other purposes. It is based on the different electrical resistance of different rocks, ores, and liquids.

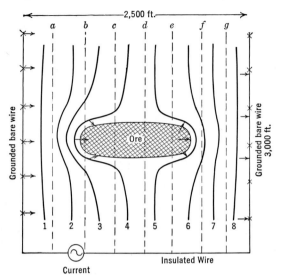

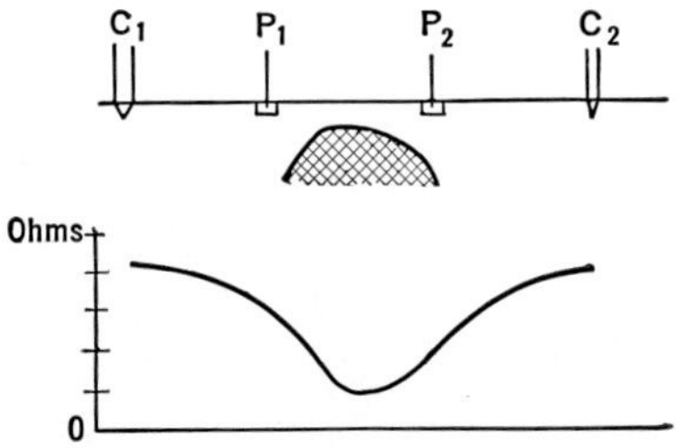

FIGURE 15–4. Resistivity measurement with four electrodes, showing drop in resistance through a good conductor.

In its simpler form, four grounded electrodes are spaced equally apart in a straight line. Current is introduced at one end electrode and taken out at the other end and measured, and the resistance offered by the ground is calculated (Fig. 15–4). The depth of penetration is about equal to the spacing between electrodes, which can be varied. Ores, being good conductors, offer less resistance than rocks, and there is a great difference between rocks. For deep work,

Electrical Methods

more complicated methods and instruments are required. The method is utilized for determining bedrock, geologic structure and strata, and ore bodies.

ELECTRICAL CORING. In oil exploration the resistivity method is used to determine the geologic strata penetrated by a drill hole, and also

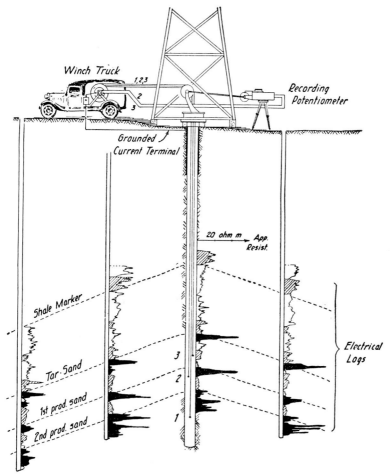

FIGURE 15-5. Electrical well-coring arrangement (schematic). (C. A. Heiland, *Geophysical Exploration.*)

beds occupied by oil, salt water, and fresh water (Fig. 15-5). The hole must be free of casing; drilling mud or water permits electrical connection with the rock walls at different levels, and the electrical resistance is read. The resistance of a porous rock containing salt water is much lower than one with oil. The results are plotted as graphic drill-hole logs, which permit correlation of oil sands and other formations from one hole to another, thus revealing structure.

Measurements of spontaneous potentials are taken at the same time and place and provide additional data. Unknown oil sands have been detected by such electrical coring.

ELECTROMAGNETIC METHODS

Electromagnetic methods are mostly confined to ore finding and are the most favored because they are more precise and yield greater information regarding the shape, size, and depth of hidden deposits. They are also slower and more costly than the methods described above. They have, however, been used successfully in many countries. The methods utilize the well-known principle that a current passing along a conductor sets up an induced current around the conductor, becoming weaker with distance from it. If a conducting ore body lies within the induced field, it sets up a secondary induced current around the ore body, and this can be measured. Two variations are common.

LOOP METHOD. A loop of insulated cable is laid on the ground, as in Figure 15–6, through which a high-frequency alternating current

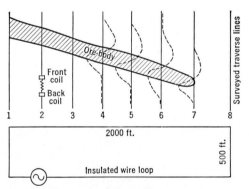

FIGURE 15–6. Electromagnetic method, utilizing loop of insulated wire on ground. Measurements are made along surveyed lines 1–8, and distortions by an underground body are plotted as curves.

is passed. This sets up an induced current in the ground which diminishes away from the loop, and measurements of the strength of the induced current are made along surveyed lines. If a conducting ore body lies hidden, the secondary induced current set up by it gives a boost to the primary induced current, and stronger readings are recorded. If the difference is large, it indicates that the conducting body either is large or is close to the surface; the approximate depth can be determined. This loop method is adaptable to almost any type of ground—rocky, sandy, mountainous, or ice, deep snow, or even underground mine workings.

SINGLE-CABLE METHOD. An alternative method utilizes a long, straight, insulated cable, well grounded at both ends, and measurements are made along straight lines at right angles to it. The method is faster than the loop for surface work, as the cable may be 1 to 2 miles long. The single cable can also be strung underground along shafts or tunnels, and readings made in workings at right angles to it, provided rails, wires, and pipes are disconnected. The single cable is less penetrating than the loop but is more sensitive and indicates many weak conductors that may not be ore bodies.

Electromagnetic methods simply record conductors, and these may be ore bodies but they may also be water-bearing strata, damp fault gouge, graphite slates, or other conducting bodies, and careful geological interpretation is essential.

GRAVIMETRIC METHODS

Gravity methods employ the well-known principle of gravitative attraction. The force of gravitation that holds the heavenly bodies in their orbits or causes an apple to fall on the ground also exerts an attraction between masses on and in the earth. A plumb bob normally points toward the center of the earth, but an adjacent mountain mass will pull it toward itself and deflect it from the vertical. Similarly, a heavy body within the earth will attract the plumb bob toward it, the deflection depending upon the size of the heavy body and its distance away. Conversely, such deflections denote the presence of a heavy or light body. Also, a given object weighs more if above a heavy mass within the earth, and less over a light mass. These principles are employed in oil exploration in three ways.

PENDULUM. A pendulum operates faster where gravity is greater, or vice versa, and the exceedingly small differences are detected by special apparatus. Two or more pendulums are employed simultaneously on a series of stations. The gravity determinations are plotted to give a gravity contour map, of which the high and low closed contours indicate heavier or lighter subsurface bodies. Anticlines or salt domes may be indicated.

TORSION BALANCE. This exceedingly delicate instrument was one of the first geophysical instruments introduced into petroleum exploration. It measures differences of gravity, or the horizontal rate of

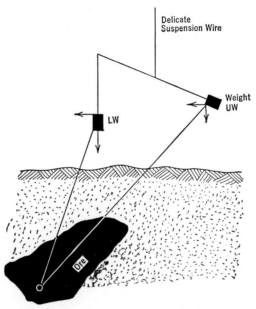

FIGURE 15-7. Schematic representation of torsion balance whose suspended weights are attracted toward a heavy body, thus twisting the suspension wire.

change of gravity, rather than the direct measurement of gravity itself. When delicately suspended weights are placed north of a heavy mass, the weights are deflected south; if the weights are south, the deflection is north; if directly above the heavy mass there is no horizontal deflection. This is the converse if the mass is a light one. The schematic arrangement of the suspended weights is shown in Figure 15-7. Because a suspended bar carries the heavy metal weights, the gravitational pull exerts a torsion on the wire, twisting it toward the heavier subsurface body, and this torsion is measured. The instrument is so sensitive that the weight of a man standing near it can be measured, and the torsion indicates his position. It can detect a weight of 15 pounds at a distance of 6 feet and detects changes in the order of 1 part in a million million.

The instrument detects the difference between limestone and sandstone, salt and rock, gravel and bedrock. Many hidden oil structures and a great many salt domes with associated oil have been discovered by its use. The drawbacks are that the method is slow, costly, and not adapted to hilly terrains. It is now largely supplanted by the gravimeter.

GRAVIMETER. The gravimeter is a relatively simple though delicate instrument by which gravity is compared with an elastic spring force. A standard object is weighed with great precision in several localities. It will weigh more over heavy rocks than over light ones and thereby indicates the density of subsurface materials. The changes are in the order of 1 part in 10,000, or 1 part in 10 million of the total value of gravity. The gravimeter is now widely used and is much faster and less costly to operate than the torsion balance.

SEISMIC METHODS

Seismic methods of prospecting utilize earthquake knowledge. An earthquake in Japan sets up waves that travel with high velocity and are measured in San Francisco, New York, or London. The exact time of arrival is automatically recorded. Dense rocks transmit waves faster than light ones, and deeper rocks are denser than shallow ones; hence the arrival time of the waves indicates the kind of material through which they travel.

In seismic prospecting the same principles are utilized. The earthquake is a charge of buried explosive, and the waves are recorded on portable seismographs. The explosion time is transmitted by radio, and the arrival times are recorded on a moving film. Two kinds of waves are transmitted, one a reflected wave or echo, the other a deeply penetrating refractive wave, and either may be used.

REFLECTION METHOD. The principle of reflection is the same as echo sounding from a ship to the ocean bottom, except that the wave is a different kind and is more complicated. By timing the interval between explosion and reflecting record, the depth to the reflecting strata can be calculated when the velocity of wave travel is known. Limestone overlain by shale makes a good reflecting surface (Fig. 15-8). With a number of receivers in use, the reflected wave can be distinguished from the refracted wave because the reflected arrives simultaneously at almost all points and the refracted arrives at different times and can be screened out. More than one reflecting surface is present, giving interfering waves, particularly in the weathered zone.

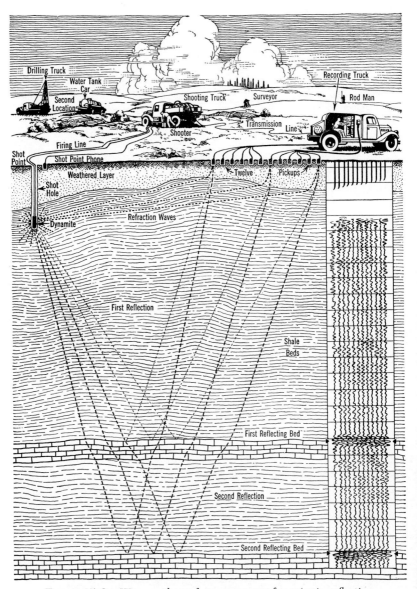

FIGURE 15–8. Wave paths and arrangements for seismic reflection.

Complicated equipment and procedure are necessary to screen out the undesired reflections. The recorder is generally less than 3,000 feet from the shot point.

This method is widely used in oil exploration by American companies and is highly successful in accurately revealing subsurface structure and salt domes.

REFRACTION METHOD. The explosion generates waves that travel along the surface (slow waves) and through deeper strata at higher speeds, which vary with the transmitting rock. A wave in overburden may travel only 1,600 feet per second, but deeper ones may be 9,000 feet per second. By spacing recorders at equal linear intervals a travel-time curve can be established. A deviation of speed from the travel-time curve gives information regarding the transmitting stratum. Thus a short-time path will not fall on the mean-time curve and indicates a salt dome or other fast-transmitting rock that lies between explosion and recorder. The recorders are placed fan-shaped, and a network of fans covers the area to be explored. In such an arrangement the shot will be recorded approximately simultaneously on each recorder, but an abnormal speed becomes immediately evident and indicates an abnormal stratum worthy of further investigation.

The refraction method has been highly successful not only in petroleum exploration but also for bedrock determinations, gold and diamond gravels, ground water, tracing coal seams through faulted blocks and under the sea, and for other purposes.

Seismic prospecting has reached an expenditure in excess of $10,000,000 a year in the United States and is the method most used in oil exploration. Vast quantities of oil have been discovered through the revelation of subsurface conditions.

RADIOMETRIC AND OTHER METHODS

Radioactivity can be detected by various devices utilizing ionization chambers. The Geiger counter can detect the presence of uranium and other radioactive minerals, although some of the emanations are blanketed by a relatively few feet of soil. Most rocks are radioactive, the acidic more so than the basic rocks, and such radioactivity can be measured by the Geiger counter and other, more sensitive equipment. A more sensitive supercounter utilizes gamma rays and readily detects many ores and differences in rocks. Counters can be made to yield continuous records by being airborne or carried in an automobile.

Modern petroleum practice records gamma rays and neutrons through the casings of oil wells and yields differences in rock strata and changes in fluid contents. From these are constructed well logs

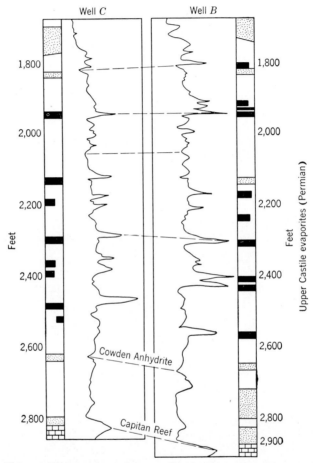

FIGURE 15–9. Radioactive logs of two wells ½ mile apart in Copper Pool, New Mexico, showing gamma-ray correlation, and sharp indications of potash (black) and anhydrite (stippled) beds; white is salt. (Courtesy of W. L. Russell.)

that resemble electric coring logs (Fig. 15–9) and are utilized for strata correlation, porosity, and oil and salt-water indications.

Soils overlying areas of high radioactivity, and the nearby ground water, may also contain radon, a gaseous radioactive emanation, which can be detected by suitable instruments. A number of faults were traced in Japan by radon emanating from them.

Valuation of Mineral Properties

Airborne equipment for detection of radioactive emanations holds high future hope for mineral exploration and geological mapping. Reflections of radio waves from subsurfaces have also been tried.

Geothermal gradients, established by careful readings in drill holes, sometimes give indication of subsurface structure. Temperatures in excess of the normal regional gradient may be due to frictional heat developed by folding, faulting, fluid flow, chemical reactions, or radioactivity.

Geochemical analyses have been used to distinguish trace elements in soils, streams, ground water, and the ashes from leaves and twigs of trees and shrubs. For example, measurable quantities of zinc have been found in certain types of shrubs growing over zinc deposits, and soil areas have been mapped on the basis of the zinc content. Analyses of saline waters encountered in oil regions are also utilized in oil finding.

VALUATION OF MINERAL PROPERTIES

After a mineral deposit has been disclosed by exploration and has been partly developed, the geologist is called upon to make an appraisal or valuation of the property. He utilizes the knowledge of the shape, occurrence, controls, and origin of the deposit to infer and predict what lies beyond the parts that can actually be seen or beyond the limits of a drill hole. Valuations may be desired to estimate the size of plant to be installed, rate of production, justifiable initial expenditures, or for purposes of sale, purchase, merger, financing, taxation, or investment.

MINING PROPERTIES

In a mine the exploration and development disclose what is termed proved or known reserves. In addition, ore that lies under one's feet and beyond the mine workings but has not been proved is called probable ore. A third category may include inferred ore, which in the judgment of the appraiser may reasonably be expected in the future. Generally, the value of a property rests more on the probable ore extensions than upon the blocked-out ore. The probable extensions in turn depend upon the kind of deposit, the behavior of the ore body, its known persistence, the ore controls, delimiting formations, and other geological features that must be evaluated.

In its simplest terms the gross value of a mining property is the amount of metal or mineral it is estimated to contain multiplied by the probable market price of the product. Such a figure, however, is meaningless and too often is the one given out in promotional mining prospectuses. It might be said, for example, that a property has a gross value of $5,000,000 arrived at by estimating 5 million tons of ore carrying $1 of copper per ton of ore. But if it would cost $10,000,-000 to win this copper, obviously the property has no value at present. What is desired is the net worth of property. This depends upon the rate of production, the life of the property, the assay value of the ore, the cost per unit for extraction, the price of the product, and the net profit per unit. An oversimplified example is as follows:

> Equipped gold mine: estimated tonnage, 1,000,000 tons.
> Annual production, 100,000 tons; life, 10 years.
> Grade, $10.76; extraction, 93%; recovery, $10 per ton.
> Costs per ton, $7.00; profit per ton, $3.00.
> Annual profit, $300,000; gross profit, $3,000,000.

However, the gross profit of $3,000,000 is not collectible until the end of 10 years and does not represent the present-day value; it must be discounted to the present. Thus, on the basis of a 6 percent return and amortization at 4 percent, the annual income is multiplied by a factor of $6.94 (from Inwood's interest tables), which equals $2,094,-000, and this represents the present value of the ultimate gross profit. If $500,000 is necessary to equip the mine for production, the present value is $1,594,000. If $100,000 shares are issued, the present value per share is $15.94, or approximately what the market price should be. Other refinements of interest or equipment cost could be included.

OIL PROPERTIES

Similar principles are applied in the valuation of oil properties. The primary data are estimates of the yield of oil in barrels per acre. This is obtained by figuring the thickness of the oil reservoir bed obtained from drilling, the acreage underlain by the oil bed, the porosity of the reservoir rock, and the recovery yield. Estimates are made of the cost of development, equipment, and extraction, which gives an estimated profit per barrel of oil. The predetermined daily or annual rate of production divided into the estimated total barrels of oil contained in the property gives the life of the pool. The annual profits are then reduced to present worth.

EXTRACTION OF METALS AND MINERALS

The mining of metals and minerals involves the following mining and metallurgical operations: mining or quarrying, milling, ore dress-

Mining Methods

ing or beneficiation, smelting, and refining. The number of necessary is determined by the nature of the ore and the fin; sired. Thus, a copper ore must be mined, milled, smelted, a to obtain pure copper, but mica need only be mined and ṛ and oil extracted and refined. Often, the mining methods ımposed because of the shape or character of the ore body and the nature of the enclosing rocks will determine whether a given deposit may have commercial value. Similarly, the metallurgical treatment of the ore made necessary because of its mineralogy or composition may be so costly as to render it profitless. Each deposit is a unit of itself, requiring its own particular mining method or metallurgical treatment. A brief outline of the different mining and metallurgical operations, omitting details, follows.

MINING METHODS

Mining methods are divided into two large groups: surface and underground. Sometimes the two are combined. Surface methods are preferred where practicable because they are generally lower in cost, can employ larger-scale operations, and are more flexible. Many low-grade ore bodies cannot be profitably mined except by large-scale, low-cost surface methods.

SURFACE METHODS. Surface methods are determined by the volume, grade, and width of the mineral body and the depth of its overburden.

Simple *pitting* and *trenching* are used for scattered surface minerals such as some residual products, placer gravels, and weathered pegmatite minerals. *Quarrying* is generally employed for removal of stone, pegmatites, and residual deposits, and it may be on a small scale or on a huge scale employing benches, power shovels, and railroad trains.

Stripping is employed for near-surface, flat or gently inclined sedimentary beds such as coal, bauxite, clay, phosphate, and other materials. The overburden must not be too great, else the value of the mineral cannot sustain the cost of stripping. In coal stripping some of the largest power shovels made excavate the overburden to expose the coal (Fig. 15–10). The coal is then excavated, and overburden from a parallel swath is dropped into the coal excavation, exposing more coal. Each new stripping swath parallels the last coal excavation. This method is low in cost, high in productivity per man, and recovers all the coal, which underground methods do not.

Open cutting is employed for near-surface deposits of large volume and low-grade or low-value minerals. It is commonly employed for iron, phosphate, clay, and "porphyry copper" deposits. The open cuts (Fig. 15-11) may be several thousand feet across, and the scale

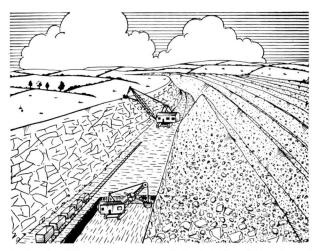

FIGURE 15-10. Stripping operation in which one shovel removes overburden, transferring it to mined-out area, while another shovel excavates a strip of newly exposed coal and loads it into cars. (*Mining Journal*, London.)

FIGURE 15-11. View of open cut of the "porphyry copper" Clay mine of the Phelps Dodge Corporation, Morenci, Arizona, taken in 1950. (Courtesy of Phelps Dodge Corp., New York.)

of operations up to tens of thousands of tons of ore per day; the Utah Copper mine has yielded 125,000 tons of ore and 200,000 tons of waste in a day. The overburden is stripped and transported to waste dumps.

Mining Methods

The ore is blasted if it is hard, and power-shoveled into railroad trains for transport to the mill or shipping point. Less commonly, ore is removed from depressed open cuts by tramways. Much waste rock has to be removed, but open-cut mining is very low in cost.

A variation of open-cut mining, combined with underground haulage, is known as *glory-holing*. In this the ore and waste are broken in open cuts and dropped into ore and waste chutes at the bottom of the glory hole from which they are drawn off by underground trains and transported to the surface. This is generally more costly than open cutting but is necessary when pits become very deep or the slopes cannot be held back. There are many deposits that yield a profit by open cutting but would become unprofitable by underground methods.

Hydraulicking is a method of excavating placer deposits. Overlying gravels are washed out by powerful jets of water, and the mineral-bearing gravels are sluiced through sluice boxes with bottom ribs that trap the heavy minerals.

Dredging is a more elaborate and larger-scale method of placer mining. Huge dredges are built in the gravel area and dig themselves into artificial ponds. Mechanical excavators dig out mineral gravels on one side and stack rejects or waste on the other side and so advance into new dredging ground. They can dig 100 feet below water line. The gravels are treated on the dredge for their desired minerals, mostly gold and tin. They can handle from 300 to 500 cubic yards of material per hour and at a cost as low as 3 cents per cubic yard.

UNDERGROUND MINING. Individual underground methods have to be devised for each mine, so details will be omitted here. In essence, entry is made into the ground by vertical or inclined shafts or tunnels. Horizontal workings or levels are driven off from the shaft at regular intervals (Fig. 15–12), from which raises are driven upward and stopes for ore removal are opened up. Ore above a level is drilled, blasted downward, and drawn off through chutes into cars, trammed to the shaft, and hoisted to the surface (Fig. 15–13). Underground mines produce up to 20,000 tons of ore per day, and the deepest mines have reached depths of over 9,200 feet. Ancient Japanese mining is shown in Figure 15–14.

In *fissure veins*, a stope is opened up a few feet above a level, the roof ore is drilled and blasted down onto a heap. Just enough of this is drawn down through chutes at the bottom of the stope to permit the miners to stand on the heap to drill the roof. If the walls are strong, the broken ore remains filling the stope until that block is mined up to the next level. The mining progresses upward to utilize gravity withdrawal. If the walls are cavey, timbering is necessary to support them, and ore is withdrawn as blasted (Fig. 15–15). The

FIGURE 15–12. An ancient method of underground mining with two levels. (Agricola.)

excavations are generally filled with waste rock or sand to prevent future caving that might endanger other workings.

Large irregular deposits are caved, in which gravity instead of blasting does the ore breaking. A large block of ground is undercut by suitable workings and penetrated and flanked by upward-branching raises, which are drilled and blasted (Fig. 15–16). The whole mass of the delimited block then caves, and the caving crushes the ore, which descends into underlying chutes in solid rock, where it is drawn off into cars. This allows the mass to cave further, and the surface is allowed to subside as the ore is withdrawn. Most of the rock breaking is thus accomplished without much blasting. An adjacent block is then started upon the completion of the first one. This method permits extremely cheap mining on a large scale with a minimum of drill-

FIGURE 15–13. Medieval method of hoisting ore. (Agricola.)

ing, blasting, and ore handling and is used in all underground "porphyry copper" mines and similar types of deposits.

In the mining of *flat coal seams and ore beds* gravity cannot be utilized, and the chief problem is supporting the roof during mining. Coal seams are mined by the *room and pillar* or the *longwall* methods. In the room and pillar method (Fig. 15–17) an area is divided into a series of rooms separated by pillars that hold up the roof while the rooms are being mined. The face of coal is undercut and the roof pressure cracks it off, until the face retreats to the end of the room.

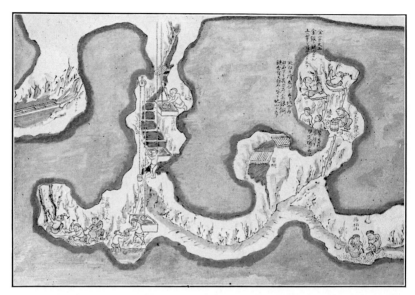

FIGURE 15–14. Ancient gold mining at Sato mine, Japan, showing underground passages, mining, ore sorting, and water disposal. Probably about 9th to 11th century. Copy from old drawing in author's possession. (Courtesy Mitsubishi Mining Co., Tokyo, Japan.)

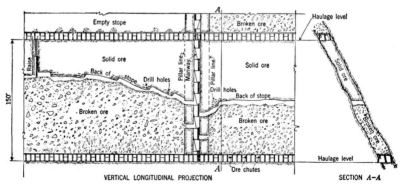

FIGURE 15–15. Underground mining method used for veins in which broken ore is left in stope to permit drilling of solid ore above (shrinkage stoping). (C. F. Jackson and J. H. Hedges, *U. S. Bur. Mines.*)

Mining Methods 303

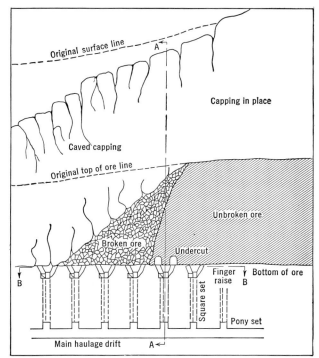

FIGURE 15–16. Underground caving method used at the Ruth mine, Nevada.
(C. A. Wright, *U. S. Bur. Mines.*)

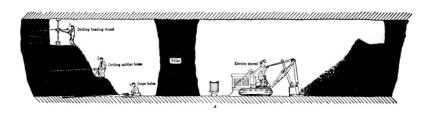

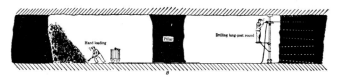

FIGURE 15–17. Mining methods used for a horizontal coal seam. (C. N. Anderson, *U. S. Bur. Mines.*)

Later, the pillars are partly recovered by using timber or stone support for the roof. Considerable coal is lost in this method.

In the longwall system (Fig. 15–18) haulage tunnels are driven to the farthest limit of mining, and from a peripheral tunnel the coal face is retreated toward the shaft where a pillar of coal is left to support it.

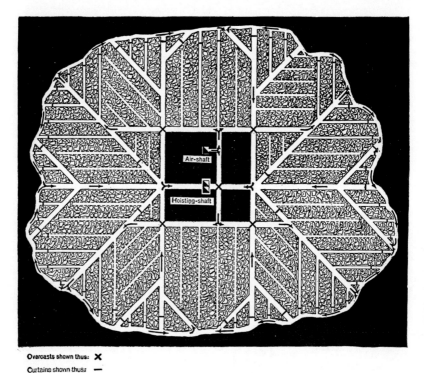

Figure 15–18. Longwall method of flat-seam coal mining by which coal is mined from periphery toward shaft. Arrows refer to ventilation control. (R. E. Swift, U. S. Bur. Mines.)

As the coal face retreats the roof slowly settles down over the excavated area. This method results in a small loss of coal.

OIL AND SULFUR. Oil is extracted by drilling oil wells, properly spaced, over the property or oil pool to the oil-producing horizons, of which there may be more than one. The depths may exceed 15,000 feet, but most wells range between 5,000 and 10,000 feet. Drilling is mostly done by a rotating bit suspended on a drilling rod (Fig. 15–19). In California a 2,000-foot hole can be drilled in 5 days, and a 9,000-foot hole in 80 days. "Drilling mud" of powdered heavy minerals

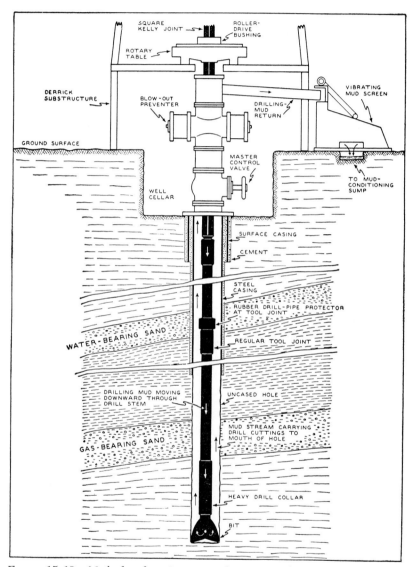

FIGURE 15–19. Method and equipment used in rotary oil-well drilling. (*Works Prog. Admin.*)

suspended in water is forced to the bottom of the hole to carry back the rock cuttings. The pressure of its weight also prevents water, gas, and generally oil from entering the hole during drilling. A tube of steel casing is driven into the hole to prevent caving and water inseep,

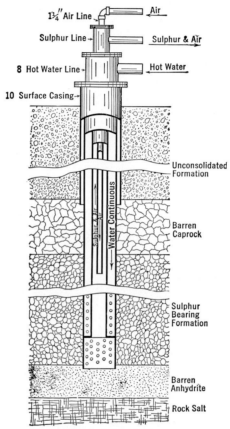

FIGURE 15–20. Method of Frasch process for extracting sulfur. (W. Lundy, Indus. Min. and Rocks.)

new lengths being added at the top as the hole is deepened (Fig. 15–19). Perforated casing permits inflow of oil and gas. The oil generally flows out naturally because of the gas pressure, which may exceed 5,000 pounds to the square inch. If gas is lost or is absent, as in old wells, the pressure drops and the well may have to be pumped. The average of some 425,000 producing wells in the United States is 10 barrels of oil a day. Generally, less than one-half of the oil originally present is recovered. After oil flow ceases, gas or water may be

Milling or Mineral Dressing

forced in some wells, made to flow through the oil sands, and be withdrawn from other wells, thereby flushing the beds and part of the residual oil. Over 1,000,000 wells, productive and nonproductive, have been drilled in the United States.

The prolific sulfur deposits on the tops of some salt domes are mined by the Frasch process (Fig. 15–20). Inside a well driven to the sulfur is a hot-water pipe, and inside this is a return sulfur pipe carrying an air pipe. The hot water melts the sulfur, which is forced by compressed air up the discharge pipe to the surface, where the melt is piped to the stockpile to solidify.

MILLING OR MINERAL DRESSING

After ore is mined it must then be milled or treated or processed to separate the desired from the undesired minerals. Various means are employed to accomplish this, depending upon the character of the ore and the desired end product. Only a few minerals do not have to be treated.

Most ores have to be milled to obtain the desired product in finished form or to yield a concentrated product that is amenable to further treatment. The nonmetallic substances are generally utilized as minerals, but the metallic ones are mostly processed to yield their metals, generally requiring several stages of treatment. Since the milling of ores always results in some loss, high-grade metallic ores are mostly shipped directly to the smelter or to the finishing plant to avoid milling losses. This, however, is not economical with low- or medium-grade ores since freight charges and smelter cost would generally be considerably greater than the value of the raw ore. Consequently, such ores are concentrated, during which the desired minerals are reserved and the waste rock or gangue minerals discarded. From 5 to 30 tons of raw ore are thus reduced to 1 ton of concentrate, thereby saving freight and smelter charges that would have had to be paid on the discarded material. Ores that contain several metals may have to be milled, even if high grade, in order to separate the metallic minerals for different subsequent treatments.

Practically all milling processes have one initial step in common, namely preliminary crushing, which yields coarse lumps that are reduced to smaller lumps, followed by grinding, generally to a powder.

AMALGAMATION. This simple, low-cost process is used only for free-milling gold and silver ores. Crushed ore is ground until the small gold particles are freed from the matrix, and the ground pulp is passed over metal plates coated with quicksilver. The gold and quicksilver form amalgam, which is scraped off, and the gold is separated from it by retorting. Quicksilver is also generally introduced into the grinding mills. In places rotating amalgamation barrels are used instead of plates. Coarse gold is commonly recovered before amalgamation by passing the ground pulp over blankets.

CONCENTRATION. The oldest and simplest type of concentration is *hand sorting*, which still has to be employed for certain minerals that cannot be ground. Many metallic ores occur in coarse chunks easily separated from the gangue minerals by coarse crushing. The broken ore travels on a moving belt or a revolving circular table from which the ore pieces are picked off by hand. In some cases the waste pieces are picked off. Hand sorting in many cases of nonmetallic minerals is the only processing step. Some minerals are fragile or too easily harmed to be treated by any other method as, for example, mica, long asbestos, and gemstones.

Gravity concentration is employed for simple ores with light and heavy minerals, and for a long time was the only mechanical method employed. Simple ores, after crushing, are separated in jigs, where the ore passes over a screen subjected to pulsating water, which jigs the heavy minerals to the bottom while the light ones are swept away for discard. Ground ores are classified as to size and passed over concentrating tables, which are tilted tables containing wooden riffles, subjected to quick pulsations. The heavy minerals gravitate to the higher side, the lighter to the lower side. The discharge is reground and passed over other tables. The concentrates are dewatered and shipped.

Oil flotation is the process now used for most metallic and a few nonmetallic ores. The method is fast, economical, and gives high extraction. Powdered ore is classified as to grain size and receives a little oil and chemicals. The pulp flows to flotation cells where a froth of air bubbles, coated by oil, is created. The metallic mineral powders adhering to the oil film are carried upward and swept into troughs, while the waste is discarded. Passage through several cells removes 90 to 95 percent of the metallic minerals. Flotation can also separate different sulfides from each other.

There are a few other concentration methods. One is *sink and float*, in which the lighter waste constituents of crushed ore are floated off by heavy solutions, leaving the heavier ore minerals. In *spiral* concentration, heavy minerals are separated from light ores by downward

Smelting and Refining

passage through a vertical "Humphrey" steel spiral. *Magnetic* concentration separates magnetic from nonmagnetic ores by means of magnets and traveling belts. Asbestos fibers are separated from matrix by *blowers,* a form of air winnowing or air concentration.

LEACHING. Some minerals not amenable to other treatment processes can be recovered by leaching. This applies to minerals soluble in common solvents, such as some of the copper minerals, nitrates, and various salts.

Acid leaching is used for oxidized copper ores that are free from carbonate. After the ore is subjected to an acid bath in acid-proof tanks the liquor is drawn off, and the copper is electrolytically precipitated. This process is used on a huge scale for the oxidized copper ores of Chuquicamata, Chile.

Ammonia leaching of oxidized copper ores has been employed where the gangue rock is carbonate. The liquor is evaporated in stills, which causes precipitation of black oxide of copper.

Natural leaching is widely employed for low-grade copper ores, particularly in large waste dumps that carry some copper. Water introduced onto the dumps carrying pyrite generates ferric sulfate and sulfuric acid, and the resulting copper sulfate undergoes precipitation of copper by scrap iron.

Water leaching is employed for water-soluble salts such as nitrates, potash, and common salt.

SMELTING AND REFINING

Smelting is the melting of ores in furnaces under high temperature to obtain metal. The smelter charge consists of concentrates, hand-sorted ore, or run-of-mine ore, along with coke or natural gas, and a flux. The last assists melting, makes a slag to remove impurities, and permits the metal to sink through the fluid melt to the bottom of the furnace, where it is tapped off. Iron ores require much limestone flux, siliceous ores require iron flux, and pyritic concentrates require siliceous flux. A smelter is really a large high-temperature, chemical furnace kept under careful chemical control.

Gold-silver ores must have a collector such as copper or lead to carry down and collect the molten droplets of precious metals, which are then recovered in a later process. Iron ores yield pig iron, later remade into steel. Copper ores yield copper matte, consisting of copper and sulfur, which is then transferred to a converter, where air is blown through, forming more slag and purifying the copper, which is poured off as blister copper, ready for further refining. Lead ores form a lead bullion, which is further refined to pure lead with separation of silver and other contained metals. Zinc is distilled in retort furnaces, and the zinc vapor is condensed to zinc or else is treated electrolytically. Mercury ores are distilled, and the mercury vapor condensed to the liquid metal. Thus, it may be seen that different processes have to be applied to different kinds of ores.

The gases are driven off through the smelter stack. If arsenic is present, it must be collected because of its poisonous nature; the other gases may be wasted to the atmosphere, but generally sulfur is recovered to form sulfuric acid, and small quantities of other volatilized metals are collected as secondary products. The dust is collected in dust chambers and further refined for its by-product metals. Most slags are discarded, but some are utilized for making mineral wool or road metal.

The last stage of metal extraction is refining, the purpose of which is threefold: (1) to recover the included precious and other metals, (2) to obtain pure metal, and (3) to get rid of deleterious ingredients.

Copper matte lacking in precious metals may be fire-refined to commercial use, and lead may be desilverized and fire-refined. The more general practice is electrolytic refining, which yields all of the included metals as by-products, the value of which generally exceeds the cost of refining. Blister copper, for example, is cast into anodes and suspended in an acid bath through which electric current is passed. Copper is dissolved from the anode and precipitated as pure electrolytic copper on the cathode, the precious metals and other impurities falling to the bottom as sludge. The sludge is further refined for its various metals. Gold always contains silver and generally some copper, and this is refined to pure gold.

SELECTED REFERENCES ON EXPLORATION AND EXPLOITATION

Ore Deposits of the Western United States—Lindgren Volume. Am. Inst. Min. Met. Eng., New York, 1933. Chap. 12. Geological methods employed by mining companies.

Handbook for Prospectors. M. W. Von Bernewitz. McGraw-Hill, New York, 1931. Prospecting and mining methods.

Mine Examination and Valuation. C. H. Baxter and R. D. Parks. Mich. Coll. Mines and Tech., Houghton, Mich., 1939. Methods of sampling and valuing mineral properties.

Oil Property Valuation. P. Paine. John Wiley & Sons, New York, 1942. Theory and methods.
The Science of Petroleum. A. E. Dunstan. Oxford Univ. Press, Oxford, 1937. Vol. I. Comprehensive treatise of petroleum geology, occurrence, and technique.
Applied Geophysics. 3d Edit. A. S. Eve and D. A. Keyes. Cambridge Univ. Press, Cambridge, 1938. A brief, readable book.
Geophysical Exploration. C. A. Heiland. Prentice-Hall, New York, 1940. A comprehensive technical textbook, with Part I, 64 pp., a summary of the subject in nontechnical language.
Exploration Geophysics. J. J. Jakosky. Times Mirror Press, Los Angeles, 1940. A comprehensive technical treatise.
"Choice of Geophysical Methods in Prospecting for Ore." H. T. F. Lundberg et al. *Min. and Met.* 26:No. 464, 1945. How to choose the most suitable method.
"Evaluation of New Geophysical Methods." W. M. Rust, Jr. *A.A.P.G. Bull.* 29:865, 1945.
Mining Geology. H. E. McKinstry. Chap. 2, "Sampling Ore and Calculating Tonnage." Prentice-Hall, New York, 1948. Methods of sampling and ore calculation.
"Some Practical Aspects of Radioactive Well Logging." W. J. Jackson and J. L. P. Campbell. *A.I.M.E. Trans.* 165:241–267, 1946. Gamma ray and neutron well logging.
"Geochemical Prospecting for Ores: a Progress Report." H. E. Hawkes. *Econ. Geol.* 44:706–712, 1949. Summary of geochemical methods.
"Aeromagnetic Survey of Allard Lake District, Quebec." (Ilmenite.) W. Bourret. *Econ. Geol.* 44:732–740, 1949. Airborne magnetometer survey of ilmenite deposit.
Mining Engineers Handbook. Edited by Robert Peele. John Wiley & Sons, New York, 1941. General reference to all phases of prospecting, mining, milling, and metallurgy.
Principles of Mining. Herbert C. Hoover. McGraw-Hill, New York, 1912. Mining practice and valuation.
Principles of Mineral Dressing. A. M. Gaudin. McGraw-Hill, New York, 1939. A standard textbook on concentration methods.
"Metal Mining Practice." C. F. Jackson and J. H. Hedges. *U. S. Bur. Mines Bull.* 419, 1939. Best general treatment of exploration, sampling, ore reserve estimates, mining methods, concentration, and ore dressing.
Economic Mineral Deposits. 2nd Edit. Alan M. Bateman. John Wiley & Sons, New York, 1950. Chaps. 10, 11, pp. 376–415.

CHAPTER SIXTEEN

Mineral Resources

Mineral resources have strikingly become almost synonymous with industrial power. Industrial power in turn is dependent upon ownership or access to liberal quantities of mineral resources. The rise of the industrial age has so accelerated the demand for minerals that the world has exploited more of its mineral resources between World War I and the present than in all preceding history. This insatiable demand for minerals to feed the hungry maw of modern industry has made sources of supply that formerly were thought adequate now look relatively trivial, and sources capable of meeting large supply are yearly becoming fewer and fewer. It is now widely realized that adequate supplies are concentrated in relatively few places on the globe and in relatively fewer hands, and this feature has become a major factor in world unrest and aggression. Those who hold the purse strings of mineral resources hold power.

The concentration of large supplies of mineral resources in few countries and under fewer owners gives rise to large international trade in minerals, which if disrupted causes industrial dislocation, unrest, and even disregard of international boundaries in the search for new supplies. It has also given rise to nationalization of mineral resources to prevent inroads upon supplies, and to political control of weak nations who own important reserves. Since the 1920's it has unquestionably been a major factor of aggression in eastern Asia, in central Europe, and in northern Africa. Thus mineral resources raise international problems of trade, industrial development, national policy, peace, conquest, and war.

Probably few people in this industrial age stop to realize the extent to which we have become dependent upon materials of the mineral kingdom. The necessities and conveniences of the home, and indeed its construction, largely depend upon mineral substances. The modern fireproof building, save for a little wood, is entirely a structure of minerals. Our various means of transportation absolutely hinge on large quantities of minerals and very little else. Our industries would be shut down without them, and our abilities to provide national security and defense would be nonexistent without minerals. Without minerals we would be back in the days of primitive agriculture and handicraft, and perhaps we would all be happier. However, the horse-and-buggy days are gone, probably forever, and whether we like it or not modern civilization is here to stay, unless the world should be largely demolished by still other energy minerals. This then means minerals and still more minerals.

A glance at the history of the development of leading industrial nations reveals that their rise coincided with the utilization of their mineral resources, notably coal and iron. Where coal and iron occurred together, industry arose. Coal was the motive power, and iron the substance. As was mentioned in Chapter 1, it is no accident that great manufacturing centers sprang up in the United States around the Great Lakes, in Pennsylvania and Alabama, and in England, the Ruhr, northern France and Belgium. There coal and iron met, and the products of their union reached the far places of the globe. Those countries attained prominence above other nations. They became the great manufacturing and trading nations, and economically and industrially aggressive. Political prominence followed and armaments were constructed for the protection of country and trade, and also, unfortunately, for the acquisition of extranational mineral resources. Leith has emphasized that self-sufficiency of minerals came to be regarded as one of the chief goals of economic nationalism. Those nations lacking an adequate supply of minerals face either a sharp curtailment of industrialism and levels of living or must find means to gain access to the minerals. This is normally accomplished by peaceful trade or investment in undeveloped foreign resources. Some strong nations, however, have exercised forcible control over the mineral resources of weaker nations, and some minor nations have found abundant mineral resources as much a liability as an asset.

MINERAL RESOURCES AND THEIR DISTRIBUTION

Scores of minerals enter into modern industry. Even a minor one may be so important a constituent in a machine, a train, or a plane, that these articles could not be constructed without them. No single nation is adequately supplied with all the minerals necessary for its industrial life or for armament or war. Some countries have much of the bulk minerals but lack those that can be flown in by airplane. Many countries are devoid of most of them. Even the United States, despite the fact that it is the most richly endowed in mineral resources of all nations, could not continue as a large industrial nation if its essential mineral imports were to be cut off. The predicament of most other nations is worse. This means, therefore, that if industrial life is to continue there must be free international trade in minerals so that all may benefit by their use or the products of their use. The distribution of mineral resources thus becomes important.

At the conclusion of the Dark Ages the chief mineral substances in use were iron, copper, lead, tin, gold, silver, mercury, precious stones, clays, and building stones. Today, more than 80 minerals enter international trade. Some are basic to essential industries, some are absolutely vital for armament or war, others contribute only to desired but unessential industries, and others are mere luxuries.

The chief minerals of industry may be grouped as follows:

Mineral group	A few examples
Mineral fuels	Coal, oil, gas
Iron ores	
Iron-alloy metals	Manganese, chrome, nickel, tungsten, molybdenum, vanadium, cobalt
Nonferrous metals	Copper, zinc, lead, tin, aluminum
Minor metals	Mercury, antimony, silver, platinum, tantalum, etc.
Metallurgical minerals	Fluorspar, barite, graphite, bauxite, refractories
Chemical minerals	Salt, sulfur, salines, borax
Structural minerals	Stones, cements, gypsum, bitumens, pigments
Fertilizer minerals	Phosphate, potash, nitrates, limes
Ceramic minerals	Clay, feldspar, silica, refractories
Abrasive minerals	Diamond, corundum, quartz, emery, garnet

Gold, because of its dominant monetary use, is not classed as an important industry mineral, but it does play a prominent part in affording means of purchasing needed mineral supplies. The chief minerals of industry and their distribution are shown in Table 11.

Minerals, unlike products of the soil, are "one-crop" materials. They are vanishing assets, most of which once removed are gone forever. Therefore their distribution and reserves command attention. In Table 11 this information has been condensed. It shows, horizontally, the major sources, and vertically, the approximate resources of this group of vital minerals. In the vertical column for each coun-

Their Distribution

TABLE 11
DISTRIBUTION OF CRITICAL MINERALS AND STATUS OF IMPORTANT COUNTRIES
[A, adequate or surplus; D, important deficiency; O, no appreciable supply.]

Important Minerals	United States	Canada	Great Britain	British Empire	France	French Empire	Belgian Colonies	Russia	China	Germany	Italy	Japan	Scandinavia	Poland	Spain	Netherlands Empire	Balkans	North Africa	South Africa	Mexico	Brazil	Venezuela	Chile	Peru	Argentina	Bolivia
Iron	A	A	D	A	A	A	D	A	A	D	D	D	A	D	A	O	D	A	A	D	A	A	A	O	D	O
Copper	A	A	O	A	O	D	A	D	O	D	O	D	A	O	A	O	A	O	A	A	O	O	A	A	O	O
Lead	D	A	D	A	O	D	D	A	D	D	D	O	D	A	A	D	D	A	D	A	O	O	O	A	A	A
Zinc	O	A	D	A	O	A	A	A	A	D	O	D	O	A	A	A	O	O	D	D	O	O	O	A	A	A
Tin	A	O	A	A	O	D	O	O	A	O	A	O	A	O	O	A	A	O	D	O	A	O	O	O	O	O
Aluminum ore	D	O	O	A	A	A	O	O	O	O	A	O	O	O	O	A	D	O	O	O	A	O	O	O	O	O
Manganese	O	O	O	A	O	D	O	A	O	O	O	O	O	O	O	O	A	O	A	O	O	O	O	O	O	O
Chromium	O	O	O	A	O	O	O	A	O	O	O	O	O	O	O	O	D	O	D	O	O	O	O	O	O	O
Nickel	A	A	O	A	O	A	O	D	O	O	O	O	A	O	O	O	A	O	O	A	O	O	O	O	O	O
Molybdenum	D	D	O	A	O	D	O	D	O	O	O	O	D	O	O	O	D	D	A	A	O	O	A	O	O	O
Tungsten	D	A	O	A	O	D	O	D	A	O	O	O	O	O	O	O	D	O	D	O	O	O	O	A	O	O
Vanadium	D	O	O	A	O	O	O	A	O	O	O	O	O	O	O	O	D	D	O	O	A	O	O	A	O	A
Cobalt	D	A	O	A	O	D	A	D	D	O	O	O	O	O	O	O	O	O	A	A	O	O	O	O	O	O
Mercury	D	O	O	A	O	O	O	D	O	O	O	O	O	O	A	O	O	O	D	A	O	O	O	O	O	D
Antimony	A	O	O	A	O	O	O	A	A	O	O	D	O	O	D	O	O	A	O	A	O	O	O	D	O	A
Gold	A	A	O	A	O	O	O	A	O	O	O	O	A	O	O	O	O	O	A	A	O	D	O	D	O	D
Platinum	D	D	O	A	O	O	D	A	O	O	O	O	D	O	O	O	A	O	A	O	O	O	O	D	D	O
Tantalum	O	O	O	A	O	O	A	A	A	O	O	D	A	O	O	O	O	O	O	O	A	O	O	O	O	O
Coal	A	A	A	A	A	A	D	A	A	A	O	A	O	A	O	A	A	O	A	O	D	O	D	A	D	D
Oil	A	A	O	D	O	O	O	A	O	D	O	O	O	D	O	A	D	O	D	D	O	A	O	O	D	O
Sulfur	A	D	O	A	A	O	D	A	O	D	A	O	A	O	A	O	A	O	D	D	O	O	D	A	O	O
Potash	A	O	O	D	O	O	A	A	O	A	D	O	O	A	O	D	O	O	O	O	O	O	O	O	O	O
Phosphates	A	A	O	A	A	A	D	A	O	D	D	A	D	D	A	D	O	A	D	D	O	O	D	D	O	A
Fluorspar	A	A	O	A	O	A	O	A	O	A	A	A	D	O	A	O	O	O	O	A	O	O	A	A	O	D
Asbestos	D	A	O	A	O	D	O	A	O	O	D	O	D	A	O	O	O	O	A	O	O	O	O	O	O	O
Graphite	O	O	O	A	O	D	O	A	O	D	O	O	D	O	O	D	O	O	D	D	O	D	O	D	D	O
Magnesite	A	A	O	A	O	D	O	A	A	A	O	A	O	O	O	O	A	O	A	O	O	O	O	O	O	O
Mica	D	O	O	A	O	O	O	A	A	D	D	O	D	O	O	D	O	O	D	D	A	O	O	O	O	O
Diamonds	O	O	O	D	O	O	A	O	O	O	O	O	O	O	O	O	O	O	A	O	A	D	O	O	O	O
Quartz Crystals	D	O	O	A	O	O	O	O	O	O	O	O	O	O	O	D	O	O	O	O	A	O	O	O	O	O
Quantities: Surplus	12	14	2	25	4	10	5	16	6	4	4	3	5	4	8	5	4	7	13	12	8	2				
Deficient	14	6	6	4	1	13	8	10	3	10	6	12	8	3	2	5	13	1	9	5	2	3				
Absent	4	10	22	1	25	7	17	3	21	16	20	15	17	23	20	20	13	22	8	13	20	25				

try, A designates an adequate or potentially adequate supply or an actual or potential surplus available for export; D designates an important use deficiency to be made up by imports, but for countries of low domestic consumption it implies minor quantities available for export; O designates unimportant supplies, although few countries are entirely lacking in these minerals, as is Brazil, where O for chrome does indicate small exports. The figures at the bottom of the vertical columns give a striking summary of the position of each country in this group of minerals.

Table 11 does not disclose the very important concentration of distribution of each mineral. It does show that copper occurs in surplus amounts in 14 countries, but it is more revealing to learn that 86 percent of the world copper comes from only 5 countries, that 55 percent of it is concentrated in 2 countries, and that these two countries represent the ownership of 82 percent of the copper production and 92 percent of world copper reserves. Further, three important nations consume 22 percent of world copper but produce only 6 percent. These striking figures have a bearing on international trade and national and foreign policies. Other, similar data are assembled in Table 12.

Another striking feature revealed by Table 11 is that many of the nonindustrial countries have large resources of critical minerals of international and perhaps political importance. These are mostly exported to countries that need them and enter international trade, except for the minerals from countries now under Russian control. There is, however, quite a falling off in the international flow of surplus minerals, which is creating a growing scarcity for those industrial countries that have to depend upon them more and more as their populations grow and their industries expand. This is because: (1) Russian mineral surpluses no longer enter world trade, (2) the former exportable surpluses of the Russian satellite countries no longer flow outward, (3) unsettled conditions in several Asian countries, (4) nationalistic tendencies in some other countries whereby their domestic resources are being conserved to help build up internal industries.

The accompanying tables have purposely been arranged to disclose the sources of primary production of minerals and therefore do not reveal that several countries are producers of finished metals but not of the ore minerals. Canada is the second largest producer of aluminum, but the ores travel from British Guiana and elsewhere to Canadian electrical power for reduction to metal. Norway and Switzer-

Their Distribution

TABLE 12

Production Distribution of Some Important Minerals

Mineral	Percent of Total	In Number of Countries	Chief Producing Countries
Oil	82	4	U. S., Venezuela, Russia, Middle East
Coal	80	4	U. S., Britain, Germany, Russia, S. Africa
Iron	64	4	U. S., France, Russia
Iron	78	5	U. S., France, Russia, Sweden, Britain
Copper	85	5	U. S., Chile, Rhodesia, Canada, Congo
Copper	82	2	Ownership: U. S., Britain
Copper	99	3	Reserves owned by U. S., Britain, Belgium
Lead	73	5	U. S., Australia, Mexico, Canada, Russia
Lead	70	2	Ownership: U. S., Britain (38)
Zinc	78	5	U. S., Belgium, Canada, Mexico, Russia, Poland
Zinc	62	2	Ownership: U. S. (44), Britain (18)
Tin	79	3	Malaya, Netherlands East Indies, Bolivia
Tin	64	2	Ownership: Britain (47), Holland (17)
Bauxite	64	5	France, Surinam, Hungary, U. S., Yugoslavia, British Guiana, Brazil
Manganese	82	5	Russia, India, S. Africa, Gold Coast, Brazil
Manganese	90	2	Russia (45), Britain (45)
Chromium	66	5	Rhodesia, Russia, Turkey, S. Africa, Cuba
Chromium	67	2	Ownership: Britain (50), Russia (17)
Nickel	97	3	Canada, New Caledonia, Russia (Finland)
Molybdenum	95	2	U. S., Chile, Mexico
Tungsten	62	3	China, Burma, U. S., Bolivia, Brazil
Vanadium	83	3	Peru, U. S., S. W. Africa
Mercury	73	4	Italy, Spain, U. S., Mexico, Canada
Antimony	81	3	China, Mexico, Bolivia
Gold	74	4	S. Africa, Russia, U. S., Canada
Platinum	85	3	Canada, Russia, S. Africa
Sulfur	76	5	U. S., Spain, Italy, Norway, Russia
Asbestos	78	3	Canada (53), Russia (18), Rhodesia (7), S. Africa
Potash	95	4	Germany, France, Russia, U. S.
Fluorspar	90	5	U. S., Germany, Russia, Britain, France
Graphite	80	5	Russia, Japan, Germany, Austria, Ceylon

land also import bauxite for the production of aluminum, utilizing their abundant water power which produces cheap electricity. Belgium and France have long been credited as important producers of zinc, which is smelted from ores that flow mainly from Mexico and South America to these well-known European smelters. Former United States tariffs have diverted this metal from American smelters. Similarly, the United States is a large producer of titanium oxide, used so abundantly in white paint, but from ores that flow mainly from India and Canada. Also, cobalt, so highly prized today for its high-temperature alloys needed in jet engines, is produced from ores that emanate chiefly from Africa and Canada. Many other examples exist of countries that produce metals but not the ores of the metals. In general, the nonindustrial countries that produce minerals export them to consuming countries. Such exports give rise to international trade and build up foreign exchange, whereby the manufactured goods of the industrial countries may be purchased.

MINERAL SUFFICIENCY AND DEFICIENCY OF LEADING AREAS

A summary of the mineral sufficiency and deficiency of the leading nations and empires is given in the bottom row of Table 11, but it is pertinent to discuss these in more detail.

UNITED STATES

The United States achieved greatness by the aid of its mineral resources. They were so bountiful as compared to those of any other individual country that they generated a smug complacency. Its people had, of course, always realized that the country was deficient in about a dozen important industry minerals, to which the term strategic and critical minerals was applied. It took the rude shock of World War II, however, to bring home the realization that former abundant supplies were no longer adequate to sustain the increased peacetime growth of industry and population, accentuated by high wartime demands for the very basis of the sinews of war. The strategic minerals of prewar times jumped from a dozen to threescore, and the term strategic almost lost significance because some of nearly every mineral of industry had to be imported. Few realize, even today, the extent to which the United States became dependent upon foreign sources of minerals during World War II. In this connection just a few figures are enlightening.

Imports of metals and minerals into the United States exceeded 60, of which 27 came exclusively from 53 foreign sources. The reliance on foreign sources for 20 of the chief wartime minerals is shown in Table 13. Of the 20 vital minerals listed, the United States was 100 percent dependent upon foreign sources for 8 of them, and 85 to 100 percent dependent for 6 of them. This table does not leave much room for complacency.

TABLE 13
Reliance on Foreign Ores during World War II in 1942-1944

Mineral	Percent Foreign	Chief Foreign Sources
Antimony	83.2	Mexico, Bolivia, Peru
Beryl	90.6	Brazil, Argentina, India
Asbestos	100	Rhodesia, South Africa
Chrome	89.6 *	South Africa, Turkey, New Caledonia, Cuba
Cobalt	85.6	Belgian Congo
Corundum	100	South Africa
Copper	37.6	Chile, Congo, Canada
Diamonds	100	Congo, Brazil, Angola, South Africa
Graphite	100	Madagascar, Ceylon
Lead	44.2	Mexico, Peru, Australia, Canada
Manganese	85.5 *	India, Africa, Cuba, Brazil
Mercury	43.2	Mexico, Canada
Mica	88 *	Brazil, India
Quartz (radio)	99 +	Brazil
Nickel	100	Canada, New Caledonia, Cuba
Tantalum	99	Brazil, Congo, Nigeria
Tin	100	Bolivia, Congo, Nigeria
Vanadium	32.4	Peru
Tungsten	61.1	China, Bolivia, Brazil
Zinc	36.7	Mexico, Peru, Australia, Canada

* Indicates that the remainder, or domestic sources, was mostly unusable; therefore these figures in reality should be higher.

Let us look at the situation of the United States today. For this we can turn most profitably to a study made by experts of the Geological Survey and the Bureau of Mines, published (1948) as the *Mineral Resources of the United States*.[1] This revealing document summarizes the self-sufficiency and reserves in two charts, reproduced here as Figures 16-1 and 16-2, portrays the ratio of production to consumption in 39 minerals during two 5-year periods, 1935-1939 and 1940-1944, one before and one during the war. Before the war production equaled or exceeded consumption in 11 out of the 39 minerals, some of which are not vital minerals. For another 12 minerals out of the 39, production yielded only 50 to 100 percent of needs; for 6 it ranged from 10 to 50 percent; for the last 10, from 0 to 10 percent.

This prewar record is factual and gives a picture of the availability of United States resources under the economic conditions existing between 1935 and 1939. It is not too happy a picture. Moreover, the

[1] Public Affairs Press, Washington, D. C., 1948.

nsumption for the period 1945–1949 was vastly greater than 1939. In the 1935–1939 period steel production, as a gauge of industry, ranged from 28 to 50 million tons. Now it is over 100 million tons. Since then there has been a population increase of 15 million people; there have been more homes, high wages, and labor short-

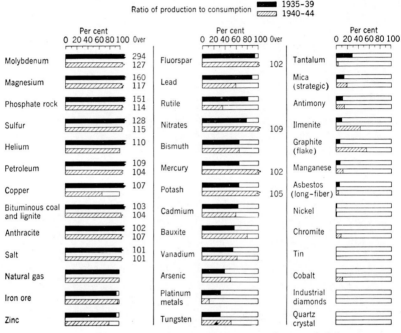

FIGURE 16–1. The United States' position in principal minerals, showing ratios of consumption to production in 1935–1939, and 1940–1944. (U. S. Dept. Interior and A. M. Bateman, *A.I.M.E.*, 1949.)

age. All these items spell demand for labor-saving and luxury articles. The home, the farm, and the factory have become more and more mechanized. This demonstrates a vastly greater future need for metals and minerals than the requirements for 1935–1939. Consequently the percentage figures given in Figure 16–1 would mean a much greater deficiency today.

Even more revealing, however, are the data disclosed by Figure 16–2, which depart from factual to conjectural and must be appraised in that light. This shows the reserves of 41 minerals expressed as years of life, based upon the 1935–1944 rate of consumption (reserves divided by average consumption). This brings out startlingly the fact that, although there is a life of half a century or more for 15 of

FIGURE 16–2. Estimated life of United States mineral reserves based upon the 1935–1944 rate of consumption. (U. S. Dept. Interior and A. M. Bateman, A.I.M.E.)

these 41 minerals, three of them would last from 33 to 39 years, and the remaining 23 have an indicated life of less than 25 years, and of these 9 have a life of less than 10 years. For 8 of them there are no reserves. These are telling estimates, and it is important to note that the long-life ones include many common substances of minor importance, such as helium, or of wide sources of supply, such as salt, magnesium, or arsenic. Note, however, that the backbone nonferrous metals such as copper, lead, and zinc are all too short, and have 19, 10, and 20 years of life, respectively. The shortage in the ferroalloy metals is painfully clear.

This picture of United States resources is the consequence of long-continued heavy drain upon them in the building of an industrial nation, and because they were freely supplied to the world. There was further heavy wartime encroachment. The estimates used as the basis for Figure 16–2 may be a little pessimistic. My opinion is a bit more optimistic; the outlook for reserves may be increased over the estimates given by increasing technology and by utilizing more scientific methods of exploration, conservation, and substitutions. However, small changes do not greatly affect the larger picture. It follows that the United States for the future will have to turn more and more to foreign sources of supply for the minerals needed to sustain its peacetime industrial maintenance and development.

The United States, although dependent upon foreign minerals, enjoys the distinct advantage of having important mineral-producing, friendly neighbors lying across unarmed boundaries to the north and south of it. From Canada there flow large tonnages of nickel, lead, zinc, copper, asbestos, silver, cobalt, platinum, aluminum, iron ore, mica, and many minor metals and minerals. From Mexico come lead, zinc, copper, mercury, silver, fluorspar, graphite, and other minerals.

THE BRITISH COMMONWEALTH

The great contrast between figures for Great Britain and the British Commonwealth, shown by Table 11, is illuminating. It forcibly brings attention to Great Britain's own paucity and her dependence on the Commonwealth for industry minerals. It indicates that her industrial existence depends upon a protected and uninterrupted flow of minerals over long ocean voyages—hence a strong navy, or strong allied navies. The Commonwealth as a whole leads the world in mineral sufficiency, since in the 30 minerals listed it lacks only 1, is deficient in 4, and has a surplus of 25. Moreover, it is the world leader in such internationally critical minerals as tin, lead, nickel, chrome, titanium, asbestos, mica, and diamonds, and is a close second in manganese, copper, zinc, and some minor minerals.

It is also striking (Table 11) to note how the British Commonwealth and the United States complement each other in mineral supplies. Of the 18 minerals in which the United States is deficient, 16 are available in abundance in the Commonwealth, and 3 of the 5 in which the Commonwealth is deficient are present in the United States in surplus quantity. Does this not mean that if the United States and Great Britain desire to maintain their industry and security, they must ever be closely associated in international trade, foreign policy, and naval strength to protect sea lanes?

The dominions and colonies of the Commonwealth, like the United States, are also facing growing depletion of their presently developed mineral resources. In addition, the separation of Burma and Egypt from the British Commonwealth has removed extensive mineral resources from British domain. On the other hand, the dominions and colonies offer exceptionally favorable opportunity for further mineral discoveries. Canada, with its great hinterland of favorable basement rocks, is a treasure-house of minerals, but the main part of this vast area has hardly been scratched. Even in recent years large deposits of iron ore, titanium ore, nickel-copper, zinc-lead, gold, cobalt, petroleum, and other minerals have been developed, and the future will unquestionably see other new mineral resources discovered. British Africa, already rich in varied mineral output, holds fruitful geologic promise for much more as the frontiers of this underdeveloped area are pushed back and more intensive exploration takes place. Australasia likewise contains large favorable areas that are every year yielding new mineral deposits to supplant ones being depleted, and even little British Guiana offers promise of continuing to pour out aluminum ore and other minerals for many years to come.

Thus areas of the British Commonwealth, next to Russia, are the greatest potential sources in the world for the development of new mineral resources.

RUSSIA

Russia is now extremely well provided with minerals and has few deficiencies. The most basic minerals, coal and iron ore, are in abundance. She has actual and potential petroleum reserves that may rival those of the United States. She leads the world in manganese and is second in iron ore, gold, chrome, platinum, and asbestos. She is probably now self-contained in copper. From the shield of

Iron-Curtain countries there flows into Russia, not to return, needed supplies of lead, zinc, nickel, tungsten, antimony, tin, bauxite, and others. Russia has now become essentially self-contained in the critical industry minerals, more so than the United States. She has fairly large potential supplies of low-grade uranium-bearing rocks but lacks high-grade ores such as are available to the United States. For the manufacture of steel and steel alloys, Russia lacks nothing nor is she now deficient in the nonferrous metals, fertilizer, or chemical minerals. It is particularly striking that she contains within her own domain every mineral essential to an industrialized nation that has to be transported in large volume and tonnage. In this Russia stands alone among nations. Moreover, the mineral resources of which Russia is apparently deficient are those that move in such small volume that they can be transported by airplane.

Russia, perhaps more than any other country of the world, includes expansive regions that are exceptionally favorable geologically for future new mineral resources. Large parts of Siberia and the southern regions are only slightly explored geologically, but indications are that new deposits of coal, oil, iron ore, gold, chemical minerals, and several nonferrous metals and nonmetallic minerals may be discovered in the future.

The actual and potential mineral resources which Russia herself controls, combined with her large population which is rapidly learning industrialization, make her potentially a colossal industrial giant of the future, and perhaps the leader of the world.

WESTERN EUROPE BLOC

The countries of western continental Europe by themselves are strikingly deficient in most of the industry minerals and have surplus in only a few. Coal in sufficient abundance is present in France, Germany, and Belgium but is deficient in Italy, Spain, Portugal, and Holland. Oil is meager in all of them. Very low-grade iron ores are utilized in France and Germany, but Spain exports good iron ore. Bauxite for aluminum is in superabundance. The essential ferroalloy metals are almost entirely lacking, and the nonferrous metals, except for aluminum, are mostly lacking, except for minor quantities of lead, zinc, and copper in Germany and Spain. Sulfur, potash, and mercury are sufficient. The industrial life of this part of Europe hinges upon extensive imports of minerals, notably the bulk minerals.

The overseas empire of France, Belgium, Holland, and Portugal do, however, contain important supplies of many of the needed industry

Atlantic Treaty Countries

minerals. French Africa yields iron ore, phosphates, mica, lead, zinc, antimony, cobalt, manganese, graphite, and diamonds. Other French colonies yield chrome, nickel, gold, and bauxite. Belgian Africa supplies copper, cobalt, tin, diamonds, manganese, and tantalum. The Netherlands Indies yield tin, gold, aluminum, nickel, oil, and fertilizers, and Portuguese Africa supplies diamonds, corundum, manganese, and mica. All these minerals must make long voyages to reach Europe, and that means available shipping and protected sea lanes are essential in event of war. All the colonies of continental European countries are lacking in sufficient quantities of the chief ferroalloy metals for the steel industry and must turn to outside sources.

The African colonial possessions of France, Belgium, and Portugal are also potential sources for new mineral resources. Mineral exploration in the Belgian Congo has been aggressive and scientific, and the results are evident in the large deposits of copper, manganese, and tin ores that have recently been developed. Additional favorable areas still await intensive geologic exploration, and, if the results of past work are any indication of what might be discovered in the future, large additions to reserves are to be expected. In the French and Portuguese African areas, however, mineral exploration has been more backward than in the British and Belgian possessions, and much ground that offers favorable possibilities for mineral deposits remains to be investigated. Geological investigations in progress may be expected to result in some new mineral deposits from these areas.

ATLANTIC TREATY COUNTRIES

The foregoing discussion of mineral sufficiency and deficiency of different areas indicates that Russia is essentially self-contained in nearly all the chief industry minerals and needs no recourse to ocean shipping to bring them in. Those minerals in which Russia is deficient are now mostly supplied by the Russian satellite countries. The only ones entirely lacking are small-tonnage items that can readily be transported by airplane.

The Atlantic Treaty nations as a group are entirely self-sufficient for all of their mineral needs. What one lacks, the others have. However, their resources are widely distributed on all the continents, and ocean voyages of several thousand miles are necessary to transport the needed minerals to the places of consumption. Moreover, the seaborne transportation involves millions of tons of minerals; the United States alone must import annually some $1\frac{1}{2}$ million tons of manganese and nearly as much chrome and bauxite. This means that the industrial existence of the Atlantic Pact nations is absolutely dependent upon seaborne traffic. Remove this traffic and industry largely ceases. It follows that to supply mineral needs, millions of tons of shipping must be available and long sea routes maintained. This spells large navies. In contrast, the mineral self-containment of Russia is superior to that of the North Atlantic industrialized nations.

EASTERN ASIA

Eastern and southeastern Asia is quite well supplied with a number of industry minerals. This section of the world is the leader in resources of tin, tungsten, and antimony and holds large resources of oil, coal, bauxite, zinc, lead, gold, copper, graphite, iron ore, chrome, sulfur, and phosphates, along with some mica, manganese, mercury, and other minerals of commerce. The tungsten, antimony, and tin of China, the tin and tungsten of Burma, the tungsten and graphite of Korea, and the varied minerals of Manchuria do not now enjoy world markets. The other minerals of this region were used formerly for its own meager industrial uses.

SOUTH AMERICA

South America is richly endowed with many minerals of commerce, and, as the continent is little industrialized, most of the minerals are surplus and are exported. Among the large resources of world importance are the great copper and nitrate deposits of Chile, the huge oil reserves of Venezuela and Colombia, the prolific bauxite deposits of Surinam and British Guiana, and the great newly explored iron and manganese deposits of Brazil, as well as its outstanding mica, beryllium, and tantalum resources. Tin is the main industry of Bolivia, and lead, zinc, and silver are won in large quantities in Peru, Bolivia, Argentina, and Chile. Manganese and iron move out of Chile, and tungsten and antimony from Bolivia and Argentina. Platinum is obtained in Colombia, and gold is mined in most of the countries. Chrome in meager quantities is recovered in Brazil. Coal is deficient on the continent. Several minor metals find their home in the Andean region.

It is to Chile that the world turns for all of its natural nitrate, and Chile will probably become the main source for copper that the United States will lack in the future. The United States east-coast iron ore needs will be largely drawn from Venezuela in the future, and considerable amounts of its manganese will probably come from Brazil. The Guianas will continue to yield the bauxite necessary for the aluminum industry of the United States and Canada, and Venezuelan oil will flow into the United States and Europe.

The South American countries are in a position to continue for a long time to supply friendly nations with the minerals in which they are deficient, providing thereby the products of international trade that can and should inure to the benefit of both the supplier and the consuming nations.

AFRICA

Africa is still the great underdeveloped continent from the standpoint of mineral resources. It contains many slightly explored or unexplored areas of great geologic promise that hold future mineral potentialities. Despite its underdevelopment, parts of it are now known to contain some of the outstanding mineral resources of the world. South Africa has long been the leading gold-producing country and has huge reserves of chrome, manganese, and asbestos. Northern Rhodesia contains the greatest single copper belt in the world, and across its border in the Belgian Congo is another great copper belt. The greatest single manganese deposit in the world, in the Gold Coast, pours out its eagerly sought mineral to Great Britain and the United States. Other manganese deposits lie in the Congo and Angola. Almost the entire world supply of diamonds comes from central and South Africa. Southern Rhodesia yields the largest supply of metallurgical chrome and the finest spinning asbestos. Algeria and Tunisia have great phosphate deposits, and important iron-ore deposits lie in the Moroccos. Large lead-zinc deposits occur in French North Africa, Southern Rhodesia, and in South-West Africa. Copper is extracted in the Transvaal and Namaqualand. Large bauxite reserves are exploited in the Gold Coast. The main corundum production of the world is in South Africa, Mozambique, and Nyasaland, and kyanite comes from Kenya. Tin is obtained in the Belgian Congo and Nigeria, and the chief supplier of tantalum is the Belgian Congo. Lesser quantities of the same minerals, and others, are found in various parts of this great lonely continent. Coal, however, is not abundant, and oil is largely lacking.

From this enumeration it will be seen that Africa is already a great contributor to the world's mineral supplies, and further exploration should yield more for the future. By far the greatest part of the known mineral resources lie in the territories of the British dominions and colonies. This is not altogether the fortuitous chance of mineral distribution but in considerable part is the result of energetic exploration, utilizing highly trained geological and engineering skills and venture capital, on the part of aggressive British, Belgian, and American developers of mineral resources. Many other parts of Africa still lack skilled mineral exploration, and vast underdeveloped areas hold geologic promise for future mineral discoveries. Although some geologic surveys have been initiated and others have been strengthened through the aid of the Marshall Plan in Portuguese and in some British colonies, much ground still remains to be geologically explored. A few scattered deposits within these areas already give indication of the presence of minerals, and further exploration should yield more, particularly within French and Portuguese Africa and within the great plateau region.

AUSTRALASIA AND INDONESIA

Australia is one of the chief mining areas of the world and has self-sufficiency and excess in several minerals. Coal is abundant and iron ore ample, but oil is almost entirely lacking. Australia provides excess supplies of gold, copper, zinc, lead, silver, tungsten, and gypsum. There are also resources of asbestos, molybdenum, and tin. In the production of lead and zinc ores Australia ranks third in the world and is a large exporter of these metals. Geological exploration continues to yield new deposits, and the area holds promise for further new mineral deposits.

New Zealand yields coal, gold, silver, and a little tungsten, and Borneo yields chiefly gold.

Indonesia is rich in petroleum, coal, tin, and aluminum ore and also yields some silver, gold, manganese, phosphate, nickel, sulfur, diamonds, and copper, most of which are exported. The chief exportable mineral wealth is in tin, petroleum, bauxite, phosphate, and nickel.

SURPLUS MINERALS FROM UNDERDEVELOPED AREAS

Many of the underdeveloped areas that are only slightly industrialized or non-industrialized have mineral production in excess of their own needs. These include most of Africa north of South Africa, Indonesia, southeastern Asia, the Middle East, and much of South America. These are mineral-bearing regions that yield surplus minerals available for export to the industrialized nations. In many of them the surplus minerals constitute the chief part of the national economy and provide most of their internal revenue. Thus in South America excess petroleum is obtained in Venezuela, Colombia, and Peru; copper and nitrates in Chile; copper, silver, lead, and zinc in Peru; tin,

Minerals in International Relations and War

silver, lead, and zinc in Bolivia; iron in Venezuela and Brazil; lead and zinc in Argentina; manganese in Brazil; and bauxite from Surinam and British Guiana. Petroleum in great abundance is yielded from the prolific wells of the Middle East. The Rhodesias and the Congo pour forth large surpluses of copper, diamonds, tin, chrome, and asbestos; manganese and bauxite reach world markets from the Gold Coast, and Nigeria yields gold and tin. North Africa sends out large tonnages of iron ore and phosphates and shortly will become an important exporter of lead and zinc to Europe. Southeastern Asia and Indonesia gain much from supplying tin to the world, as well as petroleum, bauxite, phosphate, and other minerals.

In most of these areas the production and export of minerals is the source of most or a large part of their internal revenue and foreign exchange. Elimination of their mineral production would bring internal hardship or misery and foment political unrest, and international trade is a necessity for their welfare. Their mineral products in turn supply industrialized countries with the basic materials for manufacture and security. Hence both the underdeveloped and the industrialized nations gain from interchange of minerals, a theme that is developed further in the next two subsections.

MINERALS IN INTERNATIONAL RELATIONS AND WAR

A study of Table 11 will indicate that no one nation possesses all the minerals necessary to modern industry and security. Deficient nations become absolutely dependent upon others for certain essential supplies. This gives rise to large international movements of minerals, and expanding mining companies become international in scope, with large investments in foreign lands. All this brings on political entanglements, alliances, bilateral treaties, protectorates, and even forceful domination. Some nations restrict or shut off exports of their minerals in order to conserve them or to build up national industries of their own. It is fortunate, however, that most essential and critical minerals are fairly well distributed and few countries have a monopoly on any one mineral, although quartz crystals of radio grade are largely restricted to Brazil, and nickel to Canada. National independence in minerals is impossible, but there are enough of the essential and critical minerals to supply the needs of all nations, so that no one need go without them. This means, however, that if industrial nations are to remain industrial they must have access to mineral supplies of other lands through normal trade. If access is denied to them by nationalistic policies or export or import restrictions, conditions are bred that become threats to world peace.

For defense and war armament, minerals become absolute necessities. Modern mechanized and motorized units, as well as ships and planes, are entirely made of and propelled by minerals, and there must be abundant supplies to maintain them. The flow of mineral supplies, therefore, may determine the course or cessation of war, and they may determine peace no less than they sustain industry in peacetime. Any warring nation cut off from supplies of essential and critical minerals during a long struggle must inevitably face defeat if its stocks become exhausted, because minerals cannot be grown like wood or cotton.

Minerals can also be used to enforce and preserve peace by limiting the supply available to would-be or potential aggressor nations. Armor for aggression hinges on steel, and a control of steel output determines the amount available for aggressive purposes. The output of steel can be controlled by limitation of just a few minerals in turn, as, for example, iron ore, manganese, and a few other ferroalloys. Other essential heavy equipment output can be limited by control of a few nonferrous metals such as copper and aluminum. Thus, the controlled flow of just a few of the essential and critical minerals to deficient aggressor nations can limit their industrial output to peacetime needs and prevent a surplus for warring purposes. Obviously such controls would have no effect upon nations who are self-sufficient in mineral resources.

Those minerals most essential for modern war are oil, iron ore, ferroalloy metals, nonferrous metals, mercury, antimony, tin, platinum, sulfur, asbestos, graphite, magnesite, fluorspar, mica, quartz crystals, beryl, tantalite, uranium, and industrial diamonds. These must move without interruption, many of them in large volume over long sea routes. The only alternative to such a flow during wartime is the availability and security of stockpiles of strategic minerals built up during times of peace. Stockpiles represent much more than an available tonnage of a given strategic mineral; they represent also a stockpile of manpower and of shipping.

DISTRIBUTION OF SOME KEY CRITICAL MINERALS

The bearing of mineral resources on industry, national security, and international trade and politics may be seen by considering briefly the geographic and political distribution of a few of the more critical minerals. For this let us choose as examples only coal, oil, iron ore, manganese, chrome, copper, tin, and uranium.

Distribution of Critical Minerals 331

Coal, as has been pointed out, is still the chief source of primary energy, of which the others are oil, natural gas, water power, wood, solar energy, volcanic energy, and atomic energy. Despite the inroads made upon it by petroleum, coal still produces about 70 percent of all energy units. Water power developed to the fullest can never displace coal. Wood is only a local fuel. Petroleum, however, is the preferred fuel for all mobile power units, and it and natural gas have greatly encroached on coal for stationary power units also. This encroachment in turn has created greater efficiency in the burning of coal. For example, steam railroad consumption per 1,000 ton-miles hauled dropped in the quarter-century 1920–1945 from 170 pounds of coal to 115 pounds; in electric power plants in the same period the coal consumption per kilowatt-hour dropped from 3.2 to 1.3 pounds; by 1951 it was 0.75 pounds in modern plants. In the United States, of the heat and energy units produced, coal supplies about 47 percent, oil 33 percent, gas 15 percent, and water power 4 percent. Regardless of energy, coal must be utilized to yield coke for the making of pig iron. Of the coal produced, 78 percent is used for fuel, and 22 percent for making pig iron, steel, and gas. The normal world production of coal is about 1.6 billion tons annually, of which the United States produces from 500 to 650 million tons.

Coal resources of most of the leading nations rank as follows (beginning with the highest): United States, Canada, Russia, China, Germany, Great Britain, Australia, and India. In continental Europe, France, Poland, Czechoslovakia, and Belgium have appreciable quantities, but coal is almost lacking in Italy, the Balkans, Portugal, and Scandinavia. The Middle East, except for Turkey, is woefully deficient. The Near East has large resources in China and Manchuria and minor quantities in Japan and elsewhere. Africa is lacking in coal southward as far as Southern Rhodesia, and it appears again in quantity in South Africa. South America is deficient in developed resources, but Colombia has undeveloped resources and a little occurs in Peru, Brazil, and Chile. The coal reserves of the world are thus largely confined to the North Atlantic areas. The leading industrialized nations have sufficient supplies except for Italy, Scandinavia, and Japan. Their deficiencies are capable of being supplied from European, Indian, and, eventually, Chinese and Korean coals. South America draws upon North American, South African, and British coal. The wide distribution of coal, therefore, offers no serious international or security problems except for Japan.

Iron, since the inception of the Iron Age, has become the backbone metal of all nations. To enumerate its uses would be to compile a history of the innumerable creations of modern civilization and industry. Each of the main types of iron—steel, cast iron, wrought iron, and iron alloys—has its particular sphere of use, but steel, of course, exceeds all others.

The leading iron ore-producing countries are United States, Russia, France, Sweden, Great Britain, Germany, India, Luxemburg, Spain, Australia, and Algeria. Relatively small quantities are produced in Canada, Mexico, Brazil, Chile, South Africa, Czechoslovakia, Austria, China, the Philippines, and Japan. Canada and Venezuela will before long become fairly large producers, and Brazil has large deposits only slightly developed. This wide distribution raises few international problems, as the only importantly industrialized nations deficient in iron ore are Italy and Japan. Mediterranean sources supply Italy, and Philippine ore flows to Japan although the natural supplier would be the East Asia mainland.

Manganese is the key mineral of all the ferroalloys. Not only is it important for the making of the hard, tough manganese steels, but it is an absolute essential, for which there is yet no substitute, in the making of every ton of steel, some 13 pounds being required for every ton of steel produced. About 95 percent of the consumption of manganese is for metallurgical purposes. Its purpose in steel making, an absolute requirement, is to remove oxygen and sulfur from the iron in order to produce sound, clean metal. But manganese is also used for dry batteries, in the glass industry, and in paints, pigments, dyes, and other products. No industrialized country can get along without manganese.

Usable manganese ores, unlike coal and iron, are not so widely distributed. Although minor production comes from some fifteen countries, 82 percent of it lies in five countries, and it is upon these areas of large deposits that the highly industrialized deficient nations have to depend for their sources of supply. The major part of the 4 to 5 million tons used annually emanates from Russia, the Gold Coast, India, South Africa, and Brazil, with subsidiary production from Morocco, Angola, United States, Cuba, Chile, Egypt, and a few other countries. Large deposits of manganese ore are now being explored and developed in Brazil and the Belgian Congo. It is noteworthy that, save for Russia, all the large consuming countries have to depend upon far-distant sources for their requirements of manganese to make steel. It is thus the most critical bulk mineral for all large nations except Russia. The problem of manganese supply for world use is not so much a bottleneck of deposits but of transportation, chiefly inland rail transportation of producing countries. The United States, for example, was able to obtain during World War II some 1 to 1½ million tons annually, of which Russia was able to supply only a minor amount.

Before World War II the United States supply of manganese came mostly from Russia (about one-half), which then supplied about 50 percent of the entire world's needs from her greatest of all deposits. The rest of the supply came mainly from India, the Gold Coast, South Africa, Brazil, Egypt, and Cuba. Today the United States supply from Russia is largely cut off, and it is improbable that it may be renewed on any large scale for some time to come. It is an impossible situation that United States needs would have to depend upon the whim of a presently unfriendly nation. The United States consumers, therefore, must look elsewhere for their future sources of supply. Of the wartime supply, the resources of Cuba are practically exhausted; the production of Brazil is now partly prohibited from export to conserve it for domestic use, and shipments from India to the United States are curtailed because of inland transportation and other difficulties. South African shipments are also limited because of railroad transportation. A new source of minor supply is Angola. The United States, therefore, faces a critical situation with respect to its steel industry. To meet this its nationals are exploring and developing new deposits in Brazil and, with the aid of government, are experimenting with the utilization of low-grade deposits in the United States and with secondary recovery of manganese from furnace slag. The United States steel industry for some time to come must rely on maintenance of uninterrupted international trade with India, South Africa, and the Gold Coast. Its stockpile will take care of part of an emergency. The leading steel-making nations of the world, except Russia, must program carefully for future supplies of manganese in the event of threatened emergency.

Great Britain depends upon Commonwealth sources of supply and has first call on South African, Gold Coast, and Egyptian sources, while Indian manganese, which is free to move, seeks world markets. India has indicated a desire to manufacture her manganese ore into the more valuable ferromanganese and thereby create greater domestic industry and larger foreign exchange. Large new deposits in Brazil may eventually supply United States shortages. In the event of war imports into European countries might be cut off.

Chrome is another important ferroalloy desired chiefly for making tough, hard, strong, wear-resisting steels, and stainless steel. There are three varieties of chrome: metallurgical, refractory, and chemical. Metallurgical chrome is also used for plating and making other alloys with tungsten, molybdenum, and cobalt. The refractory variety is in high demand to make linings for many kinds of high-temperature furnaces, and the chemical variety is sought for dyeing, tanning, bleaching, pigments, and oxidizing agents.

Chrome is another of the bulk alloy minerals that has little production in the countries of greatest consumption and little consumption in the countries of greatest production, save in Russia. The 1 to 2 million tons annually produced come chiefly from Russia, Southern Rhodesia, Turkey, South Africa, Cuba, the Philippines, Yugoslavia, India, Greece, Japan, and New Caledonia. Minor production comes from Brazil, Bulgaria, and Canada.

The United States chief chrome problem is its large requirement for metallurgical chrome and its almost negligible resources of this variety. It does have fairly large deposits of low-grade, high-cost chemical chrome partly developed in Montana. Its imports of about 1 million tons normally come chiefly from Rhodesia, Russia, Turkey, the Philippines, South Africa, Cuba, New Caledonia, Yugoslavia, and Turkey. The large Russian imports of the past are now practically nonexistent and cannot be relied on in the future. Other countries with good-quality metallurgical chrome are Rhodesia, Turkey, and New Caledonia; minor quantities are found in Greece, Yugoslavia, Brazil, and India. The New Caledonia supply is running low, which will leave Rhodesia and Turkey as chief sources, with the others as minor sources. If Turkey should become embraced within Russian domination, that source of supply might be denied to the United States and Europe. Chief reliance for high-quality ores would then rest upon Rhodesia, whose resources are none too great and upon which Great Britain has first call for her own needs. In the event of war European requirements might be denied to them. The United States in the future may have to turn to the utilization of refractory or chemical ores at present unfit for existing metallurgical practice. The leading consumers of chrome, except Russia, have to plan ahead for future supplies and to guard against disruption of supplies through possible war. For the United States, the supplies of chrome and manganese involve more than mere supplies of ore; they touch upon national economic life and national security.

Distribution of Critical Minerals

Copper can serve as an example of another type of mineral. It is not used as a part of the steel industry, but its use multiplies along with that of steel, as it is consumed along with steel in the manufacturing and electrical industries and in the building of all vehicles of transport. Copper is one of the most essential minerals. It is in greatest demand for all equipment that makes, transmits, and uses electricity, but brass, bronze, and other alloys consume much copper.

Although copper is widely distributed, about 80 percent of it emanates from four regions: the southwestern United States; the Andean belt; the central African belt; and the Canadian Shield. Additional copper districts are found in other parts of the United States, and in Mexico, Russia, Japan, Australia, Finland, Sweden, Spain, Portugal, Germany, Yugoslavia, and southern Africa.

The United States has long been the world's greatest producer of copper, most of which has come from the great "porphyry coppers," Butte, Montana, and the Lake Superior district. It has been the greatest consumer in the world, but now with a rapidly growing population which has become mechanically minded in the home, farm, and factory, and which is the world's greatest consumer of electricity, consumption has jumped more rapidly than production. Formerly an exporter of copper to Europe, the United States in recent years has been a net importer, to the extent of one-quarter to one-third of its primary production. It seems doubtful that the United States will ever again produce all the copper it needs. It therefore faces the new problem of establishing other dependable supplies for the future. We saw earlier in this chapter that the government estimates gave 19 years of life for United States resources, based upon the 1935–1944 rate of consumption, but the rate is now higher. Even though the estimate of reserves may be wrong by as much as 50 percent, there is still a short life for a great industrialized and electrically minded country like the United States. Moreover, sources of supply must be large to sustain the high demand. Outside of the United States, the only known large sources of supply are Chile, Northern Rhodesia, Belgian Congo, and Canada, with smaller output from Peru, Mexico, and South Africa. Of the chief sources, that of the Belgian Congo flows to Belgium and, by long arrangement, the remainder to Great Britain. The Rhodesian copper and some Canadian copper flow to Great Britain. That leaves Chile as the only large source of United States supply for the future, except for a small surplus of Canadian copper. Chile's copper resources vastly exceed those of the United States and, under normal rates of production, have a life four to five times greater than those of the United States. Copper from Chile moves also to Europe. This means that relations between the United States and Chile should remain on such friendly terms that Chilean copper will always continue to flow to the United States, and under conditions that inure to the benefit of both countries.

Tin is another metal of international and political interest because of its restricted distribution, both geographically and politically. Desired chiefly for tin-plated food containers, bearing metal, bronzes, and solders, it is widely used in most countries, but the United States, although the producer of none, is by far the largest consumer. The three chief centers of production are Malaya, the Netherlands Indies, and Bolivia, with lesser amounts in Nigeria, the Congo, China, Thailand, and Australia. About two-thirds of the production is in British domain and Indonesia. About 60 percent of the production lies in the Far East in one elongated belt extending from the Indonesian islands at the tip of Malaya to China in the north—a strictly eastern Asiatic domain where political unrest and future uncertainty prevail. Nearly one-quarter of the world production comes from resources in Bolivia, a country of notable political instability. The Chinese and much of the Thailand production flows to Russia; the remainder seeks competitive world markets. Normally, there is more than enough production of tin to supply world demand. The problem, therefore, is not one of scarcity of supply, as in the case of the metals discussed above, but largely one of distribution. Like diamonds, its allocation and distribution have been handled by a cartel that has served producers and consumers with equity and fairness. The chief problem in relation to tin is the location of its resources in areas of political uncertainty, so that in the event of emergency the supply may be denied to the large consumers of the world.

Uranium, the atomic mineral now in everyone's mind, may become one of the most important of international minerals in the future. It is widely distributed in minute quantities but is rarely concentrated in large or rich deposits. At present the large producing regions are few and the large consumers even fewer. So far, there are probably five important areas of resources. They are Katanga in the Belgian Congo, northwest Canada, the United States, Czechoslovakia-Erzgebirge region, and Russia, the first two being the chief sources of supply. The last two are presently veiled in secrecy. At present the output of the first three areas moves mostly to the United States, and that of the last two to Russia. New sources of supply are being developed, notably in Canada and South Africa and presumably in

Conservation of Mineral Resources

Russia. The South African source, in the great Witwatersrand gold deposits, may prove to be the greatest known developed source, although it is extremely low-grade as compared to the rich vein deposits of the Congo and Canada.

Large potential supplies of uranium may be won in the future from thin, black, carbonaceous marine shales, and similar black shales that are the source beds of petroleum. These have been investigated in the United States, Sweden, Russia, and other areas. The uranium-bearing varieties may prove to be fairly widespread, particularly in petroleum regions, although the richest known so far lie in Sweden, a nonpetroliferous region.

It is presumed that the large consuming areas of the future will be the United States, Russia, Great Britain, Canada, and France. The present flow of the mineral outside Russian domain is through international agreements, which presumably will be continued.

The eight examples of key minerals discussed above are chosen to disclose some of the problems raised by the distribution and international flow of mineral resources. Other strategic examples, such as lead, zinc, tungsten, diamonds, and mica, yield similar, although somewhat diversified, problems. They indicate a world challenge for alert and sound political policies in order that nations deficient in one or more of these critical and waning minerals may obtain their requirements, which are the foundation of security and future industrial prosperity.

CONSERVATION OF MINERAL RESOURCES

Conservation of mineral resources is particularly important because mineral resources are irreplaceable assets that often have been wantonly and wastefully exploited. They do not regrow like forests or wheat. There is no mineral "crop." The older concept of conservation against possible future shortages of minerals by withdrawing mineral resources from exploitation has given way to conservation by preventing waste during production; by wise use of scarce minerals; by utilizing low-grade materials; by the development of substitutes and of preservatives; and by some conserving of strategic minerals in favor of importation during times of peace. These questions concern both public interest and private ownership.

Most large organizations today propagate their mineral reserves by developing new deposits to replace exhausted ones, by utilizing research in mining and metallurgy, and by practicing conservation to extend what they already have for long-time economy. It is commonly assumed that measures of conservation must be imposed on owners of mineral lands, presuming that they are motivated only by selfish interests, and that public action alone can bring the desired results. This is unquestionably true of owners of oil and gas resources. It is also largely true of the small owner whose sole interest is concentrated in one property, without expectation of continuance elsewhere, but is not true of the larger organizations that can afford research and carry on exploration of many properties in the hope that one may prove productive. In general, conservation acts that prohibit development of mineral resources so that they may be conserved for posterity do more harm than good because they inhibit the growth of industries that give rise to other permanent industries, particularly in regions devoid of other opportunities of development. Enforced conservation of petroleum and natural gas, however, has proved beneficial to the public and the petroleum industry.

Prevention or elimination of loss or waste of ore during mining is highly desirable. Too often a thin seam of coal is lost during the mining of a thick one; or a layer of marginal ore is left behind in the mining of adjacent profitable ore; or pillars of coal or ore are left behind to support the rock roof because they are cheaper than artificial supports. These unrecoverable losses of minerals are made to permit a higher profit per ton of the minerals extracted, and such practices should not be condoned. Again, rock below economic grade is too often sent to waste dumps and intermingled with barren material, thus eliminating its possible future use, instead of segregating it against the time when higher prices or lower costs might permit its utilization.

Improvements in mining methods and mineral processing, such as oil flotation of ores, have greatly aided mineral conservation by permitting utilization of lower-grade materials and thereby increasing the ore reserves. Much still remains to be done in increasing extraction of tin from tin ores, in utilizing lower-grade chrome, manganese, and bauxite ores, in decreasing slag losses in zinc, manganese, and other metals, and in improved processing of nonmetallic minerals. Losses of zinc in slag are particularly bad. When it and its companions, lead or copper, enter the blast furnace, the lead or copper behave reasonably and the molten metal mostly sinks to the bottom where it is drawn off. But the elusive zinc seeks a home in the slag and comes to rest in the cold slag dump of most smelters. Millions upon millions of tons of smelter slag rear their black faces in smelter yards with up to 15 percent or more zinc in them. Its recovery is a problem of conservation.

For the most part metals are indestructible; even though they may change their form, they are manufactured into materials that serve their purpose and become worn out or discarded. They enter the scrap pile. But they are still metals—metals that have been eagerly prospected for, mined, processed, and have become a form of wealth. Something that once had value as a metal still has value, even in the form of scrap. Although some secondary metal is recovered, much too much is wasted, and virgin metal sought in its place. Most of the zinc is wasted when copper is recovered from brass scrap, and so are many of the scarce alloy metals. There is still need of further conservation by better and wider recovery of scrap metals.

Wise use of scarce minerals could be more widely practiced. Restrictions have been accepted for tin and atomic minerals, and this could be extended to other scarce minerals whose use for secondary purposes could be restricted; other minerals or synthetic products could be substituted in their place, thereby reserving the scarce minerals for uses for which there are no known substitutes.

Perhaps the greatest field of mineral saving is the development of substitutes for metals, where a more common metal can be used for a scarce one or, particularly, where synthetic products, such as plastics, can entirely replace metal. If, for example, the lead sheathings of electrical cables could be replaced by an entirely satisfactory substitute, some 155,000 tons of lead per year could be saved in the United States, thereby easing the drain on rapidly diminishing lead resources. Other substances, such as ceramic products, glass, spun glass, laminates, and plywoods, can be used to conserve metals.

A newer field in metal conservation is the development of inhibitors and protectors to help save metals, in the form of lacquers, special paints, plastic coatings, and other compounds. These not only protect metals against corrosion but permit the use of common corrosive metals in place of scarce noncorrosive metals. Chromium, nickel, tin, or cadmium platings can be displaced by synthetic protectors.

These and other conservation practices will help make unreplaceable minerals go further and conserve waning mineral resources. The chemical laboratory serves conservation better than man-made laws, and it should be called upon to do so to a greater extent in the future.

SOME ECONOMIC FACTORS IN THE DEVELOPMENT OF MINERAL DEPOSITS

Exploration for mineral deposits and their development are affected in large measure by various economic factors. If a new deposit, or an indication of a possible deposit, is discovered, considerable expenditure of money is necessary to determine if it has merit. Most prospects, as they are called, do not turn out to be profitable mines. Only a rare one does, and even this cannot be determined until much exploratory and development work has been expended upon it for a period of one to several years. The few prospects that turn out profitably have to carry the losses on the many that fail.

Mining operations, if successful, yield a return from the sale of the metals and minerals that exceeds the cost of extracting them. This is often erroneously called profit. Actually, the excess of sales over cost may not be profit at all, as mining represents depletion of irreplaceable assets in the ground. When these are totally removed the capital assets have vanished. Therefore the return from each ton of metal or mineral that is sold has to repay, first, the cost of extraction, freight, and marketing, and second, the proportionate part of the original investment and development costs, with interest. Any excess above these two items might then be considered as profit. In simplified terms, if a ton of mineral produced at a total cost of $20 were sold for $25, and $5 a ton represents the original investment, the operation would be profitless. It would merely be a return of original capital. Profit would occur only if the selling price exceeded $25. Obviously no one is going to invest his money in developing mineral resources merely to get his money back without profit. Hence a strong profit incentive must exist to stimulate the development of mineral resources, particularly since, as may be gleaned from the previous chapters, the many vagaries of mineral bodies so often make the mining of them a hazardous investment venture.

Because of the hazards of mining a larger profit is deemed necessary than in fabrication or manufacturing in order to justify the risk of capital. If the profit incentive is high geographic and transportation handicaps of developing new mineral deposits are generally overcome by engineering skills. If the profit incentive is low the development of new mineral resources dwindles or ceases, unless supported by government subsidy. Some of the factors that lessen incentives to develop mineral resources are the character of taxation, unreasonably high taxa-

Economic Factors in Development 341

tion, inept political interference, bad labor conditions, tariffs and export taxes, inability to export capital returns and profits, artificial exchange controls, and other trade barriers.

The deterrents of high taxation, depletion allowances, and write-offs of unsuccessful exploration apply to the United States and many other countries, and have greatly curtailed expenditures for mineral exploration and development. United States tariffs, foreign export taxes, and the other trade barriers referred to above relate to the development of mineral resources by outside capital in other lands. In one region of large metal production the rate of progressive taxation is so high that all incentive to increase production is removed. Further production of additional metal would yield no profit, and valuable future assets in the ground would thus be wasted to the owners. In many parts of the world today uncertain rates of foreign exchange, high export taxes, political controls of income allowable for development, replacements and betterments, and restrictions on export of capital returns and profits are effective deterrents to development of new mineral resources within those countries by outside capital.

In a world that is undergoing inflation, the cost of producing minerals is rising even higher. Normally, the price of minerals also rises at the same time so that a spread of profit persists. In the case of gold, however, its price, being fixed by law, cannot rise with rising costs of production; hence the spread of profit becomes squeezed smaller and smaller. Because of this, most of the gold mines in the United States and many in other countries have been forced to close.

International cartels or associations have long played a part in the international movement of many minerals. They have usually been formed and dominated by the producers of minerals, but there has also been government participation directly or indirectly. The primary purpose of cartels has been to stabilize prices, markets, and employment against wide or violent fluctuations resulting from open market conditions. This is usually accomplished by restriction or expansion of production to meet anticipated demands, by proration or allocation of production, and by research to discover new uses and new market outlets. Sometimes cartels became monopolistic and exerted undesirable influences upon consumers of minerals, but such cartels have now largely disappeared. For the most part existing cartels in minerals exert desirable stabilization, particularly in employment and in prices.

The diamond cartel, for example, draws most of its supply of diamonds from the countries of central Africa, notably Belgian Congo,

Angola, Sierra Leone, the Gold Coast, and Nigeria, South Africa, South-West Africa, and French Africa. These are distant regions of difficult transportation and sparsely scattered labor, which has no other industry to support them. Supplies have to be programmed a year or so ahead. If violent fluctuations of demand occurred, labor would become widely dispersed without adequate food and would be slow, difficult, and costly to regain. Supplies stopped during light demand could not be obtained quickly for reopening, and unemployment would persist for longer than necessary. These undesirable and costly features are eliminated through the action of the cartel which assures continuous operation. The tin cartel, drawing supplies mainly from southeast Asia and Indonesia, serves a similar purpose. This has now been reorganized into an International Tin Council, combining a Tin Study Group and a Tin Research Institute, made up of members of 15 countries, including the chief consumers as well as producers.

FUTURE OUTLOOK FOR MINERALS

With waning mineral resources and increasing demand by expanding population and industrialization, the question arises, What of the future outlook for minerals? It is a misfortune, as stated earlier, that the more we manufacture, the greater are the inroads upon the very basis of industry; the more goods that are sold abroad in world trade, the more mineral resources are depleted, and most of the metals in exported goods do not return for scrap. Unfortunately, there still exists a smug complacency that all that is needed is to go out and dig more. That is no longer easy. The readily found deposits have now been largely discovered; the obvious places to search are mostly eliminated. Investigations of the future, therefore, must be directed to the less obvious places for exploration of minerals. For this there is needed a basic knowledge of the origin, occurrence, and controls of minerals, such as has been covered in Chapters 4 to 14, and a knowledge of scientific techniques of exploration such as those summarized in Chapter 15. Not only should attention be directed to the underdeveloped and less explored regions of the earth, but newer scientific methods should be applied with concentrated vigor to the older exhausted mining districts. Still other methods and skills need to be brought to bear in an attempt to discover minerals covered by deep soils, gravels, or later rock formations.

In western North America (including Alaska), in the Andean region of South America, in the Precambrian shields of Canada and

Brazil, in the great plateau region of central and southern Africa, are vast expanses of geologically favorable areas for mineralization which are covered by late lava flows, deposits of valley wash, late sedimentary formations, glacial deposits, tundra, tropical forests, or soils. Only a very minor part of the bedrock is exposed to view—the rest lies hidden—and the bedrock is the home of most mineral deposits. In such places mineral deposits unquestionably lie hidden, awaiting the means to discover them. In Arizona, a distinctive copper province, are many notable copper districts, such as Bisbee, Morenci, Ray, Miami, Inspiration, Globe, Ajo, and Jerome. These all occur in ranges of hills or mountains, which viewed from the air look like black ships sparsely scattered in a sea of sand. Why should all these deposits be confined to hills? Obviously because the hills are places of outcropping bedrock that could be seen and prospected. There are also, of course, accompanying geologic features like faults, folds, and intrusives that favor localization of mineralization. But, in between, generally similar conditions must prevail under parts of the sea of sand that cannot be seen. If a minor part of the bedrock contains such an array of world-famous copper districts, then it is a reasonable conclusion that a much larger area hidden by overlying gravel and sand contains at least an equal number of undiscovered mineral deposits of equal size. But the problem of how to find such deposits in Arizona or Africa or Canada or Brazil has not yet been solved. The surface has only been slightly scratched. Here then lies a challenge of first magnitude to the geologist, in collaboration of the geophysicist, physicist, and chemist. Here also lies the brightest outlook for future mineral resources.

Another future source of minerals lies in the utilization of materials that today are too low-grade to be profitably exploited. We have witnessed in the past the gradual lowering of the average metal content of most ores. For example, in 1910 the average content of copper in ores mined in the United States was 1.88 percent copper; in 1920, 1.63 percent; in 1930, 1.43 percent; in 1940, 1.20 percent; in 1950, 0.90 percent. There will probably be still further reduction, and each little drop in percentage of minable grade of ore permits the utilization of previous uneconomic material, and thereby increases the ore reserves. Of course there is a limit to the amount of reduction of minable grade of copper ores. This fine example of increase in mineral resources is due to managerial skill and efficiency in which both mining and metallurgy were a part of one operation under one management. Mining improvements that permitted the mining of lower-grade ores were met immediately by metallurgical improvements that permitted metal extraction from lower-grade materials supplied. Conversely, metallurgical advances that increased efficiency and lowered the cost of operation were in turn met by mining improvements to extract lower-grade rock.

However, the same degree of reduction has not been applied to the ores of many other metals and minerals. The average grade of manganese or chrome, for example, has not dropped comparably with that of copper. Part of the reason for this is that the metallurgy and mining are separated in ownership and location. The metallurgical users have dictated the specifications of the ores and have kept them high for their own advantage, with the result that less marginal ore has been brought into the classification of ore reserves. However, we are witnessing today a change in this trend in the case of United States iron ores. With the rapid exhaustion and the short life of the high-grade open-pit iron ores of the Lake Superior region, much research is now in progress aimed to utilize by concentration the huge reserves of low-grade disseminated "taconite" material that at the moment is below the economic limit of operation. Similar progress could be applied to many other metals and minerals and thereby provide greater mineral resources for the future.

SELECTED REFERENCES ON MINERAL RESOURCES

Mineral Resources of the United States. By staffs of the U. S. Bureau of Mines and U. S. Geological Survey. Pp. 212. Public Affairs Press, Washington, D. C., 1948. Summary of domestic reserves and sufficiency; separate treatment of 39 minerals.

Ore Deposits of the Western United States—Lindgren Volume. Am. Inst. Min. Met. Eng., New York, 1933. Contributions by 44 authors relating to the various groups of mineral deposits in the western United States.

Industrial Minerals and Rocks. 2nd Edit. Edited by S. H. Dolbear and Oliver Bowles. Am. Inst. Min. Met. Eng., New York, 1949. Articles by a group of authorities covering each of the industrial nonmetallic minerals and rocks.

Mineral Raw Materials. U. S. Bureau of Mines. McGraw-Hill, New York, 1937. Treatment of 32 minerals giving use, technology, and distribution; survey of mineral position of important industrial countries.

Minerals in World Affairs. T. S. Lovering. Prentice-Hall, New York, 1943. The uses, technology, distribution, geology, country, occurrences, and political features of 16 important metals and minerals.

Seventy-five Years of Progress in the Mineral Industry—Anniversary Volume. Edited by A. B. Parsons. Am. Inst. Min. Met. Eng., New York, 1947. Authoritative contributions by group of authors on various phases of the mineral industry.

Strategic Mineral Supplies. G. A. Roush. McGraw-Hill, New York, 1939. Discussion of United States strategic minerals.

Selected References

Mineral Economics. Edited by F. G. Tryon and E. C. Eckel. McGraw-Hill, New York, 1932.

"Political Control of Mineral Resources." C. K. Leith. *Foreign Affairs*, July, 1925. Essay on mineral geography.

World Minerals and World Peace. C. K. Leith, J. W. Furness, and C. Lewis. Brookings Institute, Washington, D. C., 1943. Survey of production, resources, geographic distribution, cartels, and policy.

"Wartime Dependence on Foreign Minerals." Alan M. Bateman. *Econ. Geol.* 41:308–327, 1946. Sources, amounts, and problems of obtaining some 60 strategic minerals for wartime industry.

Our Natural Resources and Their Conservation. 2nd Edit. A. E. Parkins and J. R. Whitaker. John Wiley & Sons, New York, 1939. Good chapters on conservation of minerals, soils, and water.

"America's Stake in World Mineral Resources." Alan M. Bateman. *Min. Engr.* 1:23–27, 1949. Survey of domestic and foreign mineral resources and policy problems.

Economic Mineral Deposits. 2nd Edit. Alan M. Bateman. John Wiley & Sons, New York, 1950. Pp. 366–375; Part II, "Metallic Mineral Deposits," 419–628; Part III, "Nonmetallic Mineral Deposits," 631–875.

ADDITIONAL GENERAL REFERENCES

The following list is a group of general references, in addition to the selected references given under each chapter, which the reader may wish to refer to for further information or references.

Journal of Economic Geology. Economic Geology Publishing Co., New Haven, Conn. Eight numbers per year. From Vol. 1, 1905, to present. The outstanding journal of the world relating to economic geology and containing original articles on all phases of economic geology.

Annotated Bibliography of Economic Geology. Economic Geology Publishing Co., Urbana, Ill. Two numbers per year. From Vol. 1, 1933, to present. Gives abstracts of all important papers in the world relating to economic geology.

Bibliography and Index of North American Geology. U. S. Geological Survey, Washington, D. C. Yearly, from 1785 to present; consolidated each 10 years. Gives authors and titles of all articles on geology in North America.

Bibliography and Index of Geology Exclusive of North America. Geol. Soc. America, New York. Yearly, from 1933 to present. Covers foreign geology by authors and titles, with brief annotations.

Minerals Yearbook. Annual volumes. U. S. Bureau of Mines, Washington, D. C. Annual volumes giving statistics and reviews for each metal and nonmetal, covering production, consumption, trade, prices, and reviews by states and countries.

Transactions, American Institute of Mining and Metallurgical Engineers. New York. Yearly since 1871. Contains many papers related to economic geology.

Mineral Resources of the United States. By staffs of the U. S. Bureau of Mines and U. S. Geological Survey. Pp. 212. Public Affairs Press, Washington, D. C., 1948. Summary of domestic reserves and sufficiency; separate treatment of 39 minerals.

Ore Deposits of the Western United States—Lindgren Volume. Pp. 797. Amer. Inst. Min. Met. Engrs., New York, 1933. Contributions by 44 authors relating to the various groups of mineral deposits in the western United States.

Ore Deposits as Related to Structural Features. Edited by W. H. Newhouse. Pp. 280. Princeton Univ. Press, Princeton, N. J., 1942. Brief résumés of most important mineral deposits that exhibit localization of ore by structural features.

Industrial Minerals and Rocks. 2nd Edit. Edited by S. H. Dolbear and Oliver Bowles. Pp. 1156. Amer. Inst. Min. Met. Engrs., New York, 1949. Articles by a group of authorities covering each of the industrial nonmetallic minerals and rocks.

Mineral Raw Materials. U. S. Bureau of Mines. Pp. 342. McGraw-Hill Book Co., New York, 1937. Treatment of 32 minerals giving use, technology, and distribution; survey of mineral position of important industrial countries.

Additional General References

Minerals in World Affairs. T. S. Lovering. Pp. 394. Prentice-Hall, New York, 1943. The uses, technology, distribution, geology, country, occurrences, and political features of 16 important metals and minerals.

Seventy-five Years of Progress in the Mineral Industry—Anniversary Volume. Edited by A. B. Parsons. Am. Inst. Min. Met. Eng., New York, 1947. Authoritative contributions by group of authors on various phases of the mineral industry.

Structural Geology of Canadian Ore Deposits—A Symposium. Pp. 948. Geology Division Can. Inst. Min. & Met., Montreal, 1948. A comprehensive assemblage by belts or areas of the geology and structure of all Canadian mining districts or mines, with emphasis on their structural controls.

APPENDIX

Geological Time Table

Major Divisions Eras Duration, years		Subdivisions Periods	Major Economic Events
Cenozoic 70,000,000		Recent	Placer gold, diamonds, titanium, etc.
		Pleistocene	Glacial: placer gold, diamonds
	Tertiary	Pliocene *	Placer gold, diamonds
		Miocene	America: gold, silver, tellurium, antimony
		Oligocene	
		Eocene	Americas: copper, gold, silver, zinc, lead, molybdenum, coal
			Europe: iron, gold, silver, nickel, cobalt, uranium, lead, zinc, mercury
70,000,000		Paleocene	Asia: gold, silver, copper, lead, zinc, antimony
Mesozoic 130,000,000		Cretaceous	Americas: copper, gold, silver, zinc, lead, coal
			Asia: copper, gold, silver, zinc, lead, coal
		Jurassic	America: copper, uranium
			Europe: iron ores, coal
200,000,000		Triassic	
		Permo-Triassic	Europe: potash, salt, gypsum, copper, phosphate
		Permian	Europe: tin, copper, gold, zinc, lead, platinum, tungsten
			America: salt, gypsum, potash
Paleozoic 300,000,000	Carboniferous	Pennsylvanian	Coal
		Mississippian	Coal
		Devonian	
		Silurian	America: sedimentary iron ores, salt, gypsum
			Europe: iron, salt, gypsum, copper, nickel, titanium
		Ordovician	
500,000,000		Cambrian	
Precambrian 1,500,000,000		Late	America: iron, titanium, gold, silver, copper, zinc, nickel, cobalt
		Middle	Europe-Asia: iron, gold, copper, zinc, manganese, tungsten
2,000,000,000±		Early	Africa: iron, gold, copper, manganese, chrome, tin, asbestos

* Oil occurs in all periods from upper Cambrian to Pliocene.

Glossary

(Explanations rather than hard-and-fast definitions. Numbers in parentheses refer to the pages where definitions are given in the text, bold face representing principal ones. Mineral species are not defined.)

ALGAE. Low forms of microscopic plant life.

ALKALIES. Common ones are potassium, sodium, lithium.

ALLUVIAL DEPOSITS. Placer deposits concentrated by stream action (220).

AMYGDALOID. A basic lava whose "blow holes" have been filled by a later mineral (103).

ANAEROBIC. Pertaining to organisms that live without oxygen.

ANORTHOSITE. A granular igneous rock composed mostly of calcium feldspar.

ANTICLINE. An uparched fold in stratified rocks (179).

AQUIFER. A pervious stratum between impervious strata that conducts underground water; used mostly in connection with artesian water.

ARTESIAN WATER. Water that is confined within a permeable bed and that rises under pressure to approximately the height of the intake. If the outlet (well or spring) is well below the height of intake, water will flow out under pressure; if even or above, the water will rise in the well but will not flow out (257).

BAR THEORY. See page 191.

BARYSPHERE. A deep-seated heavy part of the earth.

BASIC MAGMAS. Those rich in iron, magnesium, and calcium.

BATHOLITH. A huge domed intrusive igneous body of at least 40 square miles in extent, whose sides slope gently outward, enlarging downward. It presumably has no bottom (36).

BAUXITE. The ore of aluminum, composed of minerals containing aluminum, oxygen, and water (209).

BEARING METAL. Metal employed for axle bearings.

BEDDING PLANES. The surfaces between individual beds of sedimentary rocks.

BIOHERM. Ancient buried reef rocks of coral, bryozoa, algae, etc. (184).

BLISTER COPPER. Copper matte that has had the sulfur removed in the first stage of refining but still contains other metals.

BONANZA. Part of a precious mineral deposit that is especially rich (228).

BOXWORK. See Limonite boxwork.

BRASS. An alloy of copper and zinc.

BRECCIA. A rock composed of fragments of rocks or minerals (**104**, 11).

BRONZE. An alloy of copper, tin, and zinc.

BUSHVELD IGNEOUS COMPLEX. A great intrusive igneous body in the Transvaal that has undergone remarkable magmatic differentiation (78).

CALICHE. Surface or near-surface deposits of soluble salts resulting from evaporation.

CAPROCK. An impervious rock (generally shale) overlying an oil reservoir rock and confining the oil beneath it (179).

CARBONATE ROCKS. Limestone, magnesian limestone, dolomite, magnesite.
CATALYTIC DEPOSITION. Deposition induced by a catalyst, which is a substance that promotes a chemical action without taking part itself (145).
CAVITY-FILLING DEPOSITS. Those formed by the deposition of minerals in cavities or rock openings (100).
CEMENT COPPER. Copper precipitated by iron from copper sulfate solution (236).
CHROME. (Not a mineral.) A term commonly used to indicate ore of chromium, consisting of the mineral chromite.
CLAY MINERALS. A family of extremely fine-grained minerals consisting chiefly of aluminum, silica, and water.
COCKADE ORE. Crusts of different minerals deposited successively around rock fragments.
COLLAPSE BRECCIA. Breccia formed by collapse of the roof of a cave (124).
COLLOIDAL. A state of subdivision of suspended matter in which the particle size ranges between 5 millionths and 200 millionths of a millimeter.
COMPLEX ORES. Ores containing several metals (9).
CONCENTRATES. Final product of ore minerals concentrated from ores during which process waste is mostly eliminated (307).
CONCENTRATION. Separation and accumulation of metal or ore mineral from gangue.
CONCORDANT. Having parallel structure.
CONGLOMERATE. Sedimentary rock of consolidated gravel (12).
CONTACT. Bounding surface between different rock units or between mineral bodies and enclosing rocks.
CONTACT METAMORPHISM. High-temperature alteration of the rocks adjacent to a deep-seated intrusive as a result of the heat and gases given off by the cooling intrusive. Little material is added from the magma (91).
CONTACT-METASOMATIC DEPOSITS. Deposits formed by high-temperature magmatic emanations around igneous contacts.
CONTACT METASOMATISM. Replacement of the contact rocks adjacent to an intrusive resulting from high-temperature emanations, from a deep-seated magma, from which constituents are carried out that combine with some of the rock constituents to form a suite of high-temperature minerals. Much material is added (91).
CRUSTIFICATION. Layering of crusts of different minerals deposited successively on the walls of a cavity (**110**, 56).
CRYSTALLIZATION. Formation of mineral crystals during cooling of a magma or by deposition from a solution.
CRYSTALLIZATION DIFFERENTIATION. Magma differentiation by separation of crystals from the melt.
DIATOMS. Microscopic plant life that lives in water.
DIFFERENTIATES. Different kinds of igneous rocks formed as result of magmatic differentiation (37).
DIFFERENTIATION. A process whereby two or more rocks of different composition, or solutions, are derived from a single body of magma (37).
DIFFUSION. Uniform spreading of a solute in a solution, or the movement of ions through liquids or crystal lattices.
DIKE. A body of igneous rock occupying a fissure (36).

Glossary

DIP. The angle between an inclined stratum or fissure and the horizontal.
DISSOCIATION. Decomposition of a chemical compound by heat (54).
DOLOMITE. Sedimentary rock composed of calcium-magnesium carbonate (12)
DRAG FOLD. Minor folding of a weak bed lying between folded strong beds.
DRUSE. An irregular opening in ore or rock containing mineral crystals.
DYNAMOTHERMAL. Pertaining to processes within the earth involving pressure and heat that bring about changes in rocks.
EARLY MAGMATIC DEPOSITS. Deposits of magmatic origin formed during the early stages of magma solidification (68).
ELUVIAL PLACERS. Placer minerals concentrated near the decomposed outcrop of the source deposit by rain wash, not by stream action (219).
EOLIAN PLACERS. Placers concentrated by wind action.
EPIGENETIC DEPOSITS. Deposits formed later than the rocks that enclose them (267).
EPITHERMAL. Term applied to hydrothermal deposits formed at low temperature and pressure (99).
EVAPORITES. The salts that result from the evaporation of ocean water or saline lakes (63, 188).
EXSOLUTION. Separation of individual minerals in solid solution when temperature is lowered (54).
EXTRUSIVE IGNEOUS BODIES. Magma extruded upon the surface, forming lava flows.
FAULT. A rupturing of rocks during which one wall has moved past the other (112).
FERRIC. Designation for iron salts in which the iron is trivalent; iron in a state of higher oxidation.
FERROUS. Designation for iron salts in which the iron is divalent; iron in a state of lower oxidation.
FILTER PRESSING. Squeezing out of a residual magma from the interstices of a mesh of crystals, a process similar to squeezing water out of a sponge (80).
FISSURE. Tabular openings or fractures formed by rupturing of rocks (104, 112).
FISSURE SYSTEM. A group of parallel fissures of the same age (113).
FISSURE VEIN. A fissure in the rocks filled with minerals (104, **112**).
FLUX. Substances added to a smelter charge to aid metal recovery in smelting operations (309).
FOLD. A bend or undulation in layered rocks caused by compression.
FORAMINIFERA. A microscopic form of shell life that lives in water.
FRACTIONAL CRYSTALLIZATION (OF SALTS). Controlled crystallization of saline waters through which different salts are crystallized out at different temperatures.
FRACTIONATION (OF OIL). The removal of certain ingredients of oil by its passage through clayey or bleaching rocks.
FUMAROLIC. Pertaining to fumaroles or vents near volcanoes from which volcanic gases escape.
GABBRO. A granular igneous rock composed of calcic feldspar and pyroxene; with olivine it is olivine gabbro (13).
GANGUE. Undesired minerals of ore, mostly nonmetallic (7).

GEIGER COUNTER. An ionization chamber that records the number of radioactive particles impinging upon it per minute, thus detecting radioactive substances (293).

GEL. A gelatinous jellylike precipitate from colloidal solution.

GEOLOGIC THERMOMETERS. Minerals that yield information of earth temperatures (54, 93).

GNEISS. A foliated or banded coarse crystalline metamorphic rock whose grains are aligned or elongated into roughly parallel arrangement (13).

GOSSAN. The rust-colored oxidized capping of an ore deposit (231).

GOUGE. Fine, puttylike or rubberlike material composed of ground-up rock along faults.

GRADE (OF ORE). The content of metal or mineral in ore expressed in percentage, ounces per ton, etc. (9).

GREENSTONE. Altered basic igneous rock.

GROUND WATER. That water of atmospheric origin which saturates rock openings beneath the water table.

GROUNDMASS (OF ROCK). Finer crystals or glass that fill in around larger crystals (11).

HYDROSOL. Colloidal solution in water.

HYDROSTATIC PRESSURE. The pressure caused by, or corresponding to, the weight of a column of water at rest.

HYDROTHERMAL SOLUTIONS. Hot waters originating within the earth carrying mineral substances in solution (14).

HYDROUS OXIDES. Oxides of the metals containing combined water.

HYPOGENE ZONE. Zone of primary ore beneath the zones affected by weathering processes (238).

HYPOTHERMAL. Term applied to hydrothermal deposits formed at high temperature and pressure (99).

IGNEOUS. Pertaining to molten matter that originated within the earth (11).

IGNEOUS ROCK. Rock formed by solidification of molten material that originated within the earth (11).

IMMISCIBLE. Pertaining to substances that do not mix, like oil and water (83).

INDIGENOUS LIMONITE. Oxidation limonite of sulfide derivation that remains fixed at the site of the parent sulfide (231).

INTERSTICES. Spaces between mineral granules.

INTRUSIVE. An igneous body formed from the consolidation of magma that was intruded into other rocks. In contrast, lavas are *extruded* upon the surface.

INVERSION POINT. A change in internal structure of a mineral at a given temperature (54).

ION. An atom or group of atoms carrying an electric charge.

KAOLIN. A type of high-grade clay (213).

KAOLINIZATION. Rock alteration causing development of kaolin.

KIMBERLITE. An ultrabasic igneous rock of deep-seated origin that occurs in pipelike masses and in which diamonds are found.

LACCOLITH. A circular dome-shaped body of igneous rock which has a flat floor and which has been intruded into stratified rocks (35).

LADDER VEIN. A series of parallel ore-filled cracks across a dike; if the dike is vertical the cracks are horizontal and resemble ladder rungs (121).

Glossary

LATE MAGMATIC DEPOSITS. Deposits of magmatic origin formed during the late stages of magma consolidation (72).

LATERAL SECRETION. A discarded theory of ore genesis by which metals were supposed to have been dissolved from the wall rocks and redeposited in nearby openings (27).

LIMESTONE. Sedimentary rock composed of calcium carbonate.

LIMONITE. A mixture of oxides of iron combined with water (232).

LIMONITE BOXWORK. Residual oxidation limonite of sulfide derivation left in former sulfide voids and displaying a characteristic meshwork (232).

LIQUID INCLUSION. Inclusions of liquid in solid crystals (54).

LODE. Mineral deposit in solid rock.

MAGMA. Molten rock matter with included gases and vapors (11, 12, 35, 36).

MAGMA RESERVOIR. A chamber within the earth that contains magma.

MAGMATIC. Related to bodies of molten rock within the earth.

MAGMATIC STOPING. Upward movement of magma by melting and prying off blocks of overlying rocks, thereby occupying their space (36).

MAGMATIC WATERS. Waters given off from a magma.

MARGINAL ORE DEPOSITS. Deposits near the lower limit of commercial workability.

MATTE. Metal with some contained sulfur as discharged from a smelting furnace; further refining yields the pure metal.

MEANDERING. Lateral migration of streams by forming large bends which become cut off, thereby shifting the channel.

MECHANICAL CONCENTRATION DEPOSITS. Deposits of heavy minerals concentrated on the surface by moving water or air (216).

MESOTHERMAL. Hydrothermal deposits formed at intermediate temperature and pressure (99).

METALLOGENETIC EPOCHS. Certain geologic periods marked by widespread mineral formation (34, **275**).

METALLOGENETIC PROVINCES. Provinces characterized particularly by one dominant type of mineralization, e.g., Arizona copper province (34, **275**).

METAMORPHIC ROCK. A rock altered from a predecessor rock by changes of texture or mineral composition induced by heat, pressure, or solution acting within the earth (12).

METAMORPHISM. Changes induced by earth agencies.

METASOMATISM. A simultaneous exchange by which a new mineral takes the form and space of an earlier mineral (14, **127**).

METEORIC WATER. Water of atmospheric origin.

MINERAL. A solid inorganic element or compound occurring naturally in the earth's crust (5).

MINERAL DEPOSIT. A body of mineral matter in or on the earth's surface which may be utilized for its industrial mineral or metal content.

MINERALIZING FLUIDS. Liquids and gases that give rise to mineralization.

MONOCLINE. A folded bed that inclines in only one direction (179).

MULLITE. Artificial product composed of aluminum oxide and silica derived by by furnacing the sillimanite group of minerals; used for high-temperature refractories (spark plugs).

NATRON. Sodium carbonate with 10 molecules of water (196).

NONMETALLIC MINERALS. Those minerals that do not have a metallic luster.

OCHER. A natural mineral substance composed of iron oxides and clay and perhaps manganese (215).
OIL SHALE. A type of shale from which oil may be obtained (187).
OÖLITE. Small spheres resembling fish roe (148).
ORE. A mixture of ore minerals and gangue from which one or more metals may be extracted at a profit (8).
ORE GENESIS. Origin of ores.
ORE MAGMA. Ore that supposedly solidified from a magma (28).
ORE MINERAL. A mineral from which one or more metals may be obtained (6).
ORE SHOOT. Concentration of primary ore along certain parts of a rock opening (117).
OUTCROP. The intersection of a rock bed or fissure with the surface.
OXIDATION. A chemical reaction that involves the addition of oxygen to a compound.
OXIDIZED DEPOSITS. Those deposits that have resulted through surficial oxidation.
PARAGENESIS. The sequence of deposition of an assemblage of minerals (110).
PAY STREAK. The areas of concentration of gold in placer deposits.
PEGMATITE. A very coarse phase of intrusive igneous rocks generally consisting mostly of quartz, feldspar, and mica (85, 40, 41).
PERIDOTITE. A granular basic igneous rock consisting of olivine and other dark minerals.
PERMEABILITY. (Geologically) the ability of a fluid or gas to flow through rock (102).
pH. Hydrogen-ion content; pH above 7 indicates alkalinity; below 7, acidity (106).
PHENOCRYST. In an igneous rock, a mineral crystal larger in size than the surrounding crystals.
PHOSPHATE. A compound of phosphorus used as fertilizer (152).
PHOSPHORITE. A sedimentary rock composed chiefly of phosphate.
PHYLLITE. A fine-grained, lustrous, foliated metamorphic rock (12).
PIEZOMETRIC SURFACE. A calculated height to which artesian water will rise in a given area.
PITCHES AND FLATS. Combined horizontal and highly inclined cracks in sedimentary beds, caused by gentle slumping (122).
PLACER DEPOSITS. Deposits of heavy minerals concentrated on the surface by moving water or air (216).
PLAYA. Dried-up lake basin in an arid region (197).
PLUTONIC. Pertaining to great depth in the earth.
PNEUMATOLYTIC. Pertaining to a high-temperature gaseous state, or to magmatic gas action (49).
PORE SPACES. Small spaces between the grains of a rock.
POROSITY. The pore space of a rock, expressed in percentage of the volume of rock (102).
PORPHYRY. Igneous rock characterized by larger crystals in a fine-grained groundmass.
"PORPHYRY COPPER." Disseminated replacement deposit in which the copper minerals occur as discrete grains and veinlets throughout a large volume of rock, which commonly is a porphyry. It is a large-tonnage, low-grade deposit (137).

Glossary

PRIMARY ZONE. Zone of primary ore below the effect of weathering processes (227).

PROBABLE ORE. Ore that is only partly indicated by mine workings.

PROTORE. Unworkable primary mineralization from which supergene enrichment ores have been formed (228).

PROVED ORE RESERVES. Ore that has been blocked out by mine workings on all sides.

PSEUDOMORPH. One mineral that has replaced another and has retained the form and size of the replaced mineral.

QUARTZITE. A metamorphosed sandstone, with sand grains tightly cemented by quartz.

RAISE. A vertical or inclined underground working that has been excavated from the bottom upward (299).

REACTION RIM. A rim of alteration products formed around an earlier crystal by reaction with a later fluid (72).

RECHARGE OF WATER. Putting water back into the body of ground water to augment the ground-water supply.

RECRYSTALLIZATION. A change from one crystal structure to another, generally by a change in temperature (54).

REPLACEMENT. Same as metasomatism.

REPLACEMENT DEPOSITS. Those formed by the replacement of rock by an equal volume of ore.

REPLACEMENT LODE. A fissure or fissures whose walls have been replaced by ore (136).

RESIDUAL CONCENTRATION DEPOSITS. Deposits concentrated at the surface by removal of undesired materials through weathering (205).

RESIDUAL LIQUIDS. Liquid of a magma that remains after most mineral constituents have solidified (46).

RIFFLES. Wooden cleats in the bottom of placer sluice boxes; natural riffles are bedrock irregularities that resemble riffles.

SADDLE REEF. An opening at the crest of a sharp fold in sedimentary rocks, occupied by ore (120).

SALT DOME. An upthrust plug of salt that punches through overlying beds (192).

SANDSTONE. Sedimentary rock of consolidated sand (12).

SCHIST. A foliated metamorphic rock whose grains have roughly parallel arrangement (12).

SCHISTOSITY. Irregular, roughly parallel cleavage induced in some metamorphic rocks.

SCORIA. A cellular lava rock, like a wasp's nest (103).

SECONDARY ENRICHMENT ZONE. See Supergene sulfide zone (227, **238**).

SEDIMENTARY ROCKS. Those formed by deposition of sediments in layers (11).

SEDIMENTATION. The deposition of sediments (142).

SEISMIC PROSPECTING. A geophysical method of prospecting, utilizing earthquake knowledge of the travel speed and reflection of waves through rocks (291).

SERPENTINE. A rock formed by alteration of a basic igneous rock (**248**).

SHALE. Sedimentary rock of consolidated mud or silt (12).

SHEAR ZONE. A zone of rock shearing consisting of many closely spaced, roughly parallel cracks (119).
SHIELDS. Large areas of ancient rocks that have remained relatively stable through most geologic time.
SILICA. A combination of silicon and oxygen (SiO_2).
SILICATES. Minerals containing silica.
SILICIC MAGMAS. Magmas rich in silica.
SILL. An intrusive igneous body injected between and parallel to the bedding planes of sedimentary rocks (36).
SKARN. Garnet, silicate rock, high in iron oxides, resulting from contact metasomatism (92).
SLICKENSIDE. A striated polished surface of a fault caused by one wall rubbing against the other (112).
SMELT. To reduce to metal in a furnace.
SODA NITER. Sodium nitrate (197).
SOLFATARAS. Volcanic gas vents (fumaroles) that give off sulfur.
SPECIFIC SURFACE. Surface area of a unit volume of a substance (105).
STEATITE TALC. A blocky variety of talc which under heat treatment becomes harder than steel.
STOCK. A sharply domed body of intrusive igneous rock (35) that presumably has no floor.
STOCKWORKS. An interlacing network of small, closely spaced, ore-bearing veinlets traversing a mass of rock (120).
STOPE. An underground excavation in ore for removing the ore.
STRATIFORM. Bedded, layered.
STRATIGRAPHIC TRAPS. Oil traps formed primarily as a result of sedimentation (179).
STRIKE. The horizontal course of a vein, fault, bed, etc.
SUBLIMATION. Change of state to a vapor through the action of heat. Sublimates are solids condensed from a vapor phase (90).
SULFATE. A salt of sulfuric acid; a compound of an element with sulfur and oxygen, e.g., lead sulfate ($PbSO_4$).
SULFIDE. A compound of sulfur and another element, e.g., lead sulfide (PbS).
SUPERGENE. Generated from above from the action of surface waters (6).
SUPERGENE ENRICHMENT. Solution of metal by surface waters from the upper part of an ore deposit and its redeposition below, causing enrichment of the underlying ore.
SUPERGENE SULFIDE ENRICHMENT ZONE. A zone of a mineral deposit beneath the oxidized zone where the metal dissolved from above by surface waters has been reprecipitated and added to that below (227, **238**).
SYNCLINE. A downarched fold in stratified rocks (179).
TACONITE. Disseminated iron ore that must be concentrated to make it usable.
TACTITE. Garnet, silicate rock resulting from contact metamorphism or contact metasomatism (92).
TECTONIC BRECCIA. Breccia formed by folding, faulting, or other earth forces (125).
TENOR. The content of metal in an ore expressed in percentage, ounces per ton, etc. (9).
TEXTURE (OF ROCK). Size and arrangement of mineral grains (11).

Glossary

TITANIFEROUS MAGNETITE. Magnetite containing titanium.
TITANOMAGNETITE. Magnetite containing titanium.
TRANSPORTED LIMONITE. Oxidation limonite that is carried away in solution and deposited outside the parent sulfide (232).
TRONA. Sodium carbonate, sodium hydrogen carbonate, and water (196).
ULTRABASIC ROCK. A dark igneous rock composed of black iron-magnesium silicates.
UMBER. A natural mineral paint composed of iron oxide, clay, and considerable manganese.
UNCONFORMITY. An eroded surface of stratified rocks (generally tilted) covered by overlying sediments (182).
VESICLES. Openings in lavas formed by expanding gas (103, 126).
VOLCANIC PIPE. A pipelike opening drilled through rock by explosive volcanic activity (102, **124**).
VUGS. Unfilled cavities remaining in the midst of cavity-filling ore deposits (110).
WEATHERING. Chemical and mechanical breakdown of rocks and minerals under the action of atmospheric agencies.
ZONAL ARRANGEMENT. A zoning of ore or minerals about a hot center, with high-temperature ores inside and low-temperature ores outside (108).
ZONE OF OXIDATION. Upper zone of a mineral deposit that has become oxidized (227).

Index

Adirondack magnetite deposits, 39, 77
Afghanistan, lapis lazuli, 17
Africa, resources, 327
African colonial possessions, 325
Agricola, Georgius, 19
  portrait of, 20
Alaska Juneau gold mine, 9
Algeria, phosphates, 153
Alkali and bitter lakes, 195
Allard Lake, Que., titanium, 283
Altenberg, Germany, stockworks, 120
Amalgamation, 307
Amosite asbestos, 250
Anaerobic bacteria, 154
Ancient shore lines, 179
Andalusite deposits, 252
Anthracite, 168
Anticlinal theory, 178
Anticline, 179
Aphanitic textures, 11
Aristotle, 18
Artesian flow in Florida, 264
Artesian water, 257
Asbestos formation, 248–250
Asia, resources, 326
Atomic energy, 331
Aurignacian, 16
Australasia, resources, 328
Avicenna, 19

Bacteria, 14
  in bauxite, 210
  in coal, 171
  in oil formation, 174, 177
Bacterial pollution in ground water, 262
Baebulo, silver mine, 19
Banat, Hungary, contact deposits of, 92
Bar theory, 191
Barite, residual, 215
Bassick breccia pipe, Colo., 124

Batholith, 36, 272
Bauxite deposits in Arkansas, 211
Bauxite formation, residual, 209–212
Bauxite from syenite, chart of, 210
Bauxite ore, 209
Beach placer formation, 224
Beaumont, Elie de, 25
Bedding planes, 103
Bendigo, Australia, saddle reefs, 121
Bentonite, 166
Bergbüchlein (*Little Book on Ores*), 19
Bicarbonate solutions, 145
Bingham Canyon, Utah, 234
Biochemical deposition, 145
Biochemical processes in weathering, 203
Bioherms, 184
Birmingham, Ala., 148
Bisbee, Ariz., 33
Bitter lakes, 195
Bituminous coal, 168
Bituminous shales, 159
Black band iron ore, 147
Black Hills, S. Dak., artesian water, 265
  gold, 138
Black sands, 225
Blow holes, 103
Blue Bell mine, Ariz., ore shoots of, 118
Bob Ingersoll Dike No. 2, Keystone, S. Dak., 86
Bonanzas, 228
Borate and bromine deposition, 194
Borax Lake, Calif., 197
Borax lakes, 197
Boulder batholith, 274
Brainerd, Minn., 145
Breccia fillings, 122
Breccias, 12, 104
Brick and tile clays, 166

British Commonwealth, 322
British Guiana, bauxite, 212
Brown coal, 168
Bull Domingo mine, Lake City, Colo., 124
Buried coral reefs, 179
Buried hills, 182
Bushveld Igneous Complex, S. Africa, 71, 78, 82
Butte, Mont., fissure systems, 113
  oxidation at, 236
  zoning at, 274

Cactus pipe, Utah, 124, 125
Calamity Gulch, uranium, 160
Calcium sulfate deposition, 190
California gold rush, 221
Cananea, Mexico, oxidation at, 234
Cannel coal, 168
Capillarity, 176
Captain Haroeris, 18
Carbonate cycle, 161–163
Carbonate deposits, 162
Carbonate rocks, kinds, 163
Carbonated waters, 144
Carboniferous coal beds, 170
Carlsbad, Czechoslovakia, 52
Carlsbad, N. Mex., potash, 194
Carnotite-bearing sandstones, 159
Carrizo sand, Tex., 261
Cartels, 341
Caspian Sea, 190
  oil-field waters, 160
Cassandra, Greece, mines of, 18
Catalytic deposition, 145
Caves, 125
Caving, 300
Cavity filling, 111–127
  resulting deposits, 111
  selected references on, 127
Cavity-filling deposits, definition of, 100
Cementation of rocks, 177
Chalk, 163
Champion Spark Plug Co., 252
Chemical controls of ore deposition, 270
Chemical evaporites, cycle of, 188–197
Chile, nitrate, 326
  nitrate deposits, 198

China clays, 166, 214
Chlorophyll porphyrin, 174
Chrome, distribution of, 334
  sands, 225
Chromite bands, figure of, 70, 71
Chrysotile asbestos, 248
Chuquicamata, Chile, 237, 309
  copper, 137
  copper deposit, 139
Clay, cycle of, 164–166
  definition of, 164
  residual formation of, 213
Clay ironstone, 147
Clay minerals, 213
Climax Molybdenum mine, Colo., 137
Clinton iron-ore beds, 148, 149
Coal, 167–172
  distribution of, 331
  occurrence, 172
  origin of, 169–172
Coal beds, to graphite, 247
Coal seams, 301
Cockade ore, 56
  Morococha, Peru, 123
Collapse breccias, 104, 124
Colloidal deposition, 146
Colorado Plateau, 269
  uranium, 160
Common salt deposition, 192
Complex ores, 9
Concentration, 308
Concordant deposits, 76
Conglomerates, 12
Conservation of mineral resources, 337
Contact-metamorphic deposits, 33
Contact metamorphism, 91
Contact-metasomatic deposits, 33, 92
Contact-metasomatic minerals, list of, 96
Contact-metasomatic ore bodies, diagram of, 94
Contact-metasomatic processes, 90–98
  examples of deposits, 97
  selected references on, 97
Contact metasomatism, 92, 272
Cooling cracks, 103
Copper, distribution of, 335–336
  sedimentary cycle of, 156–157
Copper matte, 309

Index

Copper province, Arizona, 343
Coral reefs, 184
Cornish kaolins, 214
Cornwall, England, tin, 274
 tin veins, 33
Cotta, von, 26
Criteria of replacement, 134
Critical minerals, distribution of, 330
Crocidolite asbestos, 250
Crustification, 56, 110
Crystal structure openings, 103
Crystalline schist, 12
Crystallization, 12
Crystallization differentiation, 71, 72
Crystallization-segregation, 70
Cuban iron ores, 207
Cupolas, 108, 273

Dakota sandstone artesian province, 265
Dark Ages, 314
Daubrée, 25
De Re Metallica, 19
Dead Sea, 195
Death Valley, borax, 197
Deposition of minerals, 14
Descartes, 21
Diamond cartel, 341
Diamond deposits, 224
Diamond gravels, Congo, 223
Diamond pipes, S. Africa, 68
Diamond placers, 224
Diamonds, 271
Differential thermal analyses, 164
Differentiation, by crystallization, 38
 products of, 40
Diffusion, 129
Dike, 35, 36
Diodorus, 18
Dip, definition, 112
Disseminated replacement deposits, 136
Dissociation, 54
Divining rod, 21
Dolomite, 12, 163
Dome, 179
Drag folds, 109
Dredging, 299
Drilling mud, 304
Druses, 87

Ducktown, Tenn., 242
Dwars River Bridge, S. Africa, 82

Early magmatic deposits, 68–72
Earth's crust, composition, 5
East Texas oil field, 182
Economic factors in mineral resources, 340–342
Economic geology, 3
Effluent stream, 259
Egyptian culture, 17
Electrical coring, 287
Electrical methods of geophysical prospecting, 284
Electrolytic refining, 310
Electromagnetic methods, geophysical, 288
Electron microscope picture, kaolinite, 165
Eluvial placer formation, 219
Ely, Nev., 34
Emery, 252
Emma chimney, Utah, 124
Encampment, Wyo., replacement deposits, 138
Energy sources of U. S., 168
Eolian placer deposits, 225
Eolian placers, 219
Epithermal deposits, 99
Equipotential method, 285
Estuarine clays, 165
Etna, 47
Euxine, 18
Evaporites, definition of, 63
Exploration for mineral deposits, 278–311, 340
Exsolution, 54
Extraction of metals and minerals, 296–311
 milling, 307–309
 mining methods, 297–307
 selected references on, 310–311
 smelting and refining, 309–310

Fairbanks, Alaska, placer gold, 223
Faults, 179
 as oil traps, 180
Ferric sulfate, 229
Ferroalloy metals, 322

Index

Ferrous sulfate, 229
Ferruginous cherts, 207
Filter pressing, 76, 80
Fire clays, 166, 213
Fissure intersections, 109
Fissure system, 113
  Butte, Mont., 113
Fissure veins, 103, 112–119, 299
  examples of deposits, 119
  terminations of, 115
Fissures, 103, 109, 179
Florida artesian province, 264
Flux, 309
Flying magnetometer, 283, 284
Fold openings, 104
Foraminifera, 162
Frasch process, 306
Freiberg, Germany, fissure veins of, 116
  mining academy, 23
Frood deposit, Sudbury, Ont., 85
Fuller's earth, 166
Fumaroles, 31, 47

Gamma rays, 294
Gangue minerals, definition of, 7
  list of, 8
Garnet, abrasive, 252
Gaseous emanations, 90
Gases, 42
  and vapors, character of, 47–49
Geiger counter, 293
Gemstones, 17
General references, additional, 346–347
Geochemical analyses, prospecting, 295
Geochemical investigations, 280
Geologic thermometers, 54, 93
Geological exploration, 278–282
Geological time table, 349
Geophysical exploration, 282–295
  electrical methods, 284–288
  electromagnetic methods, 288–289
  gravimetric methods, 289–291
  magnetic methods, 283–284
  radiometric and other methods, 293–295
  seismic methods, 291–293
  selected references on, 310–311

Geophysical prospecting, 278
Georgetown, Colo., fissure veins of, 116
Geothermal gradients, 295
Geyserite, 52
Glacial erratic boulders, 280
Glassy texture, 11
Glauber's salt, 196
Glory-holing, 299
Gneiss, 12
Gold Coast bauxite deposits, 212
Gold deposits, 220
Gold nuggets, 220
Gold placers, 220, 224
Golden Fleece, 18, 217
Gossans, 228, 231–233
  false, 233
Grade, definition of, 9
Graphite formation, 246
Gravimeter, 291
Gravimetric methods, geophysical, 289
Gravitative accumulation, 72
Gravitative liquid accumulation, 77
Gravity concentration, 308
Great Bear Lake, Canada, uranium, 157
Great Salt Lake, Utah, 194, 195
Greek culture, 17
Green River, oil shale, 187
Ground water, 51
  deposition from, 198
Ground-water processes, 254–266
  artesian, 257–258
  occurrence, 255
  recharge, 259
  selected references on, 266
  supplies, 261–266
Groundmass, 11
Guanajuato, Mexico, stockworks, 120
Gypsum beds, 191

Halite deposition, 192
Hand sorting, 308
Hannibal, silver mine, 19
Helium, 175
Herodotus, 18
Herzberg formula, 258
High quartz, 93
History of origin of mineral deposits, 16–29

Index

Homestake Gold mine, 133
Host rocks, 105
Hot-spring deposition, 199
Hot-spring waters, 50–52
  composition of, 52
Hot springs, 31, 50
Humic acids, 144
Humphrey spiral, 309
Hurghada oil field, Egypt, 183
Hutton, 25
Hydraulicking, 299
Hydrogen-ion concentration for uranium, 159
Hydrothermal deposits, 99
Hydrothermal mineralization, localization of, 108
Hydrothermal minerals, 14
Hydrothermal processes, 99–141
  selected references on, 127, 140
Hydrothermal solutions, 14, 42
  character of, 49–57
  evidence of, 55
  temperature and pressure, 53
Hypogene zone, 238
Hypothermal deposits, 99

Igneous breccia cavities, 103
Igneous intrusives, 179
Igneous rocks, 11, 30, 40
Immiscible liquid separation, 83
Indigenous limonite, 232, 237
Indonesia, resources, 328
Industrial minerals, 5
Influent stream, 259
Insizwa type, S. Africa, 84
Iron, carbonate deposits of, 206
  cycle of, 147–150
  distribution of, 332
  residual concentration of, 206–208
Iron bacteria, 145
Iron Curtain countries, 324
Iron-ore deposits, examples, 149
Istria, bauxite, 212

Joachimstal, Czechoslovakia, uranium, 157

Kaolin clay, 213
Kaolinite, electron microscope picture of, 165
Kaolins, 166
Karabugaz, Gulf of, 190, 196
Katmai volcano, Alaska, 48, 53
Kennecott, Alaska, 108, 235
  replacement deposits, 138
Kentucky dam site, Tennessee River, 259
Kern County, Calif., magnesite, 163
Keweenaw Peninsula, Mich., copper, 126
Kiangsi Province, China, clays, 214
Kimberley diamond pipe, figure of, 69
Kimberlite, 68
Kiruna, Sweden, magnetite deposit, 81
Knibyshev, Russia, sulfur, 155
Kolm, 160
Kozuke, Japan, sulfur, 155
Krissites district, Greece, 18
Kyanite deposits, 252
  residual, 215

Laccolith, 35
Ladder veins, 121
Lake Bonneville, 195
Lake clays, 165
Lake Sanford, N. Y., deposits, 76
Lake Shore mine, Kirkland Lake, Ont., breccias, 123
Lake Superior copper belt, 268
Lake Superior iron region, 207
Lassen Volcanic National Park, Calif., 51
Late gravitative liquid accumulation, 79
Late magmatic deposits, 72–87
  gravitative liquid accumulation, 75–83
  residual accumulation, 73
Lateral secretion, 27
Laurium, mines of, 19
Leaching, 309
Leadville, Colo., replacement deposits, 138
Lichtenburg, S. Africa, diamond deposits, 224
Limestone, 12, 162
Limonite in gossans, 232
Liquid immiscibility, 72
Liquid inclusions, 55

Liquid inclusions, minerals, 54
Liquids, 42
Lode deposits, exploration for, 279
Long Island, N. Y., ground-water supply, 263
Longwall method of flat-seam coal mining, 304
Longwall methods, 301
Loop method, geophysical, 288
Lorraine, France, iron ores, 150

Mackay, Ida., deposits of, 95
Magma, definition of, 11
Magmas and crystallization, 35
Magmatic deposits, diagram of early and late, 74
  examples of, 87
Magmatic differentiation, 37–43
Magmatic ore deposits, 38
Magmatic processes, 66–88
  classification of, 68
  mode of formation, 67
  resulting deposits, 87
  selected references on, 88
Magmatic stoping, 36
Magmatic vapors and gases, 51
Magnesite, 163
Magnesium salts, 162
Magnetic concentration, 309
Magnetic methods of exploration, 282
Magnetometer, 283
Manganese, cycle of, 150–152
  deposits of, 151–152
  distribution of, 332–333
  residual concentration of, 208
Manganese carbonate, 208
Mansfeld, Germany, copper, 156
Marine clays, 165
Marine deposition, 147
Marl, 162
Marysville, Mont., 95
Mazarron, Spain, fissure veins of, 116
Mercury ores, 310
Merensky Reef, S. Africa, 72
Mesothermal deposits, 99
Metallic minerals, definition of, 6
Metallogenetic provinces and epochs, 34, 275–276
Metallurgy, 3

Metals, list of, 10
Metamorphic processes, 246, 253
  selected references on, 253
Metamorphic rocks, 12
Metamorphism, definition of, 15
Metasomatism, 14
Miami-Globe, Ariz., 234
Middle East oil fields, 185
Mineral deposits, formation of, 59–65
  list of, 61
  selected references on, 65
  materials of, 5–15
  selected references on, 15
  related to igneous activity, 30–43
  selected references on, 43
Mineral localization, controls of, 267–277
  selected references on, 276
Mineral paints, 215
Mineral resources, 1, 312–345
  Africa, 327–328
  Atlantic Treaty countries, 324–326
  British Commonwealth, 322–323
  chief minerals of industry, 314
  conservation of, 337, 339
  distribution of, 313–318
  distribution of critical, 331–337
  economic factors of, 340–342
  future outlook for, 342, 344
  in international relations, 329, 331
  production distribution, 317
  Russia, 323, 324
  selected references on, 344–345
  South America, 326, 327
  sufficiency and deficiency, 318–330
  surplus minerals, 328–329
  table of distribution, 315
  U. S. imports, table, 319
  U. S., reserves, 318–322
Mineral sequence, 110
Mineral sufficiency and deficiency, 318
Mineral zoning, 32
Mineralizing solutions, 42
  composition of, 52
  primary, 45–57
  selected references on, 57
Minerals, outlook for, 342
Mining geology, 3
Mississippi Valley, zinc deposits, 125

Monocline, 179
Monteponi, Sardinia, lead deposit, 270
Mooihoek, S. Africa, 83
Moravia, 16
Morenci, Ariz., clay mine, 298
   contact metasomatism at, 93
Morocco, phosphates, 153
Mt. Bischoff, Australia, tin ore, 105
Multiple fissures, 109

Namaqualand, S. Africa, diamonds, 225
Natural gas, 172–188
Nebona, stele of, 17
Neolithic man, 17
Neptunists, 25
Neutrons, 294
Nevada, gold gravels, 223
New Caledonia, chromite dike, 82
   nickel, 215
Newfoundland, iron ores, 149
Nicaro, Cuba, nickel, 215
Nickel concentration, residual, 214–215
   New Caledonia, 215
   Nicaro, Cuba, 215
Nickel-copper ores, 271
Nickel-sulfide deposits, 84
Nikopol, Ukraine, Russia, 151
Nitrate deposits, 198
Nitrates, 197
Nome, Alaska, placer gold, 225
Nonmetallic minerals, 3
   definition of, 6
Nonmetallics, 10
Noranda, Quebec, Horne mine, 139
Northern Rhodesia, copper, 157
   Copper Belt, 137
Nuggets, gold, 223

Oceanic water, composition, 189
Ochers, 215
Oil and gas, exploration, 280
Oil and gas fields, in U. S., map of, 186
   of world, map of, 186
Oil extraction, 304
Oil flotation, 308
Oil shale, 187

Oölites, 148
Open cut, 298
Open cutting, 297
Ore beneficiation, 3
Ore magmas, 27, 28
Ore minerals, 9
   definition of, 6
   list of, 7
Ore shoots, definition of, 117
Ores, list of, 10
Organic acids, 144
Organic solutions, 145
Origin of mineral deposits, 16–29
   selected references on, 29
Otago, New Zealand, shear zones, 119
Ouray, Colo., fissure veins of, 116
Outcrop, 117
Overlaps, 179, 182
Oxidation and secondary enrichment, 226–245
   causes of deposition, 236–238
   factors controlling oxide ore deposition, 235–238
   factors influencing enrichment, 242
   gossans, 231–233
   oxidation and solution, 228–234
   secondary sulfide enrichment, 238–244
   selected references on, 245
   sulfide enrichment deposits, 243
Oxide ore deposition, 235–238

Paleolithic pottery, 16
Palisades of the Hudson, 70
   differentiation, 77
Paragenesis, 110
Pay streaks, 222
Peat, 168
Pegmatite dikes, 62
Pegmatites, 40, 85
Pendulum, 289
Permeability, 102
Petrified logs, 160
Petroleum, 172–188
   and natural gas, 172–188
   migration and accumulation, 176
   oil fields, 185–187
   origin, 173
   pressures, 184

368 Index

Petroleum, traps, 178–184
pH, 106, 150, 211
Philipsburg, Mont., 95
Phosphate, residual, 215
Phosphate beds, 160
Phosphate deposition, 158
Phosphate deposits, 153
Phosphoria formation, 154
Phosphorites, 158
Phosphorus, cycle of, 152–154
Phyllite, 12
Pinches and swells, 112
Pitches and flats, 104, 122
Pittsburgh seam, coal, 172
Placers, 216–226
  beach, 225
  deposits, 222–224
  diamond, 224
  eluvial, 219–220
  eolian, 219, 225–226
  gold, 223
  minerals of, 217
  principles of, 217–218
  requisites for, 221–222
  selected references on, 226
  stream, 220–224
Platinum pipe, Mooihoek, S. Africa, 83
Platinum pipes, S. Africa, 82
Plutonists, 25
Pneumatolytic, definition of, 49
Pneumatolytic action, 91
Pore-space fillings, 125
Pore spaces, 102
Porosity, 102
Porphyritic texture, 11
Porphyry coppers, 33, 107, 335
  deposits, 243, 298
Potash deposition, 193
Primary minerals, 6
Protore, 238
  definition of, 228
Pyramids, 16
Pyritologia, print from, 22
Pyrophyllite, 251

Quarrying, 297
Quartzite, 12
Quebec, asbestos deposits, 249

Radioactivity, 175
Radiometric method, geophysical, 293
Radium in logs, 160
Radon in natural gas, 175
Recrystallization, 54
Redruth, Cornwall, England, crustified vein, 110
Refining, 309
Reflected buried hills, 179
Reflection method, seismic, 291
Refraction method, seismic, 293
Refractory clays, 166
Rektor ore body, diagram of, 81
Replacement, 1, 127–140
  agencies of, 131–132
  criteria of, 134–135
  deposits, 135–140
  disseminated deposits, 136–137
  examples of, 140
  localization of, 133–134
  process, 128–135
  selected references on, 140
  veins, 136
Replacement deposits, 100
Replacement lodes, 132, 139
Residual concentration, 205–216
  clay, 213–214
  iron, 206–208
  manganese, 208
  nickel, 214–215
Residual formation, bauxite, 208–210
  clay, 213–214
Residual liquid injection, 80
Resistivity, 286
Rhodesia, chrome, 334
  copper belt, 234
Rio Tinto, Spain, 145
  copper, 271
Rock alteration openings, 104
Rock-forming minerals, 6
Rock openings, 101–104
Rocks, list of, 13
Room and pillar methods, 301
Rotary oil-well drilling, 305
Round Mountain oil field, Calif., 181
Russian mineral surpluses, 316

Saddle reef openings, 104
Saddle reefs, 120

Index 369

Salt domes, 179, 181, 192, 293
Salt lakes, 195
Salts in ocean, 189
San Miguel River, uranium, 160
Sandstone, 12
Sandstone lenses, 179
Santa Eulalia, Mexico, 105
Saones, 18
Sato mine, Japan, 302
Schist, 12
Searles Lake, salts of, 196
Secondary enrichment, 227
Secondary minerals, 6
Secondary sulfide enrichment, 238–244
    deposits, 243
Sedimentary breccias, 104
Sedimentary controls of mineral localization, 268
Sedimentary cycles and deposits, 147–199
    carbonate, 161–163
    chemical evaporites, 188–197
    clay, 164–166
    coal, 167–172
    copper, 156–157
    iron, 147–150
    manganese, 150–152
    petroleum and gas, 172–188
    phosphorus, 152–154
    silica, 166–167
    sulfur, 154–156
    uranium, vanadium, 157–160
Sedimentary processes, 142–201
    deposition, 144–147
    principles, 143–147
    selected references on, 199
    solution and transportation, 144
    sources of materials, 143
Sedimentary rocks, 11
Seismic methods, geophysical, 291
Seismic prospecting, 293
Selected references, on cavity filling, 127
    on contact metasomatism, 97
    on exploration and exploitation, 310
    on ground water, 266
    on history of theories of origin, 29
    on magmatic concentration, 88
    on materials of mineral deposits, 15

Selected references, on metamorphism, 253
    on mineral controls, 276
    on mineral resources, 344
    on mineralizing solutions, 57
    on oxidation and enrichment, 245
    on placer deposits, 226
    on processes of formation of mineral deposits, 65
    on relation of deposits to igneous activity, 43
    on replacement, 140
    on residual concentration, 216
    on sedimentation, 199
Self-potential prospecting, 284
Sericite alteration, 107
Seven Devils, Ida., 92
Shale, 12
Shear-zone cavities, 103
Shear-zone deposits, 119
Shear zones, 109
Shinkolobwe, Belgian Congo, uranium, 157
Shoestring sands, 179, 182
Shore lines, 182
Sicily, sulfur, 155
Sierra Mojada, Mexico, ores of, 231
Silica cycle, 166–167
Silicic magmas, 43
Sill, 35, 36
Sillimanite group minerals, 251–252
Silver, oxide ores of, 230
Silver chloride, 237
Sinai Peninsula, turquoise, 18
Sink and float, 308
Skarn, 92
Slag, 309
Smelters, 317
Smelting and refining, 309
Snake River, gold, 223
Soapstone formation, 251
Soda Lakes, Nev., 196
Solfataras, definition of, 48
Solution cavity deposits, 125
Solution openings, 104
Solutions, movement of, 104
South America, resources, 326
Specific surface, 105
Spiral concentration, 308

370 Index

Stassfurt, Germany, potash salts, 193
Steamboat Springs, Nev., 32, 51, 52
Stockworks, 120
Stope, 299
Strabo, 18
Stratigraphic traps, 182
   for oil, 179
Stream placer formation, 220
Strike, definition, 112
Stripping, 297
Stromboli, 47
Structural control of veins by igneous contacts, 272
Structural controls of mineral localization, 267
Structural traps for oil, 179
Substitutes for metals, 339
Substitution in replacement, 128
Sulfate solutions in sedimentation, 144, 145
Sulfur, cycle of, 154–156
   extraction, 304
Supergene minerals, 6
   sulfide enrichment, 57, 227
   sulfide zone, 238
   sulfides, 241
Surplus minerals, 328
Swamp clays, 166
Swedish kolm, 160
Synclines, 179

Tactite, 92
Talc formation, 251
Tchiaturi, Caucasus, Russia, 151
Tectonic breccias, 104, 125
   Collins mine, Ariz., 126
Tenor, definition of, 9
Tension cracks, 122
Terraces, 179
Theophrastus, 18
Tin, distribution of, 336
Tin ore, Mt. Bischoff, Australia, 105
Titaniferous magnetite, picture of, 78
Titaniferous magnetite bodies, 76
Tomboy mine, Telluride, Colo., 117
Torsion balance, 290
Transported limonite, 233
Travancore, India, placers, 225
Travertine, 52

Trenching, 297
Tripoli, residual, 215
Tri-State district, 108
Tropical weathering, 204, 205, 209–212
Tunisia, phosphates, 153
Turkey, chrome, 334

Unconformities, 179, 182
Underground caving, 303
Underground mining, 299
United States oil fields, 185
United Verde Extension mine, Ariz., 235
Up-dip porosity diminution, 179
Up-dip wedging of sands, 179
Uraninite, S. Africa, 161
Uranium, distribution of, 336
   with petroleum, 176
Uranium content of earth, 158
Uranium deposits, 159–160
Uranium and vanadium, cycle of, 157
Uranium and vanadium ores, 126
Utah Copper mine, Bingham, Utah, 137, 274

Valley of Ten Thousand Smokes, 31, 43, 48
   fumaroles of, 13
Valuation of mineral properties, 295–296
Vanadium, cycle of, 157
Vapors, 42
Vein dikes, 28
Ventura Avenue oil field, Calif., section, 180
Vesicles, 103
Vesicular fillings, 126
Vesuvius, 47
Victoria, Australia, ladder veins, 121
Volcanic breccias, 11
Volcanic flow drains, 103
Volcanic pipe breccia deposits, 124
Volcanic pipes, 104
Volcanic tuff, 11
Volcanoes, relation of ores to, 31

Wabana basin, Newfoundland, 149
Wall rock alteration, 107
   at Butte, Mont., 107

Index

Waste, 10
Water power, 331
Water table, 255
Weathering, 14
Weathering processes, 202–245
  other residual products, 215
  principles of, 202–204
  products of, 204
  residual bauxite formation, 209–212
  residual clay formation, 213–214
  residual iron concentration, 206–208
  residual manganese concentration, 208–209
  residual nickel concentration, 215

Weathering processes, selected references on, 216, 226, 245
  tropical, 204
Werner, Abraham Gottlob, 23, 100
  portrait of, 24
Western Europe bloc, 324
Witwatersrand, uranium, 161, 337
World War II, 318

Yellowstone National Park, 32, 51
  waters of, 52

Zonal arrangement of minerals, 108
Zonal distribution of deposits, 273
Zone of oxidation, 228